Contractor's Pricing Guide: Residential Repair & Remodeling Costs
with RSMeans™ Data

Sam Babbitt, Senior Editor

GORDIAN®

2025

Chief Product Officer
Chris Gaudreau

Chief Technology Officer
Seth Gregorie

Director, Data Operations
Sam Giffin

Engineering and Research
Matthew Doheny, Sr. Manager

Engineers: Architectural Divisions
1, 3, 4, 5, 6, 7, 8, 9, 10, 11, 12, 13, 41
Sam Babbitt
Richard Goldin
Scott Keller
Chris Pedersen
Melissa Stewart

Engineers: Civil Divisions
2, 31, 32, 33, 34, 35, 44, 46
Derrick Hale, PE
Michael Lynch
Giles Munyard

Engineers: Mechanical, Electrical, Plumbing & Conveying Divisions
14, 21, 22, 23, 25, 26, 27, 28, 48
Brian Adams
Michelle Curran
Antonio D'Aulerio
Chris Clough

Contributing Engineers
John Melin, PE, Manager
Paul Cowan
Thomas Lyons
Matthew Sorrentino

Production
Debra Panarelli, Manager
Sheryl Rose
Janice Thalin

Data Analysis
Yonique McNeil
Ellen D'Amico

Cover Design
Blaire Collins

Data Engineering
Monique Law
Kedar Gaikwad
Mike Bruesch
Jonathan Forgit

City Cost Index
Adam Raimond

Labor Research
Elisa Mello
David Yazbek

GORDIAN®

Contractor's Pricing Guide:
Residential Repair & Remodeling Costs
with RSMeans™ Data

2025

RSMeans Data from Gordian
Construction Publishers & Consultants
30 Patewood Dr., Suite 350
Greenville, SC 29615
United States of America
1.800.448.8182
rsmeans.com

Copyright 2024 by The Gordian Group Inc.
All rights reserved.
Cover photo © iStock.com/quackersnaps

Printed in the United States of America
ISSN 1542-9156
ISBN 978-1-961006-69-0

Gordian's authors, editors, and engineers apply diligence and judgment in locating and using reliable sources for the information published. However, Gordian makes no express or implied warranty or guarantee in connection with the content of the information contained herein, including the accuracy, correctness, value, sufficiency, or completeness of the data, methods, and other information contained herein. Gordian makes no express or implied warranty of merchantability or fitness for a particular purpose. Gordian shall have no liability to any customer or third party for any loss, expense, or damage, including consequential, incidental, special, or punitive damage, including lost profits or lost revenue, caused directly or indirectly by any error or omission, or arising out of, or in connection with, the information contained herein. For the purposes of this paragraph, "Gordian" shall include The Gordian Group, Inc., and its divisions, subsidiaries, successors, parent companies, and their employees, partners, principals, agents and representatives, and any third-party providers or sources of information or data. Gordian grants the purchaser of this publication limited license to use the cost data

contained herein for purchaser's internal business purposes in connection with construction estimating and related work. The publication, and all cost data contained herein, may not be reproduced, integrated into any software or computer program, developed into a database or other electronic compilation, stored in an information storage or retrieval system, or transmitted or distributed to anyone in any form or by any means, electronic or mechanical, including photocopying or scanning, without prior written permission of Gordian. This publication is subject to protection under copyright law, trade secret law and other intellectual property laws of the United States, Canada and other jurisdictions. Gordian and its affiliates exclusively own and retain all rights, title and interest in and to this publication and the cost data contained herein including, without limitation, all copyright, patent, trademark and trade secret rights. Except for the limited license contained herein, your purchase of this publication does not grant you any intellectual property rights in the publication or the cost data.

3425 | $110.00 per copy (in United States)
Price is subject to change without prior notice.

Table of Contents

Foreword	vii
How the Cost Data Is Built: An Overview	ix
How to Use the Cost Data: The Details	xi
Unit Price Section	xiii
RSMeans™ Data: Unit Prices—How They Work	xv
Reference Section	
Location Factors	295
Abbreviations	301
Index	302
Other Data and Services	318
Labor Trade Rates including Overhead & Profit	Inside Back Cover

Foreword

The Value of RSMeans™ Data from Gordian

Since 1942, RSMeans Data has been the industry–standard materials, labor, and equipment cost information database for contractors, facility owners and managers, architects, engineers and anyone else that requires the latest localized construction cost information. More than 80 years later, the objective remains the same: to provide facility and construction professionals with the most current and comprehensive construction cost database possible.

With the constant influx of new construction methods and materials, in addition to ever-changing labor and material costs, last year's cost data is not reliable for today's designs, estimates or budgets. Gordian's cost engineers apply real-world construction experience to identify and quantify new building products and methodologies, adjust productivity rates and adjust costs to local market conditions across the nation. This adds up to more than 30,000 hours in cost research annually. This unparalleled construction cost expertise is why so many facility and construction professionals rely on RSMeans Data year over year.

About Gordian

Gordian is the leading provider of Building Intelligence™ Solutions, delivering unrivaled insights, robust technology and comprehensive expertise that fuel customers' success during every phase of the building lifecycle. Gordian created Job Order Contracting (JOC) and the industry standard RSMeans Data. We empower organizations to optimize capital investments, improve project performance and minimize long-term operating expenses. We develop and maintain the largest collection of labor, material, and equipment data and associated costs for all areas of construction. Gordian's solutions are accessed through our innovative software platforms and supported by a team of industry experts proven to help clients maximize efficiency, optimize cost savings, and increase building quality.

Our Commitment

At Gordian, we do more than talk about the quality of our data and the usefulness of its application. We stand behind all of our RSMeans Data — from historical cost indexes to construction materials and techniques — to craft current costs and predict future trends. If you have any questions about our products or services, please call us toll-free at 800.448.8182 or visit our website at rsmeans.com.

How the Cost Data Is Built: An Overview

A Powerful Construction Tool

You have in your hands one of the most powerful construction tools available today. A successful project is built on the foundation of an accurate and dependable estimate. This data will enable you to construct such an estimate.

For the casual user the data is designed to be:

- quickly and easily understood so you can get right to your estimate.
- filled with valuable information so you can understand the necessary factors that go into the cost estimate.

For the regular user, the data is designed to be:

- a handy reference that can be quickly referred to for key costs.
- a comprehensive, fully reliable source of current construction costs, so you'll be prepared to estimate any project.
- a source for project costs, product selections, and alternate materials and methods.

To meet all of these requirements we have organized the data into the following clearly defined sections.

Unit Price Section

The cost data has been divided into 23 sections representing the order of construction. Within each section is a listing of components and variations of those components that are applicable to that section. Costs are shown for the various actions that must be done to that component.

Reference Section

This section includes information on Location Factors and a listing of abbreviations.

Location Factors

You can adjust total project costs to over 900 locations throughout the U.S. and Canada by using the data in this section.

Abbreviations

A listing of the abbreviations used throughout, along with the terms they represent, is included.

Index

A comprehensive listing of all terms and subjects will help you quickly find what you need when you are not sure where it falls in the order of construction.

The Scope of this Data

This data is designed to be as comprehensive and easy to use as possible. To that end we have made certain assumptions and limited its scope in two key ways:

1. We have established material prices based on a national average.
2. We have computed labor costs based on residential wage rates.

Project Size

This data is intended for use by those involved primarily in residential repair and remodeling construction costing between $10,000 and $100,000.

With reasonable exercise of judgment, the figures can be used for any building work. For other types of projects, such as new home construction or commercial buildings, consult the appropriate RSMeans dataset for more information.

How to Use the Cost Data: The Details

What's Behind the Numbers? The Development of Cost Data

RSMeans™ Data engineers continually monitor developments in the construction industry in order to ensure reliable, thorough, and up-to-date cost information. While overall construction costs may vary relative to general economic conditions, price fluctuations within the industry are dependent upon many factors. Individual price variations may, in fact, be opposite to overall economic trends. Therefore, costs are constantly tracked and complete updates are performed yearly. Also, new items are frequently added in response to changes in materials and methods.

Costs in U.S. Dollars

All costs represent U.S. national averages and are given in U.S. dollars. The City Cost Index (CCI) with RSMeans Data or Location Factors can be used to adjust costs to a particular location. The CCI or Location Factors for Canada can be used to adjust U.S. national averages to local costs in Canadian dollars. No exchange rate conversion is necessary because it has already been factored in.

Material Costs

RSMeans Data engineers contact manufacturers, dealers, distributors, and contractors all across the U.S. and Canada to determine national average material costs. If you have access to current material costs for your specific location, you may wish to make adjustments to reflect differences from the national average. Included within material costs are fasteners for a normal installation. RSMeans Data engineers use manufacturers' recommendations, written specifications, and/or standard construction practices for the sizing and spacing of fasteners. Adjustments to material costs may be required for your specific application or location. The manufacturer's warranty is assumed. Extended warranties are not included in the material costs. **Material costs do not include sales tax.**

Labor Costs

Labor costs are based upon a mathematical average of trade-specific wages in 30 major U.S. cities. The type of wage (union, open shop, or residential) is identified on the inside back cover of printed publications or selected by the estimator when using the electronic products. Markups for the wages can also be found on the inside back cover of printed publications and/or under the labor references found in the electronic products.

- If wage rates in your area vary from those used, or if rate increases are expected within a given year, labor costs should be adjusted accordingly.

Labor costs reflect productivity based on actual working conditions. In addition to actual installation, these figures include time spent during a normal weekday on tasks, such as material receiving and handling, mobilization at the site, site movement, breaks, and cleanup.

Productivity data is developed over an extended period so as not to be influenced by abnormal variations and reflects a typical average.

Equipment Costs

Equipment costs include not only rental but also operating costs for equipment under normal use. The operating costs include parts and labor for routine servicing, such as the repair and replacement of pumps, filters, and worn lines. Normal operating expendables, such as fuel, lubricants, tires, and electricity (where applicable), are also included. Extraordinary operating expendables with highly variable wear patterns, such as diamond bits and blades, are excluded. These costs are included under materials. Equipment rental rates are obtained from industry sources throughout North America—contractors, suppliers, dealers, manufacturers, and distributors.

Rental rates can also be treated as reimbursement costs for contractor-owned equipment. Owned equipment costs include depreciation, loan payments, interest, taxes, insurance, storage, and major repairs.

Equipment costs do not include operators' wages.

Factors Affecting Costs

Costs can vary depending upon a number of variables. Here's a listing of some factors that affect costs and points to consider.

Quality—The prices for materials and the workmanship upon which productivity is based represent sound construction work. They are also in line with industry standards and manufacturer specifications and are frequently used by federal, state, and local governments.

Overtime—We have made no allowance for overtime. If you anticipate premium time or work beyond normal working hours, be sure to make an appropriate adjustment to your labor costs.

Productivity—The productivity, daily output, and labor-hour figures for each line item are based on an eight-hour work day in daylight hours in moderate temperatures and up to a 14' working height unless otherwise indicated. For work that extends beyond normal work hours or is performed under adverse conditions, productivity may decrease.

Size of Project—The size, scope of work, and type of construction project will have a significant impact on cost. Economies of scale can reduce costs for large projects. Unit costs can often run higher for small projects.

Location—Material prices are for metropolitan areas. However, in dense urban areas, traffic and site storage limitations may increase costs. Beyond a 20-mile radius of metropolitan areas, extra trucking or transportation charges may also increase the material costs slightly. On the other hand, lower wage rates may be in effect. Be sure to consider both of these factors when preparing an estimate, particularly if the job site is located in a central city or remote rural location. In addition, highly specialized subcontract items may require travel and per-diem expenses for mechanics.

Other Factors—

- season of year
- contractor management
- weather conditions
- local union restrictions
- building code requirements
- availability of:
  - adequate energy
  - skilled labor
  - building materials
- owner's special requirements/restrictions
- safety requirements
- environmental considerations
- access

Unpredictable Factors—General business conditions influence "in-place" costs of all items. Substitute materials and construction methods may have to be employed. These may affect the installed cost and/or life cycle costs. Such factors may be difficult to evaluate and cannot necessarily be predicted on the basis of the job's location in a particular section of the country. Thus, where these factors apply, you may find significant but unavoidable cost variations for which you will have to apply a measure of judgment to your estimate.

Final Checklist

Estimating can be a straightforward process provided you remember the basics. Here's a checklist of some of the steps you should remember to complete before finalizing your estimate.

Did you remember to:

- Factor in the City Cost Index or Location Factors for your locale?
- Take into consideration which items have been marked up and by how much?

- Mark up the entire estimate sufficiently for your purposes?
- Read the background information on techniques and technical matters that could impact your project time span and cost?
- Include all components of your project in the final estimate?
- Double check your figures for accuracy?
- Call RSMeans Data engineers if you have any questions about your estimate or the data you've used? Remember, Gordian stands behind all of our products, including our extensive RSMeans Data solutions. If you have any questions about your estimate, about the costs you've used from our data, or even about the technical aspects of the job that may affect your estimate, feel free to call the Gordian RSMeans editors at 1.800.448.8182.

Unit Price Section

Table of Contents

Job Costs

	Page
Construction Fees	1
Construction Permits	1
Trade Labor	2
Temporary Utilities	3
Equipment Rental	3
Move/Reset Contents	6
Cover and Protect	6
Temporary Bracing	6
Jack and Relevel	7
Dumpster Rental	7
Debris Hauling	7
Dump Fees	7
Scaffolding	7
Construction Clean-up	8

Job Preparation

Building Demolition	9
General Clean-up	10
Flood Clean-up	10
Asbestos Removal	12
Room Demolition	13
Selective Demolition	14

Excavation

Excavation	17
Backfill	17

Foundation

Piles	19
Foundation Post	23
Concrete Footing	24
Concrete Slab	25

Rough Frame/Structure

Beam/Girder	27
Column	29
Floor Framing System	29
Subflooring	33
Wall Framing System	34
Rough-in Opening	38
Glue-Laminated Beams	42
Metal Stud Framing	44
Metal Joist	45
Exterior Sheathing	45
Plywood Sheathing	46
Stairs	47
Ceiling Framing	47
Roof Framing	49
Residential Trusses	54
Commercial Trusses	55
Roof Sheathing	56
Wood Deck	57
Porch	63

Exterior Trim

	Page
Exterior Trim	65
Fascia	67
Soffit	69

Roofing

Metal Flashing	71
Roof Shingles	73
Rolled Roofing	73
Composition Shingles	73
Wood Shingles	75
Slate Tile Roofing	76
Flat Clay Tile Roofing	77
Mission Tile Roofing	77
Spanish Tile Roofing	78
Concrete Tile Roofing	80
Aluminum Sheet	81
Fiberglass Sheet	83
Galvanized Steel	83
Standing Seam	84
Elastomeric Roofing	85
Modified Bitumen Roofing	87
Built-up Roofing	87

Exterior Doors and Windows

Exterior Type Door	89
Residential Metal Door	91
Metal Door	92
Solid Core Door	93
Components	96
Sliding Patio Door	99
Screen Door	101
Storm Door	102
Garage Door	102
Commercial Metal Door	104
Aluminum Window	106
Vinyl Window	113
Wood Window	114
Window Screens	118
Window Glass	119
Storefront System	120
Curtain Wall System	122
Skylight	123
Industrial Steel Window	124

Masonry

Masonry Wall	125
Chimney	129
Fireplace	129

Siding

Exterior Siding	133

Site Work

Grading	139
Asphalt Paving	139
Masonry Pavers	139
Fencing	140

Rough Mechanical

	Page
Plumbing	147
Water Heater	148
Ductwork	150
Heating	153
Furnace Unit	154
Boiler	156
Heat Pump	157
Heating	158
Heating Control	159
Air Conditioning	159
Sprinkler System	163

Rough Electrical

Electrical Per S.F.	165
Main Service	165

Insulation

Floor Insulation	167
Wall Insulation	168
Ceiling Insulation	170
Exterior Insulation	172
Roof Insulation Board	172

Walls/Ceilings

Wall Plaster	175
Ceiling Plaster	176
Gypsum Wallboard	177
Ceramic Wall Tile	180
Wall Tile	181
Wall Paneling	182
Wallcovering	184
Acoustical Ceiling	187

Finish Carpentry

Hollow Core Door	191
Solid Core Door	195
Bifold Door	196
Louver Door	199
Bypass Sliding Door	200
Base Molding	201
Molding	204
Window Trim Set	205
Window Sill	206
Shelving	206
Closet Shelving	207
Stair Assemblies	208
Stairs	208
Stair Components	209

Cabinets and Countertops

Wall Cabinets	211
Tall Cabinets	218
Base Cabinets	222
Countertop	229
Vanity Cabinets	231
Vanity Cabinet Tops	234

xiii

Table of Contents (cont.)

Painting
Page

Paint Preparation, Walls.....235
Paint Preparation, Ceilings.....236
Paint/Texture, Walls.....236
Paint/Texture, Ceilings.....237

Flooring

Wood Plank Flooring.....239
Wood Parquet Tile Floor.....241
Ceramic Tile Flooring.....243
Hard Tile Flooring.....243
Sheet Vinyl Per S.F......245
Sheet Vinyl.....245
Vinyl Tile.....245
Carpeting Per S.F......246
Carpeting.....247
Carpet Pad Per S.F......248
Carpet Pad.....249
Stair Components.....249

Finish Mechanical

Commode.....251
Sink (assembly).....252
Sink Only.....253

Finish Mechanical (cont.)
Page

Bathtub.....255
Shower.....257
Ductwork.....258

Finish Electrical

Electrical Per S.F......261
Circuits.....261
Light Switch.....262
Cover for Outlet/Switch.....264
Residential Light Fixture.....264
Commercial Light Fixture.....266
Fan.....268
Low Voltage Systems.....269

Improvements/Appliances/Treatments

Window Treatment.....273
Stairs.....275
Elevator.....275
Rangetop.....276
Rangehood.....276
Oven.....277
Dryer.....278
Washer.....279

Improvements/Appliances/ Treatments (cont.)
Page

Refrigerator.....279
Freezer.....280
Dishwasher.....281
Garbage Disposal.....281
Trash Compactor.....281
Kitchenette Unit.....281
Bath Accessories.....282
Fire Prevention/Protection.....283
Gutters.....284
Downspouts.....284
Shutters.....285
Awning.....287
Vent.....288
Cupolas.....289
Aluminum Carport.....290
Storage Shed.....290
Gazebo.....291
Pool.....291

Close Out

Construction Clean-up.....293

RSMeans™ Data: Unit Prices—How They Work

The prices in this publication include all overhead and profit mark-ups for the installing contractor. If a general contractor is involved, an additional 10% should be added to the cost shown in the "Total" column.

The costs for material, labor, and any equipment are shown here. Material costs include delivery and a 10% mark-up. No sales taxes are included. Labor costs include all mark-ups as shown on the inside back cover of this dataset. Equipment costs also include a 10% mark-up.

The cost in the Total column should be multiplied by the quantity of each item to determine the total cost for that item. The sum of all items' total costs becomes the project's total cost.

The dataset is arranged into 23 categories that follow the order of construction for residential projects. The category is shown as white letters on a black background at the top of each page.

A brief specification of the cost item is shown here.

Graphics representing some of the cost items are found at the beginning of each category.

The cost data is arranged in a tree structure format. Major cost sections are left justified and shown as bold headings within a grey background. Cost items and sub-cost items are found under the bold headings. The complete description should read as "bold heading, cost item, sub-cost item."

For example, the identified item's description is "Beam/Girder, Solid Wood Beam, 2" x 8"."

Each action performed to the cost item is conveniently listed to the right of the description. See the explanation below for the scope of work included as part of the action.

The unit of measure upon which all the costs are based is shown here. An abbreviations list is included within the Reference Section found at the back of this dataset.

Rough Frame / Structure

Rough Framing

Beam / Girder		Unit	Material	Labor	Equip.	Total	Specification
Solid Wood Beam							
2" x 8"							
	Demolish	L.F.		1.08		1.08	Cost includes material and labor to install 2" x 8" beam.
	Install	L.F.	1.46	1.60		3.06	
	Demolish and Install	L.F.	1.46	2.68		4.14	
	Clean	L.F.	.11	.36		.47	
	Paint	L.F.	.36	2.33		2.69	
	Minimum Charge	Job		260		260	
2" x 10"							
	Demolish	L.F.		1.34		1.34	Cost includes material and labor to install 2" x 10" beam.
	Install	L.F.	3.88	1.74		5.62	
	Demolish and Install	L.F.	3.88	3.08		6.96	
	Clean	L.F.	.13	.43		.56	
	Paint	L.F.	.25	1.44		1.69	
	Minimum Charge	Job		260		260	
2" x 12"							
	Demolish	L.F.		1.61		1.61	Cost includes material and labor to install 2" x 12" beam.
	Install	L.F.	4.09	1.89		5.98	
	Demolish and Install	L.F.	4.09	3.50		7.59	
	Clean	L.F.	.14	.50		.64	
	Paint	L.F.	.55	3.45		4	
	Minimum Charge	Job		260		260	
4" x 8"							
	Demolish	L.F.		3.29		3.29	Cost includes material and labor to install 4" x 8" beam.
	Install	L.F.	12.25	2.42	2.55	17.22	
	Demolish and Install	L.F.	12.25	5.71	2.55	20.51	
	Clean	L.F.	.13	.43		.56	
	Paint	L.F.	.25	1.44		1.69	
	Minimum Charge	Job		260		260	
4" x 10"							
	Demolish	L.F.		4.12		4.12	Cost includes material and labor to install 4" x 10" beam.
	Install	L.F.	8.70	2.55	2.69	13.94	
	Demolish and Install	L.F.	8.70	6.67	2.69	18.06	
	Clean	L.F.	.14	.50		.64	
	Paint	L.F.	.29	1.68		1.97	
	Minimum Charge	Job		260		260	
4" x 12"							
	Demolish	L.F.		4.95		4.95	Cost includes material and labor to install 4" x 12" beam.
	Install	L.F.	11.85	2.69	2.84	17.38	
	Demolish and Install	L.F.	11.85	7.64	2.84	22.33	
	Clean	L.F.	.17	.57		.74	
	Paint	L.F.	.55	3.45		4	
	Minimum Charge	Job		260		260	

Action	Scope
Demolish	Includes removing the item and hauling the debris to a truck, dumpster, or storage area.
Install	Includes the installation of the item with normal placement and spacing of fasteners.
Demolish and Install	Includes the demolish action and the install action.
Reinstall	Includes the reinstallation of a removed item.
Clean	Includes cleaning of the item.
Paint	Includes up to two coats of finish on the item (unless otherwise indicated).
Minimum Charge	Includes the minimum labor or the minimum labor and equipment cost. Use this charge for small quantities of material.

Job Costs

Scaffolding

Temporary Electrical Service

Air Compressor

Construction Fees

		Unit	Material	Labor	Equip.	Total	Specification
Asbestos Test	Minimum Charge	Job		310		310	Minimum charge for testing.
Architectural	Minimum Charge	Day	2950			2950	Minimum labor charge for architect.
Engineering	Minimum Charge	Day	2500			2500	Minimum labor charge for field engineer.

Construction Permits

		Unit	Material	Labor	Equip.	Total	Specification
Job (less than 5K)	Minimum Charge	Job	76			76	Permit fees, job less than 5K.
Job (5K to 10K)	Minimum Charge	Job	185			185	Permit fees, job more than 5K, but less than 10K.
Job (10K to 25K)	Minimum Charge	Job	360			360	Permit fees, job more than 10K, but less than 25K.
Job (25K to 50K)	Minimum Charge	Job	695			695	Permit fees, job more than 25K, but less than 50K.
Job (50K to 100K)	Minimum Charge	Job	1300			1300	Permit fees, job more than 50K, but less than 100K.
Job (100K to 250K)	Minimum Charge	Job	2675			2675	Permit fees, job more than 100K, but less than 250K.
Job (250K to 500K)	Minimum Charge	Job	5175			5175	Permit fees, job more than 250K, but less than 500K.
Job (500K to 1M)	Minimum Charge	Job	9325			9325	Permit fees, job more than 500K, but less than 1 million.

For customer support on your Contractor's Pricing Guide: Residential Repair & Remodeling, call 800.448.8182.

Job Costs

Trade Labor		Unit	Material	Labor	Equip.	Total	Specification
Carpenter							
Daily (8 hours)							
	Install	Day		520		520	Daily labor rate for one carpenter.
	Minimum Charge	Job		260		260	
Weekly (40 hours)							
	Install	Week		2600		2600	Weekly labor rate for one carpenter.
	Minimum Charge	Job		260		260	
Drywaller							
Daily (8 hours)							
	Install	Day		520		520	Daily labor rate for one carpenter.
	Minimum Charge	Job		260		260	
Weekly (40 hours)							
	Install	Week		2600		2600	Weekly labor rate for one carpenter.
	Minimum Charge	Job		260		260	
Roofer							
Daily (8 hours)							
	Install	Day		490		490	Daily labor rate for one skilled roofer.
	Minimum Charge	Job		244		244	
Weekly (40 hours)							
	Install	Week		2450		2450	Weekly labor rate for one skilled roofer.
	Minimum Charge	Job		244		244	
Painter							
Daily (8 hours)							
	Install	Day		430		430	Daily labor rate for one painter.
	Minimum Charge	Job		216		216	
Weekly (40 hours)							
	Install	Week		2150		2150	Weekly labor rate for one painter.
	Minimum Charge	Job		216		216	
Mason							
Daily (8 hours)							
	Install	Day		515		515	Daily labor rate for one skilled mason.
	Minimum Charge	Job		259		259	
Weekly (40 hours)							
	Install	Week		2575		2575	Weekly labor rate for one skilled mason.
	Minimum Charge	Job		259		259	
Electrician							
Daily (8 hours)							
	Install	Day		620		620	Daily labor rate for an electrician.
	Minimum Charge	Job		310		310	
Weekly (40 hours)							
	Install	Week		3100		3100	Weekly labor rate for an electrician.
	Minimum Charge	Job		310		310	
Plumber							
Daily (8 hours)							
	Install	Day		585		585	Daily labor rate for a plumber.
	Minimum Charge	Job		294		294	
Weekly (40 hours)							
	Install	Week		2925		2925	Weekly labor rate for a plumber.
	Minimum Charge	Job		294		294	
Common Laborer							
Daily (8 hours)							
	Install	Day		400		400	Daily labor rate for a common laborer.
	Minimum Charge	Job		200		200	

For customer support on your Contractor's Pricing Guide: Residential Repair & Remodeling, call 800.448.8182.

Job Costs

Trade Labor

Trade Labor		Unit	Material	Labor	Equip.	Total	Specification
Weekly (40 hours)							
	Install	Week		2000		2000	Weekly labor rate for a common
	Minimum Charge	Job		200		200	laborer.

Temporary Utilities

Temporary Utilities		Unit	Material	Labor	Equip.	Total	Specification
Power Pole							
Job Site Electricity							
	Install	Month	76.50			76.50	Includes average cost of electricity per
	Minimum Charge	Job		310		310	month.
Water Service							
Job Site Water							
	Install	Month	139			139	Includes average costs for water for
	Minimum Charge	Month	139			139	one month.
Telephone							
	Install	Month	109			109	Includes average cost for local and
	Minimum Charge	Month	109			109	long distance telephone service for
							one month.
Chemical Toilet							
	Install	Month			395	395	Monthly rental of portable toilet.
	Minimum Charge	Day			19.70	19.70	
Office Trailer							
Place and Remove							
	Install	Job	485			485	Includes cost to a contractor to deliver,
	Minimum Charge	Job		247		247	place and level an office trailer and
							remove at end of job.
Storage Container							
Place and Remove							
	Install	Job	278			278	Includes cost to a contractor to deliver
	Minimum Charge	Job		247		247	and place a storage container and
							remove at end of job.
Fencing							
	Install	L.F.	4.51			4.51	Includes material costs for temporary
	Minimum Charge	Job		200		200	fencing.
Railing							
	Install	L.F.	22	13.55	1.07	36.62	Includes material and labor to install
	Minimum Charge	Job		200		200	primed steel pipe.

Equipment Rental

Equipment Rental		Unit	Material	Labor	Equip.	Total	Specification
Flatbed Truck							
Daily Rental							
	Install	Day			465	465	Daily rental of flatbed truck.
	Minimum Charge	Job			465	465	
Weekly Rental							
	Install	Week			2325	2325	Weekly rental of flatbed truck.
	Minimum Charge	Job			465	465	
Monthly Rental							
	Install	Month			9275	9275	Monthly rental of flatbed truck.
	Minimum Charge	Job			465	465	
Dump Truck							
Daily Rental							
	Install	Day			575	575	Daily rental of dump truck.
	Minimum Charge	Day			575	575	

For customer support on your Contractor's Pricing Guide: Residential Repair & Remodeling, call 800.448.8182.

Job Costs

Equipment Rental		Unit	Material	Labor	Equip.	Total	Specification
Weekly Rental							
	Install	Week			2875	2875	Weekly rental of dump truck.
	Minimum Charge	Day			575	575	
Monthly Rental							
	Install	Month			11500	11500	Monthly rental of dump truck.
	Minimum Charge	Day			575	575	
Forklift							
Daily Rental							
	Install	Day			490	490	Daily rental of forklift.
	Minimum Charge	Day			490	490	
Weekly Rental							
	Install	Week			2450	2450	Weekly rental of forklift.
	Minimum Charge	Day			490	490	
Monthly Rental							
	Install	Month			9800	9800	Monthly rental of forklift.
	Minimum Charge	Day			490	490	
Bobcat							
Daily Rental							
	Install	Day			305	305	Daily rental of wheeled, skid steer loader.
	Minimum Charge	Day			305	305	
Weekly Rental							
	Install	Week			1525	1525	Weekly rental of wheeled, skid steer loader.
	Minimum Charge	Day			305	305	
Monthly Rental							
	Install	Month			6050	6050	Monthly rental of wheeled, skid steer loader.
	Minimum Charge	Day			305	305	
Air Compressor							
Daily Rental							
	Install	Day			234	234	Daily rental of air compressor.
	Minimum Charge	Day			234	234	
Weekly Rental							
	Install	Week			1175	1175	Weekly rental of air compressor.
	Minimum Charge	Day			234	234	
Monthly Rental							
	Install	Month			4675	4675	Monthly rental of air compressor.
	Minimum Charge	Day			234	234	
Jackhammer							
Daily Rental							
	Install	Day			45.50	45.50	Daily rental of air tool, jackhammer.
	Minimum Charge	Day			45.50	45.50	
Weekly Rental							
	Install	Week			227	227	Weekly rental of air tool, jackhammer.
	Minimum Charge	Day			45.50	45.50	
Monthly Rental							
	Install	Month			905	905	Monthly rental of air tool, jackhammer.
	Minimum Charge	Day			45.50	45.50	
Concrete Mixer							
Daily Rental							
	Install	Day			133	133	Daily rental of concrete mixer.
	Minimum Charge	Day			133	133	
Weekly Rental							
	Install	Week			665	665	Weekly rental of concrete mixer.
	Minimum Charge	Day			133	133	

For customer support on your Contractor's Pricing Guide: Residential Repair & Remodeling, call 800.448.8182.

Job Costs

Equipment Rental

Equipment Rental		Unit	Material	Labor	Equip.	Total	Specification
Monthly Rental							
	Install	Month			2675	2675	Monthly rental of concrete mixer.
	Minimum Charge	Day			133	133	
Concrete Bucket							
Daily Rental							
	Install	Day			50	50	Daily rental of concrete bucket.
	Minimum Charge	Day			50	50	
Weekly Rental							
	Install	Week			249	249	Weekly rental of concrete bucket.
	Minimum Charge	Day			50	50	
Monthly Rental							
	Install	Month			995	995	Monthly rental of concrete bucket.
	Minimum Charge	Day			50	50	
Concrete Pump							
Daily Rental							
	Install	Day			990	990	Daily rental of concrete pump.
	Minimum Charge	Day			990	990	
Weekly Rental							
	Install	Week			4950	4950	Weekly rental of concrete pump.
	Minimum Charge	Day			990	990	
Monthly Rental							
	Install	Month			19800	19800	Monthly rental of concrete pump.
	Minimum Charge	Day			990	990	
Generator							
Daily Rental							
	Install	Day			86	86	Daily rental of generator.
	Minimum Charge	Day			86	86	
Weekly Rental							
	Install	Week			430	430	Weekly rental of generator.
	Minimum Charge	Day			86	86	
Monthly Rental							
	Install	Month			1725	1725	Monthly rental of generator.
	Minimum Charge	Day			86	86	
Welder							
Daily Rental							
	Install	Day			64.50	64.50	Daily rental of welder.
	Minimum Charge	Day			64.50	64.50	
Weekly Rental							
	Install	Week			325	325	Weekly rental of welder.
	Minimum Charge	Day			64.50	64.50	
Monthly Rental							
	Install	Month			1300	1300	Monthly rental of welder.
	Minimum Charge	Day			64.50	64.50	
Sandblaster							
Daily Rental							
	Install	Day			100	100	Daily rental of sandblaster.
	Minimum Charge	Day			100	100	
Weekly Rental							
	Install	Week			500	500	Weekly rental of sandblaster.
	Minimum Charge	Day			100	100	
Monthly Rental							
	Install	Month			2000	2000	Monthly rental of sandblaster.
	Minimum Charge	Day			100	100	

For customer support on your Contractor's Pricing Guide: Residential Repair & Remodeling, call 800.448.8182.

Job Costs

Equipment Rental

Equipment Rental		Unit	Material	Labor	Equip.	Total	Specification
Space Heater							
Daily Rental							
	Install	Day			76	76	Daily rental of space heater.
	Minimum Charge	Day			53	53	
Weekly Rental							
	Install	Day			380	380	Weekly rental of space heater.
	Minimum Charge	Day			53	53	
Monthly Rental							
	Install	Day			1525	1525	Monthly rental of space heater.
	Minimum Charge	Day			53	53	

Move / Reset Contents

Move / Reset Contents		Unit	Material	Labor	Equip.	Total	Specification
Small Room							
	Install	Room		48.50		48.50	Includes cost to remove and reset the
	Minimum Charge	Job		200		200	contents of a small room.
Average Room							
	Install	Room		58.50		58.50	Includes cost to remove and reset the
	Minimum Charge	Job		200		200	contents of an average room.
Large Room							
	Install	Room		81		81	Includes cost to remove and reset the
	Minimum Charge	Job		200		200	contents of a large room.
Extra Large Room							
	Install	Room		128		128	Includes cost to remove and reset the
	Minimum Charge	Job		200		200	contents of an extra large room.
Store Contents							
Weekly							
	Install	Week	365			365	Storage fee, per week.
Monthly							
	Install	Month	1100			1100	Storage fee, per month.

Cover and Protect

Cover and Protect		Unit	Material	Labor	Equip.	Total	Specification
Cover / Protect Walls							
	Install	SF Wall		.15		.15	Includes labor costs to cover and
	Minimum Charge	Job		200		200	protect walls.
Cover / Protect Floors							
	Install	SF Flr.		.15		.15	Includes labor costs to cover and
	Minimum Charge	Job		200		200	protect floors.

Temporary Bracing

Temporary Bracing		Unit	Material	Labor	Equip.	Total	Specification
Shim Support Piers							
	Install	Ea.	6.95	155		161.95	Includes labor and material to install
	Minimum Charge	Job		465		465	shims support for piers.

For customer support on your Contractor's Pricing Guide: Residential Repair & Remodeling, call 800.448.8182.

Job Costs

Jack and Relevel

		Unit	Material	Labor	Equip.	Total	Specification
Single Story							
	Install	S.F.	1.84	1.86		3.70	Includes labor and equipment to jack
	Minimum Charge	Job		465		465	and relevel single-story building.
Two-story							
	Install	S.F.	1.84	3.09		4.93	Includes labor and equipment to jack
	Minimum Charge	Job		465		465	and relevel two-story building.

Dumpster Rental

		Unit	Material	Labor	Equip.	Total	Specification
Small Load							
	Install	Week			735	735	Dumpster, weekly rental, 2 dumps, 5
	Minimum Charge	Week			735	735	yard container.
Medium Load							
	Install	Week			845	845	Dumpster, weekly rental, 2 dumps, 10
	Minimum Charge	Week			845	845	yard container.
Large Load							
	Install	Week			1300	1300	Dumpster, weekly rental, 2 dumps, 30
	Minimum Charge	Week			1300	1300	yard container.

Debris Hauling

		Unit	Material	Labor	Equip.	Total	Specification
Per Ton							
	Install	Ton		127	82	209	Includes hauling of debris by trailer or
	Minimum Charge	Ton		955	615	1570	dump truck.
Per Cubic Yard							
	Install	C.Y.		42.50	27.50	70	Includes hauling of debris by trailer or
	Minimum Charge	Job		955	615	1570	dump truck.
Per Pick-up Truck Load							
	Install	Ea.		222		222	Includes hauling of debris by pick-up
	Minimum Charge	Ea.		222		222	truck.

Dump Fees

		Unit	Material	Labor	Equip.	Total	Specification
Per Cubic Yard							
	Install	C.Y.	70			70	Dump charges, building materials.
	Minimum Charge	C.Y.	70			70	
Per Ton							
	Install	Ton	166			166	Dump charges, building materials.
	Minimum Charge	Ton	166			166	

Scaffolding

		Unit	Material	Labor	Equip.	Total	Specification
4' to 6' High							
	Install	Month	168			168	Includes monthly rental of scaffolding,
	Minimum Charge	Month	168			168	steel tubular, 30" wide, 7' long, 5' high.
7' to 11' High							
	Install	Month	212			212	Includes monthly rental of scaffolding,
	Minimum Charge	Month	212			212	steel tubular, 30" wide, 7' long, 10' high.

For customer support on your Contractor's Pricing Guide: Residential Repair & Remodeling, call 800.448.8182.

Job Costs

Scaffolding

Scaffolding	Unit	Material	Labor	Equip.	Total	Specification
12' to 16' High						
Install	Month	256			256	Includes monthly rental of scaffolding, steel tubular, 30" wide, 7' long, 15' high.
Minimum Charge	Month	256			256	
17' to 21' High						
Install	Month	370			370	Includes monthly rental of scaffolding, steel tubular, 30" wide, 7' long, 20' high.
Minimum Charge	Month	370			370	
22' to 26' High						
Install	Month	415			415	Includes monthly rental of scaffolding, steel tubular, 30" wide, 7' long, 25' high.
Minimum Charge	Month	415			415	
27' to 30' High						
Install	Month	455			455	Includes monthly rental of scaffolding, steel tubular, 30" wide, 7' long, 30' high.
Minimum Charge	Month	455			455	

Construction Clean-up

Construction Clean-up	Unit	Material	Labor	Equip.	Total	Specification
Initial						
Clean	S.F.		.50		.50	Includes labor for general clean-up.
Minimum Charge	Job		200		200	
Progressive						
Clean	S.F.		.25		.25	Includes labor for progressive site/job clean-up on a weekly basis.
Minimum Charge	Job		200		200	

For customer support on your Contractor's Pricing Guide: Residential Repair & Remodeling, call 800.448.8182.

Job Preparation

Building Demolition		Unit	Material	Labor	Equip.	Total	Specification
Wood-framed Building							
1 Story Demolition							
	Demolish	S.F.		1.94	2.61	4.55	Includes demolition of 1 floor only light wood-framed building, including hauling of debris. Dump fees not included.
	Minimum Charge	Job		3075	4125	7200	
2nd Story Demolition							
	Demolish	S.F.		1.98	2.66	4.64	Includes demolition of 2nd floor only light wood-framed building, including hauling of debris. Dump fees not included.
	Minimum Charge	Job		3075	4125	7200	
3rd Story Demolition							
	Demolish	S.F.		2.58	3.46	6.04	Includes demolition of 3rd floor only light wood-framed building, including hauling of debris. Dump fees not included.
	Minimum Charge	Job		3075	4125	7200	
Masonry Building							
	Demolish	S.F.		2.77	3.72	6.49	Includes demolition of one story masonry building, including hauling of debris. Dump fees not included.
	Minimum Charge	Job		3075	4125	7200	
Concrete Building							
Unreinforced							
	Demolish	S.F.		5.60	1.76	7.36	Includes demolition of a one story non-reinforced concrete building, including hauling of debris. Dump fees not included.
	Minimum Charge	Job		3075	4125	7200	
Reinforced							
	Demolish	S.F.		6.40	2.01	8.41	Includes demolition of a one story reinforced concrete building, including hauling of debris. Dump fees not included.
	Minimum Charge	Job		3075	4125	7200	
Steel Building							
	Demolish	S.F.		3.69	4.96	8.65	Includes demolition of a metal building, including hauling of debris. Dump fees not included. No salvage value assumed for scrap metal.
	Minimum Charge	Job		3075	4125	7200	

Job Preparation

General Clean-up

General Clean-up		Unit	Material	Labor	Equip.	Total	Specification
Remove Debris							Includes removal of debris caused by the loss including contents items and loose debris.
	Clean	S.F.		1.29	.46	1.75	
	Minimum Charge	Job		645	232	877	
Muck-out							Includes labor to remove 2" of muck and mud after a flood loss.
	Clean	S.F.		.78		.78	
	Minimum Charge	Job		200		200	
Deodorize / Disinfect							Includes deodorizing building with application of mildicide and disinfectant to floors and walls - will vary depending on the extent of deodorization required.
	Clean	SF Flr.	.01	.27		.28	
	Minimum Charge	Job		200		200	
Clean Walls							Includes labor and material for heavy cleaning (multiple applications) with detergent and solvent.
Heavy							
	Clean	S.F.		.17		.17	
	Minimum Charge	Job		200		200	
Clean Ceiling							Includes labor and material to clean light to moderate smoke and soot.
Light							
	Clean	S.F.		.29		.29	
	Minimum Charge	Job		200		200	
Heavy							Includes labor and material to clean heavy smoke and soot.
	Clean	S.F.		.43		.43	
	Minimum Charge	Job		200		200	

Flood Clean-up

Flood Clean-up		Unit	Material	Labor	Equip.	Total	Specification
Emergency Service Call							
After Hours / Weekend							Includes minimum charges for deflooding clean-up after hours or on weekends.
	Minimum Charge	Job		266		266	
Grey Water / Sewage							Includes minimum charges for deflooding clean-up.
	Minimum Charge	Job		200		200	
Water Extraction							
Clean (non-grey)							Includes labor to remove clean uncontaminated water from the loss.
	Clean	S.F.	.08	.80		.88	
	Minimum Charge	Job		200		200	
Grey (non-solids)							Includes labor to remove non-solid grey water from the loss.
	Clean	S.F.	.15	1.60		1.75	
	Minimum Charge	Job		200		200	
Sewage / Solids							Includes labor to remove water from the loss that may contain sewage or solids.
	Clean	SF Flr.	.25	2		2.25	
	Minimum Charge	Job		200		200	

For customer support on your Contractor's Pricing Guide: Residential Repair & Remodeling, call 800.448.8182.

Job Preparation

Flood Clean-up		Unit	Material	Labor	Equip.	Total	Specification
Mildicide Walls							
Topographical							
	Clean	SF Wall		.36		.36	Clean-up of deflooding including
	Minimum Charge	Job		200		200	topographical mildicide application.
Injection Treatments							
	Clean	SF Flr.	65	6.15		71.15	Clean-up of deflooding including
	Minimum Charge	Job		200		200	mildicide injection.
Carpet Cleaning							
Uncontaminated							
	Clean	SF Flr.		.25		.25	Includes labor to clean carpet.
	Minimum Charge	Job		200		200	
Contaminated							
	Clean	SF Flr.		.40		.40	Includes labor to clean contaminated
	Minimum Charge	Job		200		200	carpet.
Stairway							
	Clean	Ea.	.13	4.70		4.83	Includes labor to clean contaminated
	Minimum Charge	Job		200		200	carpet on a stairway.
Carpet Treatment							
Disinfect / Deodorize							
	Clean	SF Flr.		.25		.25	Includes disinfecting and deodorizing
	Minimum Charge	Job		200		200	of carpet.
Disinfect For Sewage							
	Clean	SF Flr.		.64		.64	Includes labor to disinfect and
	Minimum Charge	Job		200		200	deodorize carpet exposed to sewage.
Mildicide							
	Clean	SF Flr.	.03	.25		.28	Includes labor and materials to apply
	Minimum Charge	Job		200		200	mildicide to carpet.
Thermal Fog Area							
	Clean	SF Flr.	.02	.61		.63	Includes labor and materials to
	Minimum Charge	Job		200		200	perform thermal fogging during
							deflooding clean-up.
Wet Fog / ULV Area							
	Clean	SF Flr.	.03	.35		.38	Includes labor and materials to
	Minimum Charge	Job		200		200	perform wet fogging / ULV during
							deflooding clean-up.
Steam Clean Fixtures							
	Clean	Ea.	.01	40		40.01	Includes labor and materials to steam
	Minimum Charge	Job		200		200	clean fixtures during deflooding
							clean-up.
Pressure Wash Sewage							
	Clean	SF Flr.		3.63		3.63	Includes labor to pressure wash
	Minimum Charge	Job		200		200	sewage contamination.
Air Mover							
	Minimum Charge	Day			106	106	Daily rental of air mover.
Dehumidifier							
Medium Size							
	Minimum Charge	Day			54	54	Daily rental of dehumidifier, medium.
Large Size							
	Minimum Charge	Day			310	310	Daily rental of dehumidifier, large.

For customer support on your Contractor's Pricing Guide: Residential Repair & Remodeling, call 800.448.8182.

Job Preparation

Flood Clean-up

Flood Clean-up	Unit	Material	Labor	Equip.	Total	Specification
Air Purifier						
HEPA Filter-Bacteria						
Minimum Charge	Ea.	455			455	HEPA filter for air purifier.

Asbestos Removal

Asbestos Removal	Unit	Material	Labor	Equip.	Total	Specification
Asbestos Analysis						
Minimum Charge	Job		310		310	Minimum charge for testing.
Asbestos Pipe Insulation						
1/2″ to 3/4″ Diameter						
Demolish	L.F.	.47	8		8.47	Minimum charge to remove asbestos. Includes full Tyvek suits for workers changed four times per eight-hour shift with respirators, air monitoring and supervision by qualified professionals.
Minimum Charge	Job		4950		4950	
1″ to 3″ Diameter						
Demolish	L.F.	1.46	24.50		25.96	Minimum charge to remove asbestos. Includes full Tyvek suits for workers changed four times per eight-hour shift with respirators, air monitoring and supervision by qualified professionals.
Minimum Charge	Job		4950		4950	
Asbestos-based Siding						
Demolish	S.F.	3.25	55		58.25	Minimum charge to remove asbestos. Includes full Tyvek suits for workers changed four times per eight-hour shift with respirators, air monitoring and supervision by qualified professionals.
Minimum Charge	Job		4950		4950	
Asbestos-based Plaster						
Demolish	S.F.	2.50	42.50		45	Minimum charge to remove asbestos. Includes full Tyvek suits for workers changed four times per eight-hour shift with respirators, air monitoring and supervision by qualified professionals.
Minimum Charge	Job		4950		4950	
Encapsulate Asbestos Ceiling						
Install	S.F.	.64	.24		.88	Includes encapsulation with penetrating sealant sprayed onto ceiling and walls with airless sprayer.
Minimum Charge	Job		4950		4950	
Scrape Acoustical Ceiling						
Demolish	S.F.	.09	1.39		1.48	Minimum charge to remove asbestos. Includes full Tyvek suits for workers changed four times per eight-hour shift with respirators, air monitoring and supervision by qualified professionals.
Minimum Charge	Job		4950		4950	

For customer support on your Contractor's Pricing Guide: Residential Repair & Remodeling, call 800.448.8182.

Job Preparation

Asbestos Removal

Asbestos Removal		Unit	Material	Labor	Equip.	Total	Specification
Seal Ceiling Asbestos							
	Install	S.F.	2.83	4.65		7.48	Includes attaching plastic cover to ceiling area and sealing where asbestos exists, taping all seams, caulk or foam insulation in cracks and negative air system.
	Minimum Charge	Job		4950		4950	
Seal Wall Asbestos							
	Install	S.F.	2.83	4.13		6.96	Includes attaching plastic cover to wall area and sealing where asbestos exists, taping all seams, caulk or foam insulation in cracks and negative air system.
	Minimum Charge	Job		4950		4950	
Seal Floor Asbestos							
	Install	S.F.	4.52	8.25		12.77	Includes attaching plastic cover to floor area and sealing where asbestos exists, taping all seams, caulk or foam insulation in cracks and negative air system.
	Minimum Charge	Job		4950		4950	
Negative Air Vent System							
	Install	Day			132	132	Daily rental of negative air vent system.
	Minimum Charge	Week			660	660	
HEPA Vacuum Cleaner							
	Install	Day			95	95	Daily rental of HEPA vacuum, 16 gallon.
	Minimum Charge	Week			475	475	
Airless Sprayer							
	Install	Day			99.50	99.50	Daily rental of airless sprayer.
	Minimum Charge	Week			500	500	
Decontamination Unit							
	Install	Day			171	171	Daily rental of decontamination unit.
	Minimum Charge	Week			855	855	
Light Stand							
	Install	Day			30	30	Daily rental of floodlight w/tripod.
	Minimum Charge	Week			149	149	

Room Demolition

Room Demolition		Unit	Material	Labor	Equip.	Total	Specification
Strip Typical Room							
	Demolish	S.F.		4		4	Includes 4 labor hours for general demolition.
	Minimum Charge	Job		222		222	
Strip Typical Bathroom							
	Demolish	S.F.		5.35		5.35	Includes 4 labor hours for general demolition.
	Minimum Charge	Job		222		222	
Strip Typical Kitchen							
	Demolish	S.F.		5.10		5.10	Includes 4 labor hours for general demolition.
	Minimum Charge	Job		222		222	
Strip Typical Laundry							
	Demolish	S.F.		4.57		4.57	Includes 4 labor hours for general demolition.
	Minimum Charge	Job		222		222	

Job Preparation

Selective Demolition

Selective Demolition		Unit	Material	Labor	Equip.	Total	Specification
Concrete Slab							
4″ Unreinforced							
	Demolish	S.F.		.28	.79	1.07	Includes labor and equipment to remove unreinforced concrete slab and haul debris to truck or dumpster.
	Minimum Charge	Job		630	1775	2405	
6″ Unreinforced							
	Demolish	S.F.		.31	.89	1.20	Includes labor and equipment to remove unreinforced concrete slab and haul debris to truck or dumpster.
	Minimum Charge	Job		630	1775	2405	
8″ Unreinforced							
	Demolish	S.F.		.36	1.01	1.37	Includes labor and equipment to remove unreinforced concrete slab and haul debris to truck or dumpster.
	Minimum Charge	Job		630	1775	2405	
4″ Reinforced							
	Demolish	S.F.		.28	.79	1.07	Includes labor and equipment to remove reinforced concrete slab and haul debris to truck or dumpster.
	Minimum Charge	Job		630	1775	2405	
6″ Reinforced							
	Demolish	S.F.		.31	.89	1.20	Includes labor and equipment to remove reinforced concrete slab and haul debris to truck or dumpster.
	Minimum Charge	Job		630	1775	2405	
8″ Reinforced							
	Demolish	S.F.		.36	1.01	1.37	Includes labor and equipment to remove reinforced concrete slab and haul debris to truck or dumpster.
	Minimum Charge	Job		630	1775	2405	
Brick Wall							
4″ Thick							
	Demolish	S.F.		3.68	.75	4.43	Includes labor and equipment to remove brick wall and haul debris to a central location for disposal.
	Minimum Charge	Job		1125	229	1354	
8″ Thick							
	Demolish	S.F.		5.05	1.03	6.08	Includes labor and equipment to remove brick wall and haul debris to a central location for disposal.
	Minimum Charge	Job		1125	229	1354	
12″ Thick							
	Demolish	S.F.		6.75	1.38	8.13	Includes labor and equipment to remove brick wall and haul debris to a central location for disposal.
	Minimum Charge	Job		1125	229	1354	
Concrete Masonry Wall							
4″ Thick							
	Demolish	S.F.		4.09	.83	4.92	Includes minimum labor and equipment to remove 4″ block wall and haul debris to a central location for disposal.
	Minimum Charge	Job		1125	229	1354	
6″ Thick							
	Demolish	S.F.		5.05	1.03	6.08	Includes minimum labor and equipment to remove 6″ block wall and haul debris to a central location for disposal.
	Minimum Charge	Job		1125	229	1354	

For customer support on your Contractor's Pricing Guide: Residential Repair & Remodeling, call 800.448.8182.

Job Preparation

Selective Demolition		Unit	Material	Labor	Equip.	Total	Specification
8" Thick							
	Demolish	S.F.		6.25	1.27	7.52	Includes minimum labor and
	Minimum Charge	Job		1125	229	1354	equipment to remove 8" block wall and haul debris to a central location for disposal.
12" Thick							
	Demolish	S.F.		9	1.83	10.83	Includes minimum labor and
	Minimum Charge	Job		1125	229	1354	equipment to remove 12" block wall and haul debris to a central location for disposal.
Tree Removal							
Small, 4" - 8"							
	Demolish	Ea.		143	123	266	Minimum charge, common labor
	Minimum Charge	Job		200		200	trade.
Medium, 9" - 12"							
	Demolish	Ea.		215	185	400	Minimum charge, common labor
	Minimum Charge	Job		200		200	trade.
Large, 13" - 18"							
	Demolish	Ea.		258	222	480	Minimum charge, tree removal.
	Minimum Charge	Job		665	209	874	
Very Large, 19" - 24"							
	Demolish	Ea.		320	277	597	Minimum charge, tree removal.
	Minimum Charge	Job		665	209	874	
Stump Removal							
	Demolish	Ea.		96.50	221	317.50	Includes minimum charges for a site
	Minimum Charge	Job		370	90	460	demolition crew and backhoe / loader.

Excavation

Backhoe/Loader - Wheel Type   Tractor Loader - Wheel Type   Backhoe - Crawler Type

Excavation

		Unit	Material	Labor	Equip.	Total	Specification
Footings							
Hand Trenching	Install	L.F.		2.50		2.50	Includes hand excavation for footing or trench up to 10" deep and 12" wide.
	Minimum Charge	Job		100		100	
Machine Trenching	Install	L.F.		.95	.32	1.27	Includes machine excavation for footing or trench up to 10" deep and 12" wide.
	Minimum Charge	Job		475	162	637	
Trenching w / Backfill							
Hand	Install	L.F.		4		4	Includes hand excavation, backfill and compaction.
	Minimum Charge	Job		100		100	
Machine	Install	L.F.		4.77	12.45	17.22	Includes machine excavation, backfill and compaction.
	Minimum Charge	Job		200	117	317	

Backfill

		Unit	Material	Labor	Equip.	Total	Specification
Hand (no compaction)	Install	L.C.Y.		33.50		33.50	Includes backfill by hand in average soil without compaction.
	Minimum Charge	Job		200		200	
Hand (w / compaction)	Install	B.C.Y.		47		47	Includes backfill by hand in average soil with compaction.
	Minimum Charge	Job		200		200	
Machine (no compaction)	Install	L.C.Y.		4.77	1.62	6.39	Includes backfill, by 55 HP wheel loader, of trenches from loose material piled adjacent without compaction.
	Minimum Charge	Job		475	162	637	
Machine (w / compaction)	Install	L.C.Y.		12.70	4.32	17.02	Includes backfill, by 55 HP wheel loader, of trenches from loose material piled adjacent with compaction by vibrating tampers.
	Minimum Charge	Job		475	162	637	

Foundation

Concrete Footing and Formwork

Wood Pile

Piles	Unit	Material	Labor	Equip.	Total	Specification
Treated Wood, 12" Butt						
To 30'						
Demolish	L.F.		6.40	1.57	7.97	Includes material, labor and
Install	V.L.F.	20	6	4.32	30.32	equipment to install up to 30' long
Demolish and Install	V.L.F.	20	12.40	5.89	38.29	treated wood piling.
Paint	L.F.	.57	2.47		3.04	
Minimum Charge	Job		3750	2700	6450	
30' to 40'						
Demolish	L.F.		6.40	1.57	7.97	Includes material, labor and
Install	V.L.F.	27.50	5.35	3.86	36.71	equipment to install 30' to 40' long
Demolish and Install	V.L.F.	27.50	11.75	5.43	44.68	treated wood piling.
Paint	L.F.	.57	2.47		3.04	
Minimum Charge	Job		3750	2700	6450	
40' to 50'						
Demolish	L.F.		6.40	1.57	7.97	Includes material, labor and
Install	V.L.F.	29.50	5.20	3.75	38.45	equipment to install 40' to 50' long
Demolish and Install	V.L.F.	29.50	11.60	5.32	46.42	treated wood piling.
Paint	L.F.	.57	2.47		3.04	
Minimum Charge	Job		3750	2700	6450	
50' to 60'						
Demolish	L.F.		6.40	1.57	7.97	Includes material, labor and
Install	V.L.F.	31	4.70	3.38	39.08	equipment to install 50' to 60' long
Demolish and Install	V.L.F.	31	11.10	4.95	47.05	treated wood piling.
Paint	L.F.	.57	2.47		3.04	
Minimum Charge	Job		3750	2700	6450	
60' to 80'						
Demolish	L.F.		6.40	1.57	7.97	Includes material, labor and
Install	V.L.F.	37.50	4.48	4.60	46.58	equipment to install 60' to 80' long
Demolish and Install	V.L.F.	37.50	10.88	6.17	54.55	treated wood piling.
Paint	L.F.	.57	2.47		3.04	
Minimum Charge	Job		3750	2700	6450	
Untreated Wood, 12" Butt						
To 30'						
Demolish	L.F.		6.40	1.57	7.97	Includes material, labor and
Install	V.L.F.	17	6	4.32	27.32	equipment to install up to 30' long
Demolish and Install	V.L.F.	17	12.40	5.89	35.29	untreated wood piling.
Paint	L.F.	.57	2.47		3.04	
Minimum Charge	Job		3750	2700	6450	

For customer support on your Contractor's Pricing Guide: Residential Repair & Remodeling, call 800.448.8182.

Foundation

Piles

	Unit	Material	Labor	Equip.	Total	Specification
30' to 40'						
Demolish	L.F.		6.40	1.57	7.97	Includes material, labor and
Install	V.L.F.	17.95	5.35	3.86	27.16	equipment to install 30' to 40' long
Demolish and Install	V.L.F.	17.95	11.75	5.43	35.13	untreated wood piling.
Paint	L.F.	.57	2.47		3.04	
Minimum Charge	Job		3750	2700	6450	
40' to 50'						
Demolish	L.F.		6.40	1.57	7.97	Includes material, labor and
Install	V.L.F.	18.85	5.20	3.75	27.80	equipment to install 40' to 50' long
Demolish and Install	V.L.F.	18.85	11.60	5.32	35.77	untreated wood piling.
Paint	L.F.	.57	2.47		3.04	
Minimum Charge	Job		3750	2700	6450	
50' to 60'						
Demolish	L.F.		6.40	1.57	7.97	Includes material, labor and
Install	V.L.F.	26.50	4.70	3.38	34.58	equipment to install 50' to 60' long
Demolish and Install	V.L.F.	26.50	11.10	4.95	42.55	untreated wood piling.
Paint	L.F.	.57	2.47		3.04	
Minimum Charge	Job		3750	2700	6450	
60' to 80'						
Demolish	L.F.		6.40	1.57	7.97	Includes material, labor and
Install	V.L.F.	31	4.48	3.22	38.70	equipment to install 60' to 80' long
Demolish and Install	V.L.F.	31	10.88	4.79	46.67	untreated wood piling.
Paint	L.F.	.57	2.47		3.04	
Minimum Charge	Job		3750	2700	6450	

Precast Concrete

	Unit	Material	Labor	Equip.	Total	Specification
10" Square						
Demolish	L.F.		6.55	1.60	8.15	Includes material, labor and
Install	V.L.F.	16.40	5.35	3.86	25.61	equipment to install a precast,
Demolish and Install	V.L.F.	16.40	11.90	5.46	33.76	prestressed, 40' long, 10" square
Paint	L.F.	.57	2.47		3.04	concrete pile.
Minimum Charge	Job		3750	2700	6450	
12" Square						
Demolish	L.F.		7	1.71	8.71	Includes material, labor and
Install	V.L.F.	29	5.55	3.97	38.52	equipment to install a precast,
Demolish and Install	V.L.F.	29	12.55	5.68	47.23	prestressed, 40' long, 12" square
Paint	L.F.	.57	2.47		3.04	concrete pile.
Minimum Charge	Job		3750	2700	6450	
14" Square						
Demolish	L.F.		7.20	1.76	8.96	Includes material, labor and
Install	V.L.F.	33	6.25	4.50	43.75	equipment to install a precast,
Demolish and Install	V.L.F.	33	13.45	6.26	52.71	prestressed, 40' long, 14" square
Paint	L.F.	.57	2.47		3.04	concrete pile.
Minimum Charge	Job		3750	2700	6450	
16" Square						
Demolish	L.F.		7.55	1.85	9.40	Includes material, labor and
Install	V.L.F.	34.50	6.70	4.83	46.03	equipment to install a precast,
Demolish and Install	V.L.F.	34.50	14.25	6.68	55.43	prestressed, 40' long, 16" square
Paint	L.F.	.57	2.47		3.04	concrete pile.
Minimum Charge	Job		3750	2700	6450	
18" Square						
Demolish	L.F.		8.20	2	10.20	Includes material, labor and
Install	V.L.F.	60	7.25	7.45	74.70	equipment to install a precast,
Demolish and Install	V.L.F.	60	15.45	9.45	84.90	prestressed, 40' long, 18" square
Paint	L.F.	.57	2.47		3.04	concrete pile.
Minimum Charge	Job		3750	2700	6450	

Foundation

Piles

	Unit	Material	Labor	Equip.	Total	Specification
Cross Bracing						
Treated 2″ x 6″						
Demolish	L.F.		1.31		1.31	Cost includes material and labor to
Install	L.F.	1.32	7.50		8.82	install 2″ x 6″ lumber for cross-bracing
Demolish and Install	L.F.	1.32	8.81		10.13	of foundation piling or posts.
Reinstall	L.F.		5.99		5.99	
Clean	S.F.		.50		.50	
Paint	L.F.	.46	1.66		2.12	
Minimum Charge	Job		930		930	
Treated 2″ x 8″						
Demolish	L.F.		1.31		1.31	Cost includes material and labor to
Install	L.F.	1.56	7.75		9.31	install 2″ x 8″ lumber for cross-bracing
Demolish and Install	L.F.	1.56	9.06		10.62	of foundation piling or posts.
Reinstall	L.F.		6.18		6.18	
Clean	S.F.		.50		.50	
Paint	L.F.	.62	1.73		2.35	
Minimum Charge	Job		930		930	
Treated 2″ x 10″						
Demolish	L.F.		1.35		1.35	Cost includes material and labor to
Install	L.F.	2.19	8		10.19	install 2″ x 10″ lumber for
Demolish and Install	L.F.	2.19	9.35		11.54	cross-bracing of foundation piling or
Reinstall	L.F.		6.40		6.40	posts.
Clean	S.F.		.50		.50	
Paint	L.F.	.75	1.76		2.51	
Minimum Charge	Job		930		930	
Treated 2″ x 12″						
Demolish	L.F.		1.61		1.61	Cost includes material and labor to
Install	L.F.	3.19	8.30		11.49	install 2″ x 12″ lumber for
Demolish and Install	L.F.	3.19	9.91		13.10	cross-bracing of foundation piling or
Reinstall	L.F.		6.63		6.63	posts.
Clean	S.F.		.50		.50	
Paint	L.F.	.88	1.80		2.68	
Minimum Charge	Job		930		930	
Metal Pipe						
6″ Concrete Filled						
Demolish	L.F.		9.10	2.22	11.32	Includes material, labor and
Install	V.L.F.	25	7.85	5.65	38.50	equipment to install concrete filled
Demolish and Install	V.L.F.	25	16.95	7.87	49.82	steel pipe pile.
Paint	L.F.	.28	2.47		2.75	
Minimum Charge	Job		3750	2700	6450	
6″ Unfilled						
Demolish	L.F.		9.10	2.22	11.32	Includes material, labor and
Install	V.L.F.	21	7.15	5.15	33.30	equipment to install unfilled steel pipe
Demolish and Install	V.L.F.	21	16.25	7.37	44.62	pile.
Paint	L.F.	.28	2.47		2.75	
Minimum Charge	Job		3750	2700	6450	
8″ Concrete Filled						
Demolish	L.F.		9.10	2.22	11.32	Includes material, labor and
Install	V.L.F.	27.50	8.15	5.90	41.55	equipment to install concrete filled
Demolish and Install	V.L.F.	27.50	17.25	8.12	52.87	steel pipe pile.
Paint	L.F.	.28	2.47		2.75	
Minimum Charge	Job		3750	2700	6450	
8″ Unfilled						
Demolish	L.F.		9.10	2.22	11.32	Includes material, labor and
Install	V.L.F.	22	7.50	5.40	34.90	equipment to install unfilled steel pipe
Demolish and Install	V.L.F.	22	16.60	7.62	46.22	pile.
Paint	L.F.	.28	2.47		2.75	
Minimum Charge	Job		3750	2700	6450	

For customer support on your Contractor's Pricing Guide: Residential Repair & Remodeling, call 800.448.8182.

Foundation

Piles

	Unit	Material	Labor	Equip.	Total	Specification
10″ Concrete Filled						
Demolish	L.F.		9.40	2.29	11.69	Includes material, labor and
Install	V.L.F.	60.50	8.35	6	74.85	equipment to install concrete filled
Demolish and Install	V.L.F.	60.50	17.75	8.29	86.54	steel pipe pile.
Paint	L.F.	.28	2.47		2.75	
Minimum Charge	Job		3750	2700	6450	
10″ Unfilled						
Demolish	L.F.		9.40	2.29	11.69	Includes material, labor and
Install	V.L.F.	49.50	7.50	5.40	62.40	equipment to install unfilled steel pipe
Demolish and Install	V.L.F.	49.50	16.90	7.69	74.09	pile.
Paint	L.F.	.28	2.47		2.75	
Minimum Charge	Job		3750	2700	6450	
12″ Concrete Filled						
Demolish	L.F.		9.40	2.29	11.69	Includes material, labor and
Install	V.L.F.	41.50	9.05	6.50	57.05	equipment to install concrete filled
Demolish and Install	V.L.F.	41.50	18.45	8.79	68.74	steel pipe pile.
Paint	L.F.	.28	2.47		2.75	
Minimum Charge	Job		3750	2700	6450	
12″ Unfilled						
Demolish	L.F.		9.40	2.29	11.69	Includes material, labor and
Install	V.L.F.	41.50	7.90	5.70	55.10	equipment to install unfilled steel pipe
Demolish and Install	V.L.F.	41.50	17.30	7.99	66.79	pile.
Paint	L.F.	.28	2.47		2.75	
Minimum Charge	Job		3750	2700	6450	

Steel H Section

	Unit	Material	Labor	Equip.	Total	Specification
H Section 8 x 8 x 36#						
Demolish	L.F.		6.35	1.55	7.90	Includes material, labor and
Install	V.L.F.	20	5.90	4.22	30.12	equipment to install steel H section
Demolish and Install	V.L.F.	20	12.25	5.77	38.02	pile.
Paint	S.F.	.30	.35		.65	
Minimum Charge	Job		3750	2700	6450	
H Section 10 x 10 x 57#						
Demolish	L.F.		6.95	1.70	8.65	Includes material, labor and
Install	V.L.F.	31	6.15	4.43	41.58	equipment to install steel H section
Demolish and Install	V.L.F.	31	13.10	6.13	50.23	pile.
Paint	S.F.	.30	.35		.65	
Minimum Charge	Job		3750	2700	6450	
H Section 12 x 12 x 74#						
Demolish	L.F.		7.20	1.76	8.96	Includes material, labor and
Install	V.L.F.	39	6.35	6.55	51.90	equipment to install steel H section
Demolish and Install	V.L.F.	39	13.55	8.31	60.86	pile.
Paint	S.F.	.30	.35		.65	
Minimum Charge	Job		3750	2700	6450	
H Section 14 x 14 x 89#						
Demolish	L.F.		7.55	1.85	9.40	Includes material, labor and
Install	V.L.F.	53	6.95	7.15	67.10	equipment to install steel H section
Demolish and Install	V.L.F.	53	14.50	9	76.50	pile.
Paint	S.F.	.30	.35		.65	
Minimum Charge	Job		3750	2700	6450	

For customer support on your Contractor's Pricing Guide: Residential Repair & Remodeling, call 800.448.8182.

Foundation

Foundation Post		Unit	Material	Labor	Equip.	Total	Specification
Wood Foundation Post							
4" x 4"							
	Demolish	L.F.		1.03	.25	1.28	Cost includes material and labor to
	Install	L.F.	1.72	6.65		8.37	install 4" x 4" treated wood post by
	Demolish and Install	L.F.	1.72	7.68	.25	9.65	hand in normal soil conditions.
	Reinstall	L.F.		5.33		5.33	
	Minimum Charge	Job		930		930	
4" x 6"							
	Demolish	L.F.		1.03	.25	1.28	Cost includes material and labor to
	Install	L.F.	2.59	6.65		9.24	install 4" x 6" treated wood post by
	Demolish and Install	L.F.	2.59	7.68	.25	10.52	hand in normal soil conditions.
	Reinstall	L.F.		5.33		5.33	
	Minimum Charge	Job		930		930	
6" x 6"							
	Demolish	L.F.		1.03	.25	1.28	Cost includes material and labor to
	Install	L.F.	4.08	6.65		10.73	install 6" x 6" treated wood post by
	Demolish and Install	L.F.	4.08	7.68	.25	12.01	hand in normal soil conditions.
	Reinstall	L.F.		5.33		5.33	
	Minimum Charge	Job		930		930	
Water Jet							
	Demolish	Ea.		118		118	Includes material, labor and
	Install	Ea.	85.50	915		1000.50	equipment to water jet a 6" x 6" wood
	Demolish and Install	Ea.	85.50	1033		1118.50	post under an existing building. Cost
	Reinstall	Ea.		913.37		913.37	may vary depending upon location
	Minimum Charge	Job		800		800	and access.
Shim							
	Install	Ea.	1.62	93		94.62	Includes labor and material to trim or
	Minimum Charge	Job		930		930	readjust existing foundation posts
							under an existing building. Access of
							2 - 4 foot assumed.
Cross-bracing							
Treated 2" x 6"							
	Demolish	L.F.		1.31		1.31	Cost includes material and labor to
	Install	L.F.	1.32	7.50		8.82	install 2" x 6" lumber for cross-bracing
	Demolish and Install	L.F.	1.32	8.81		10.13	of foundation piling or posts.
	Reinstall	L.F.		5.99		5.99	
	Clean	S.F.		.50		.50	
	Paint	L.F.	.46	1.66		2.12	
	Minimum Charge	Job		930		930	
Treated 2" x 8"							
	Demolish	L.F.		1.31		1.31	Cost includes material and labor to
	Install	L.F.	1.56	7.75		9.31	install 2" x 8" lumber for cross-bracing
	Demolish and Install	L.F.	1.56	9.06		10.62	of foundation piling or posts.
	Reinstall	L.F.		6.18		6.18	
	Clean	S.F.		.50		.50	
	Paint	L.F.	.62	1.73		2.35	
	Minimum Charge	Job		930		930	
Treated 2" x 10"							
	Demolish	L.F.		1.35		1.35	Cost includes material and labor to
	Install	L.F.	2.19	8		10.19	install 2" x 10" lumber for
	Demolish and Install	L.F.	2.19	9.35		11.54	cross-bracing of foundation piling or
	Reinstall	L.F.		6.40		6.40	posts.
	Clean	S.F.		.50		.50	
	Paint	L.F.	.75	1.76		2.51	
	Minimum Charge	Job		930		930	

For customer support on your Contractor's Pricing Guide: Residential Repair & Remodeling, call 800.448.8182.

Foundation

Foundation Post

Foundation Post		Unit	Material	Labor	Equip.	Total	Specification
Treated 2″ x 12″							
	Demolish	L.F.		1.61		1.61	Cost includes material and labor to
	Install	L.F.	3.19	8.30		11.49	install 2″ x 12″ lumber for
	Demolish and Install	L.F.	3.19	9.91		13.10	cross-bracing of foundation piling or
	Reinstall	L.F.		6.63		6.63	posts.
	Clean	S.F.		.50		.50	
	Paint	L.F.	.88	1.80		2.68	
	Minimum Charge	Job		930		930	

Concrete Footing

Concrete Footing		Unit	Material	Labor	Equip.	Total	Specification
Continuous							
12″ w x 6″ d							
	Demolish	L.F.		.70	1.97	2.67	Includes material and labor to install
	Install	L.F.	8.30	4.01	.05	12.36	continuous reinforced concrete footing
	Demolish and Install	L.F.	8.30	4.71	2.02	15.03	poured by chute cast against earth
	Minimum Charge	Job		1050		1050	including reinforcing, forming and
							finishing.
12″ w x 12″ d							
	Demolish	L.F.		.84	2.36	3.20	Includes material and labor to install
	Install	L.F.	19.20	8.10	.09	27.39	continuous reinforced concrete footing
	Demolish and Install	L.F.	19.20	8.94	2.45	30.59	poured by chute cast against earth
	Minimum Charge	Job		1050		1050	including reinforcing, forming and
							finishing.
18″ w x 10″ d							
	Demolish	L.F.		.84	2.36	3.20	Includes material and labor to install
	Install	L.F.	17.75	10.30	.12	28.17	continuous reinforced concrete footing
	Demolish and Install	L.F.	17.75	11.14	2.48	31.37	poured by chute cast against earth
	Minimum Charge	Job		1050		1050	including reinforcing, forming and
							finishing.
24″ w x 24″ d							
	Demolish	L.F.		1.26	3.55	4.81	Includes material and labor to install
	Install	L.F.	51.50	31	.36	82.86	continuous reinforced concrete footing
	Demolish and Install	L.F.	51.50	32.26	3.91	87.67	poured by chute cast against earth
	Minimum Charge	Job		1050		1050	including reinforcing, forming and
							finishing.
Stem Wall							
Single Story							
	Demolish	L.F.		17.95	15.80	33.75	Includes material and labor to install
	Install	L.F.	43.50	134	1.57	179.07	stem wall 24″ above and 18″ below
	Demolish and Install	L.F.	43.50	151.95	17.37	212.82	grade for a single story structure.
	Minimum Charge	Job		1050		1050	Includes 12″ wide and 8″ deep
							footing.
Two Story							
	Demolish	L.F.		17.95	15.80	33.75	Includes material and labor to install
	Install	L.F.	52.50	193	2.25	247.75	stem wall 24″ above and 18″ below
	Demolish and Install	L.F.	52.50	210.95	18.05	281.50	grade typical for a two story structure,
	Minimum Charge	Job		1050		1050	includes 18″ wide and 10″ deep
							footing.

Foundation

Concrete Slab		Unit	Material	Labor	Equip.	Total	Specification
Reinforced							
4"							
	Demolish	S.F.		.28	.79	1.07	Cost includes material and labor to
	Install	S.F.	4.33	.52		4.85	install 4" reinforced concrete
	Demolish and Install	S.F.	4.33	.80	.79	5.92	slab-on-grade poured by chute
	Clean	S.F.	.12	.44		.56	including forms, vapor barrier, wire
	Paint	S.F.	.25	.98		1.23	mesh, 3000 PSI concrete, float finish,
	Minimum Charge	Job		256		256	and curing.
6"							
	Demolish	S.F.		.31	.89	1.20	Cost includes material and labor to
	Install	S.F.	5.65	.78	.01	6.44	install 6" reinforced concrete
	Demolish and Install	S.F.	5.65	1.09	.90	7.64	slab-on-grade poured by chute
	Clean	S.F.	.12	.44		.56	including forms, vapor barrier, wire
	Paint	S.F.	.25	.98		1.23	mesh, 3000 PSI concrete, float finish,
	Minimum Charge	Job		256		256	and curing.
8"							
	Demolish	S.F.		.36	1.01	1.37	Cost includes material and labor to
	Install	S.F.	7.90	1.02	.01	8.93	install 8" reinforced concrete
	Demolish and Install	S.F.	7.90	1.38	1.02	10.30	slab-on-grade poured by chute
	Clean	S.F.	.12	.44		.56	including forms, vapor barrier, wire
	Paint	S.F.	.25	.98		1.23	mesh, 3000 PSI concrete, float finish,
	Minimum Charge	Job		256		256	and curing.
Unreinforced							
4"							
	Demolish	S.F.		.28	.79	1.07	Cost includes material and labor to
	Install	S.F.	2.70	1.26	.01	3.97	install 4" concrete slab-on-grade
	Demolish and Install	S.F.	2.70	1.54	.80	5.04	poured by chute, including forms,
	Clean	S.F.	.12	.44		.56	vapor barrier, 3000 PSI concrete, float
	Paint	S.F.	.25	.98		1.23	finish, and curing.
	Minimum Charge	Job		256		256	
6"							
	Demolish	S.F.		.31	.89	1.20	Cost includes material and labor to
	Install	S.F.	3.94	1.28	.01	5.23	install 6" concrete slab-on-grade
	Demolish and Install	S.F.	3.94	1.59	.90	6.43	poured by chute, including forms,
	Clean	S.F.	.12	.44		.56	vapor barrier, 3000 PSI concrete, float
	Paint	S.F.	.25	.98		1.23	finish, and curing.
	Minimum Charge	Job		256		256	
8"							
	Demolish	S.F.		.36	1.01	1.37	Cost includes material and labor to
	Install	S.F.	5.40	1.35	.01	6.76	install 8" concrete slab-on-grade
	Demolish and Install	S.F.	5.40	1.71	1.02	8.13	poured by chute, including forms,
	Clean	S.F.	.12	.44		.56	vapor barrier, 3000 PSI concrete, float
	Paint	S.F.	.25	.98		1.23	finish, and curing.
	Minimum Charge	Job		256		256	
Lightweight							
	Demolish	S.F.		1.18	.42	1.60	Includes material and labor to install
	Install	S.F.	2.78	3.80		6.58	lightweight concrete, 3" - 4" thick.
	Demolish and Install	S.F.	2.78	4.98	.42	8.18	
	Clean	S.F.	.12	.44		.56	
	Paint	S.F.	.25	.98		1.23	
	Minimum Charge	Job		256		256	

For customer support on your Contractor's Pricing Guide: Residential Repair & Remodeling, call 800.448.8182.

Foundation

Concrete Slab

	Unit	Material	Labor	Equip.	Total	Specification
Gunite						Includes material and labor to install
Demolish	S.F.		1.48	.36	1.84	gunite on flat plane per inch of
Install	S.F.	.61	1.60	.25	2.46	thickness using 1″ as base price. No
Demolish and Install	S.F.	.61	3.08	.61	4.30	forms or reinforcing are included.
Clean	S.F.	.12	.44		.56	
Paint	S.F.	.25	.98		1.23	
Minimum Charge	Job		1425		1425	
Epoxy Inject						
Install	L.F.	3.74	8.55		12.29	Includes labor and material to inject
Minimum Charge	Job		256		256	epoxy into cracks.
Slabjacking						
Install	S.F.	.86	1.49	.16	2.51	Includes labor, material and
Minimum Charge	Job		256		256	equipment to install slabjacking under
						an existing building.
Scrape and Paint						
Paint	S.F.	.62	1.08		1.70	Includes labor and materials to scrape
Minimum Charge	Job		216		216	and paint a concrete floor.
Vapor Barrier						
Install	S.F.	.08	.14		.22	Cost includes material and labor to
Minimum Charge	Job		260		260	install 4 mil polyethylene vapor barrier
						10′ wide sheets with 6″ overlaps.

For customer support on your Contractor's Pricing Guide: Residential Repair & Remodeling, call 800.448.8182.

Rough Frame / Structure

Rough Framing

Beam / Girder

	Unit	Material	Labor	Equip.	Total	Specification
Solid Wood Beam						
2" x 8"						
Demolish	L.F.		1.08		1.08	Cost includes material and labor to install 2" x 8" beam.
Install	L.F.	1.46	1.60		3.06	
Demolish and Install	L.F.	1.46	2.68		4.14	
Clean	L.F.	.11	.36		.47	
Paint	L.F.	.36	2.33		2.69	
Minimum Charge	Job		260		260	
2" x 10"						
Demolish	L.F.		1.34		1.34	Cost includes material and labor to install 2" x 10" beam.
Install	L.F.	3.88	1.74		5.62	
Demolish and Install	L.F.	3.88	3.08		6.96	
Clean	L.F.	.13	.43		.56	
Paint	L.F.	.25	1.44		1.69	
Minimum Charge	Job		260		260	
2" x 12"						
Demolish	L.F.		1.61		1.61	Cost includes material and labor to install 2" x 12" beam.
Install	L.F.	4.09	1.89		5.98	
Demolish and Install	L.F.	4.09	3.50		7.59	
Clean	L.F.	.14	.50		.64	
Paint	L.F.	.55	3.45		4	
Minimum Charge	Job		260		260	
4" x 8"						
Demolish	L.F.		3.29		3.29	Cost includes material and labor to install 4" x 8" beam.
Install	L.F.	12.25	2.42	2.55	17.22	
Demolish and Install	L.F.	12.25	5.71	2.55	20.51	
Clean	L.F.	.13	.43		.56	
Paint	L.F.	.25	1.44		1.69	
Minimum Charge	Job		260		260	
4" x 10"						
Demolish	L.F.		4.12		4.12	Cost includes material and labor to install 4" x 10" beam.
Install	L.F.	8.70	2.55	2.69	13.94	
Demolish and Install	L.F.	8.70	6.67	2.69	18.06	
Clean	L.F.	.14	.50		.64	
Paint	L.F.	.29	1.68		1.97	
Minimum Charge	Job		260		260	
4" x 12"						
Demolish	L.F.		4.95		4.95	Cost includes material and labor to install 4" x 12" beam.
Install	L.F.	11.85	2.69	2.84	17.38	
Demolish and Install	L.F.	11.85	7.64	2.84	22.33	
Clean	L.F.	.17	.57		.74	
Paint	L.F.	.55	3.45		4	
Minimum Charge	Job		260		260	

For customer support on your Contractor's Pricing Guide: Residential Repair & Remodeling, call 800.448.8182.

Rough Frame / Structure

Beam / Girder		Unit	Material	Labor	Equip.	Total	Specification
6" x 8"							
	Demolish	L.F.		4.95		4.95	Cost includes material and labor to
	Install	L.F.	12.55	4.61	4.87	22.03	install 6" x 8" beam.
	Demolish and Install	L.F.	12.55	9.56	4.87	26.98	
	Clean	L.F.	.14	.50		.64	
	Paint	L.F.	.29	1.68		1.97	
	Minimum Charge	Job		260		260	
6" x 10"							
	Demolish	L.F.		6.20		6.20	Cost includes material and labor to
	Install	L.F.	11.65	2.08		13.73	install 6" x 10" triple beam.
	Demolish and Install	L.F.	11.65	8.28		19.93	
	Clean	L.F.	.17	.57		.74	
	Paint	L.F.	.34	1.92		2.26	
	Minimum Charge	Job		260		260	
6" x 12"							
	Demolish	L.F.		7.35		7.35	Cost includes material and labor to
	Install	L.F.	12.25	2.19		14.44	install 6" x 12" triple beam.
	Demolish and Install	L.F.	12.25	9.54		21.79	
	Clean	L.F.	.19	.64		.83	
	Paint	L.F.	.37	2.16		2.53	
	Minimum Charge	Job		260		260	
Steel I-type							
W8 x 31							
	Demolish	L.F.		11.50	11.80	23.30	Includes material and labor to install
	Install	L.F.	60	3.27	2.97	66.24	W shaped steel beam / girder.
	Demolish and Install	L.F.	60	14.77	14.77	89.54	
	Minimum Charge	Job		590		590	
W8 x 48							
	Demolish	L.F.		11.50	11.80	23.30	Includes material and labor to install
	Install	L.F.	92.50	3.43	3.11	99.04	W shaped steel beam / girder.
	Demolish and Install	L.F.	92.50	14.93	14.91	122.34	
	Minimum Charge	Job		590		590	
W8 x 67							
	Demolish	L.F.		11.50	11.80	23.30	Includes material and labor to install
	Install	L.F.	130	3.59	3.26	136.85	W shaped steel beam / girder.
	Demolish and Install	L.F.	130	15.09	15.06	160.15	
	Minimum Charge	Job		590		590	
W10 x 45							
	Demolish	L.F.		11.50	11.80	23.30	Includes material and labor to install
	Install	L.F.	87	3.43	3.11	93.54	W shaped steel beam / girder.
	Demolish and Install	L.F.	87	14.93	14.91	116.84	
	Minimum Charge	Job		590		590	
W10 x 68							
	Demolish	L.F.		11.50	11.80	23.30	Includes material and labor to install
	Install	L.F.	131	3.59	3.26	137.85	W shaped steel beam / girder.
	Demolish and Install	L.F.	131	15.09	15.06	161.15	
	Minimum Charge	Job		590		590	
W12 x 50							
	Demolish	L.F.		11.50	11.80	23.30	Includes material and labor to install
	Install	L.F.	96.50	3.43	3.11	103.04	W shaped steel beam / girder.
	Demolish and Install	L.F.	96.50	14.93	14.91	126.34	
	Minimum Charge	Job		590		590	
W12 x 87							
	Demolish	L.F.		16.95	17.30	34.25	Includes material and labor to install
	Install	L.F.	168	3.59	3.26	174.85	W shaped steel beam / girder.
	Demolish and Install	L.F.	168	20.54	20.56	209.10	
	Minimum Charge	Job		590		590	

For customer support on your Contractor's Pricing Guide: Residential Repair & Remodeling, call 800.448.8182.

Rough Frame / Structure

Beam / Girder

Beam / Girder	Unit	Material	Labor	Equip.	Total	Specification
W12 x 120						
Demolish	L.F.		16.95	17.30	34.25	Includes material and labor to install
Install	L.F.	232	3.68	3.34	239.02	W shaped steel beam / girder.
Demolish and Install	L.F.	232	20.63	20.64	273.27	
Minimum Charge	Job		590		590	
W14 x 74						
Demolish	L.F.		16.95	17.30	34.25	Includes material and labor to install
Install	L.F.	143	3.59	3.26	149.85	W shaped steel beam / girder.
Demolish and Install	L.F.	143	20.54	20.56	184.10	
Minimum Charge	Job		590		590	
W14 x 120						
Demolish	L.F.		16.95	17.30	34.25	Includes material and labor to install
Install	L.F.	232	3.68	3.34	239.02	W shaped steel beam / girder.
Demolish and Install	L.F.	232	20.63	20.64	273.27	
Minimum Charge	Job		590		590	

Column

Column	Unit	Material	Labor	Equip.	Total	Specification
Concrete						
Small Diameter						
Demolish	L.F.		18.75	3.82	22.57	Includes material and labor to install
Install	V.L.F.	74.50	40.50	10.25	125.25	small diameter concrete column.
Demolish and Install	V.L.F.	74.50	59.25	14.07	147.82	
Minimum Charge	Job		865	325	1190	
Large Diameter						
Demolish	L.F.		56	11.45	67.45	Includes material and labor to install
Install	V.L.F.	108	47.50	12.05	167.55	large diameter concrete column.
Demolish and Install	V.L.F.	108	103.50	23.50	235	
Minimum Charge	Job		865	325	1190	

Floor Framing System

Floor Framing System	Unit	Material	Labor	Equip.	Total	Specification
12″ O.C.						
2″ x 6″ Joists						
Demolish	S.F.		.92		.92	Cost includes material and labor to
Install	S.F.	1.56	.83		2.39	install 2″ x 6″ joists, 12″ O.C.
Demolish and Install	S.F.	1.56	1.75		3.31	including box or band joist. Blocking
Reinstall	S.F.		.67		.67	or bridging not included.
Minimum Charge	Job		260		260	
2″ x 8″ Joists						
Demolish	S.F.		.95		.95	Cost includes material and labor to
Install	S.F.	1.71	.95		2.66	install 2″ x 8″ joists, 12″ O.C.
Demolish and Install	S.F.	1.71	1.90		3.61	including box or band joist. Blocking
Reinstall	S.F.		.76		.76	or bridging not included.
Minimum Charge	Job		260		260	
2″ x 10″ Joists						
Demolish	S.F.		.97		.97	Cost includes material and labor to
Install	S.F.	4.53	1.16		5.69	install 2″ x 10″ joists, 12″ O.C.
Demolish and Install	S.F.	4.53	2.13		6.66	including box or band joist. Blocking
Reinstall	S.F.		.93		.93	or bridging not included.
Minimum Charge	Job		260		260	

Rough Frame / Structure

Floor Framing System		Unit	Material	Labor	Equip.	Total	Specification
2" x 12" Joists							
	Demolish	S.F.		1.01		1.01	Cost includes material and labor to
	Install	S.F.	4.77	1.19		5.96	install 2" x 12" joists, 12" O.C.
	Demolish and Install	S.F.	4.77	2.20		6.97	including box or band joist. Blocking
	Reinstall	S.F.		.95		.95	or bridging not included.
	Minimum Charge	Job		260		260	
Block / Bridge							
	Demolish	Ea.		1.39		1.39	Includes material and labor to install
	Install	Ea.	5.20	2.93		8.13	set of cross bridging or per block of
	Demolish and Install	Ea.	5.20	4.32		9.52	solid bridging for 2" x 10" joists cut to
	Minimum Charge	Job		260		260	size on site.

16" O.C.

		Unit	Material	Labor	Equip.	Total	Specification
2" x 6" Joists							
	Demolish	S.F.		.69		.69	Cost includes material and labor to
	Install	S.F.	1.20	.62		1.82	install 2" x 6" joists including box or
	Demolish and Install	S.F.	1.20	1.31		2.51	band joist installed 16" O.C. Does not
	Reinstall	S.F.		.50		.50	include beams, blocking or bridging.
	Minimum Charge	Job		260		260	
2" x 8" Joists							
	Demolish	S.F.		.71		.71	Cost includes material and labor to
	Install	S.F.	1.31	.71		2.02	install 2" x 8" joists including box or
	Demolish and Install	S.F.	1.31	1.42		2.73	band joist installed 16" O.C. Does not
	Reinstall	S.F.		.57		.57	include beams, blocking or bridging.
	Minimum Charge	Job		260		260	
2" x 10" Joists							
	Demolish	S.F.		.73		.73	Cost includes material and labor to
	Install	S.F.	3.49	.87		4.36	install 2" x 10" joists including box or
	Demolish and Install	S.F.	3.49	1.60		5.09	band joist 16" O.C. Does not include
	Reinstall	S.F.		.69		.69	beams, blocking or bridging.
	Minimum Charge	Job		260		260	
2" x 12" Joists							
	Demolish	S.F.		.76		.76	Cost includes material and labor to
	Install	S.F.	3.67	.89		4.56	install 2" x 12" joists including box or
	Demolish and Install	S.F.	3.67	1.65		5.32	band joist installed 16" O.C. Does not
	Reinstall	S.F.		.72		.72	include beams, blocking or bridging.
	Minimum Charge	Job		260		260	
Block / Bridge							
	Demolish	Ea.		1.39		1.39	Includes material and labor to install
	Install	Ea.	5.20	2.93		8.13	set of cross bridging or per block of
	Demolish and Install	Ea.	5.20	4.32		9.52	solid bridging for 2" x 10" joists cut to
	Minimum Charge	Job		260		260	size on site.

24" O.C.

		Unit	Material	Labor	Equip.	Total	Specification
2" x 6" Joists							
	Demolish	S.F.		.46		.46	Cost includes material and labor to
	Install	S.F.	.85	.42		1.27	install 2" x 6" joists including box or
	Demolish and Install	S.F.	.85	.88		1.73	band joist installed 24" O.C. Does not
	Reinstall	S.F.		.33		.33	include beams, blocking or bridging.
	Minimum Charge	Job		260		260	
2" x 8" Joists							
	Demolish	S.F.		.47		.47	Cost includes material and labor to
	Install	S.F.	.92	.47		1.39	install 2" x 8" joists including box or
	Demolish and Install	S.F.	.92	.94		1.86	band joist installed 24" O.C. Does not
	Reinstall	S.F.		.38		.38	include beams, blocking or bridging.
	Minimum Charge	Job		260		260	

For customer support on your Contractor's Pricing Guide: Residential Repair & Remodeling, call 800.448.8182.

Rough Frame / Structure

Floor Framing System

Floor Framing System		Unit	Material	Labor	Equip.	Total	Specification
2" x 10" Joists							
	Demolish	S.F.		.49		.49	Cost includes material and labor to
	Install	S.F.	2.48	.58		3.06	install 2" x 10" joists including box or
	Demolish and Install	S.F.	2.48	1.07		3.55	band joist installed 24" O.C. Does not
	Reinstall	S.F.		.46		.46	include beams, blocking or bridging.
	Minimum Charge	Job		260		260	
2" x 12" Joists							
	Demolish	S.F.		.50		.50	Cost includes material and labor to
	Install	S.F.	2.61	.60		3.21	install 2" x 12" joists including box or
	Demolish and Install	S.F.	2.61	1.10		3.71	band joist installed 24" O.C. Does not
	Reinstall	S.F.		.48		.48	include beams, blocking or bridging.
	Minimum Charge	Job		260		260	
Block / Bridge							
	Demolish	Ea.		1.39		1.39	Includes material and labor to install
	Install	Ea.	5.20	2.93		8.13	set of cross bridging or per block of
	Demolish and Install	Ea.	5.20	4.32		9.52	solid bridging for 2" x 10" joists cut to
	Minimum Charge	Job		260		260	size on site.

Engineered Lumber, Joist

		Unit	Material	Labor	Equip.	Total	Specification
9-1/2"							
	Demolish	S.F.		.73		.73	Includes material and labor to install
	Install	S.F.	3.74	.52		4.26	engineered lumber truss / joists per
	Demolish and Install	S.F.	3.74	1.25		4.99	S.F. of floor area based on joists 16"
	Reinstall	S.F.		.42		.42	O.C. Beams, supports and bridging
	Minimum Charge	Job		260		260	are not included.
11-7/8"							
	Demolish	S.F.		.76		.76	Includes material and labor to install
	Install	S.F.	1.65	.53		2.18	engineered lumber truss / joists per
	Demolish and Install	S.F.	1.65	1.29		2.94	S.F. of floor area based on joists 16"
	Reinstall	S.F.		.43		.43	O.C. Beams, supports and bridging
	Minimum Charge	Job		260		260	are not included.
14"							
	Demolish	S.F.		.78		.78	Includes material and labor to install
	Install	S.F.	8.95	.57		9.52	engineered lumber truss / joists per
	Demolish and Install	S.F.	8.95	1.35		10.30	S.F. of floor area based on joists 16"
	Reinstall	S.F.		.46		.46	O.C. Beams, supports and bridging
	Minimum Charge	Job		260		260	are not included.
16"							
	Demolish	S.F.		.81		.81	Includes material and labor to install
	Install	S.F.	9.35	.60		9.95	engineered lumber truss / joists per
	Demolish and Install	S.F.	9.35	1.41		10.76	S.F. of floor area based on joists 16"
	Reinstall	S.F.		.48		.48	O.C. Beams, supports and bridging
	Minimum Charge	Job		260		260	are not included.
Block / Bridge							
	Demolish	Pair		1.39		1.39	Includes material and labor to install
	Install	Pair	1.22	3.86		5.08	set of steel, one-nail type cross
	Demolish and Install	Pair	1.22	5.25		6.47	bridging for trusses placed 16" O.C.
	Minimum Charge	Job		260		260	

Block / Bridge

		Unit	Material	Labor	Equip.	Total	Specification
2" x 6"							
	Demolish	Ea.		1.39		1.39	Includes material and labor to install
	Install	Ea.	1.78	2.35		4.13	set of cross bridging or per block of
	Demolish and Install	Ea.	1.78	3.74		5.52	solid bridging for 2" x 6" joists cut to
	Minimum Charge	Job		260		260	size on site.

For customer support on your Contractor's Pricing Guide: Residential Repair & Remodeling, call 800.448.8182.

Rough Frame / Structure

Floor Framing System		Unit	Material	Labor	Equip.	Total	Specification
2″ x 8″							
	Demolish	Ea.		1.39		1.39	Includes material and labor to install
	Install	Ea.	1.94	2.60		4.54	set of cross bridging or per block of
	Demolish and Install	Ea.	1.94	3.99		5.93	solid bridging for 2″ x 8″ joists cut to
	Minimum Charge	Job		260		260	size on site.
2″ x 10″							
	Demolish	Ea.		1.39		1.39	Includes material and labor to install
	Install	Ea.	5.20	2.93		8.13	set of cross bridging or per block of
	Demolish and Install	Ea.	5.20	4.32		9.52	solid bridging for 2″ x 10″ joists cut to
	Minimum Charge	Job		260		260	size on site.
2″ x 12″							
	Demolish	Ea.		1.39		1.39	Includes material and labor to install
	Install	Ea.	5.45	3.45		8.90	set of cross bridging or per block of
	Demolish and Install	Ea.	5.45	4.84		10.29	solid bridging for 2″ x 12″ joists cut to
	Minimum Charge	Job		260		260	size on site.
Ledger Strips							
1″ x 2″							
	Demolish	L.F.		.37		.37	Cost includes material and labor to
	Install	L.F.	.18	1.51		1.69	install 1″ x 2″ ledger strip nailed to the
	Demolish and Install	L.F.	.18	1.88		2.06	face of studs, beams or joist.
	Minimum Charge	Job		260		260	
1″ x 3″							
	Demolish	L.F.		.37		.37	Includes material and labor to install
	Install	L.F.	.78	1.74		2.52	up to 1″ x 4″ ledger strip nailed to the
	Demolish and Install	L.F.	.78	2.11		2.89	face of studs, beams or joist.
	Minimum Charge	Job		260		260	
1″ x 4″							
	Demolish	L.F.		.37		.37	Includes material and labor to install
	Install	L.F.	.78	1.74		2.52	up to 1″ x 4″ ledger strip nailed to the
	Demolish and Install	L.F.	.78	2.11		2.89	face of studs, beams or joist.
	Minimum Charge	Job		260		260	
2″ x 2″							
	Demolish	L.F.		.40		.40	Cost includes material and labor to
	Install	L.F.	.50	1.58		2.08	install 2″ x 2″ ledger strip nailed to the
	Demolish and Install	L.F.	.50	1.98		2.48	face of studs, beams or joist.
	Minimum Charge	Job		260		260	
2″ x 4″							
	Demolish	L.F.		.44		.44	Cost includes material and labor to
	Install	L.F.	.99	2.08		3.07	install 2″ x 4″ ledger strip nailed to the
	Demolish and Install	L.F.	.99	2.52		3.51	face of studs, beams or joist.
	Reinstall	L.F.		1.67		1.67	
	Minimum Charge	Job		260		260	
Ledger Boards							
2″ x 6″							
	Demolish	L.F.		.44		.44	Cost includes material and labor to
	Install	L.F.	5.30	3.26		8.56	install 2″ x 6″ ledger board fastened to
	Demolish and Install	L.F.	5.30	3.70		9	a wall, joists or studs.
	Reinstall	L.F.		2.60		2.60	
	Minimum Charge	Job		260		260	
4″ x 6″							
	Demolish	L.F.		.74		.74	Cost includes material and labor to
	Install	L.F.	10.60	3.72		14.32	install 4″ x 6″ ledger board fastened to
	Demolish and Install	L.F.	10.60	4.46		15.06	a wall, joists or studs.
	Reinstall	L.F.		2.98		2.98	
	Minimum Charge	Job		260		260	

For customer support on your Contractor's Pricing Guide: Residential Repair & Remodeling, call 800.448.8182.

Rough Frame / Structure

Floor Framing System

Floor Framing System	Unit	Material	Labor	Equip.	Total	Specification
4" x 8"						
Demolish	L.F.		.99		.99	Cost includes material and labor to
Install	L.F.	17.20	4.34		21.54	install 4" x 8" ledger board fastened to
Demolish and Install	L.F.	17.20	5.33		22.53	a wall, joists or studs.
Reinstall	L.F.		3.47		3.47	
Minimum Charge	Job		260		260	

Sill Plate Per L.F.

	Unit	Material	Labor	Equip.	Total	Specification
2" x 4"						
Demolish	L.F.		.38		.38	Cost includes material and labor to
Install	L.F.	.92	1.89		2.81	install 2" x 4" pressure treated lumber,
Demolish and Install	L.F.	.92	2.27		3.19	drilled and installed on foundation
Reinstall	L.F.		1.52		1.52	bolts 48" O.C. Bolts, nuts and washers
Minimum Charge	Job		260		260	are not included.
2" x 6"						
Demolish	L.F.		.57		.57	Cost includes material and labor to
Install	L.F.	1.29	2.08		3.37	install 2" x 6" pressure treated lumber,
Demolish and Install	L.F.	1.29	2.65		3.94	drilled and installed on foundation
Reinstall	L.F.		1.67		1.67	bolts 48" O.C. Bolts, nuts and washers
Minimum Charge	Job		260		260	not included.
2" x 8"						
Demolish	L.F.		.76		.76	Cost includes material and labor to
Install	L.F.	1.51	2.31		3.82	install 2" x 8" pressure treated lumber,
Demolish and Install	L.F.	1.51	3.07		4.58	drilled and installed on foundation
Reinstall	L.F.		1.85		1.85	bolts 48" O.C. Bolts, nuts and washers
Minimum Charge	Job		260		260	not included.

Earthquake Strapping

	Unit	Material	Labor	Equip.	Total	Specification
Install	Ea.	2.76	3.26		6.02	Includes labor and material to replace
Minimum Charge	Job		260		260	an earthquake strap.

Subflooring

Subflooring	Unit	Material	Labor	Equip.	Total	Specification
Plywood						
1/2"						
Demolish	S.F.		1.16		1.16	Cost includes material and labor to
Install	SF Flr.	.83	.56		1.39	install 1/2" plywood subfloor, CD
Demolish and Install	SF Flr.	.83	1.72		2.55	standard interior grade, plugged and
Minimum Charge	Job		260		260	touch sanded.
5/8"						
Demolish	S.F.		1.17		1.17	Cost includes material and labor to
Install	SF Flr.	1.10	.62		1.72	install 5/8" plywood subfloor, CD
Demolish and Install	SF Flr.	1.10	1.79		2.89	standard interior grade, plugged and
Reinstall	SF Flr.		.50		.50	touch sanded.
Minimum Charge	Job		260		260	
3/4"						
Demolish	S.F.		1.18		1.18	Cost includes material and labor to
Install	SF Flr.	1.10	.67		1.77	install 3/4" plywood subfloor, CD
Demolish and Install	SF Flr.	1.10	1.85		2.95	standard interior grade, plugged and
Reinstall	SF Flr.		.54		.54	touch sanded.
Minimum Charge	Job		260		260	

For customer support on your Contractor's Pricing Guide: Residential Repair & Remodeling, call 800.448.8182.

Rough Frame / Structure

Subflooring

Subflooring		Unit	Material	Labor	Equip.	Total	Specification
Particle Board							
1/2"							
	Demolish	S.F.		.58		.58	Cost includes material and labor to
	Install	S.F.	.65	.72		1.37	install 1/2" particle board subfloor.
	Demolish and Install	S.F.	.65	1.30		1.95	
	Minimum Charge	Job		260		260	
5/8"							
	Demolish	S.F.		.58		.58	Cost includes material and labor to
	Install	SF Flr.	2.10	.76		2.86	install 5/8" particle board subfloor.
	Demolish and Install	SF Flr.	2.10	1.34		3.44	
	Minimum Charge	Job		260		260	
3/4"							
	Demolish	S.F.		.59		.59	Cost includes material and labor to
	Install	SF Flr.	.86	.83		1.69	install 3/4" particle board subfloor.
	Demolish and Install	SF Flr.	.86	1.42		2.28	
	Minimum Charge	Job		260		260	
Plank Board							
	Demolish	S.F.		1.78		1.78	Cost includes material and labor to
	Install	SF Flr.	2.38	1.16		3.54	install 1" x 6" standard grade plank
	Demolish and Install	SF Flr.	2.38	2.94		5.32	flooring.
	Reinstall	SF Flr.		1.16		1.16	
	Minimum Charge	Job		260		260	
Felt Underlay							
	Demolish	S.F.		.10		.10	Includes material and labor to install
	Install	S.F.	.06	.14		.20	#15 felt building paper.
	Demolish and Install	S.F.	.06	.24		.30	
	Minimum Charge	Job		260		260	
Prep (for flooring)							
	Install	S.F.		.48		.48	Preparation of subflooring for
	Minimum Charge	Job		254		254	installation of new finished flooring. Cost reflects average time to prep area.

Wall Framing System

Wall Framing System		Unit	Material	Labor	Equip.	Total	Specification
2" x 4"							
8' High							
	Demolish	L.F.		5.90		5.90	Cost includes material and labor to
	Install	L.F.	9.90	8.70		18.60	install 2" x 4" x 8' high wall system,
	Demolish and Install	L.F.	9.90	14.60		24.50	including studs, treated bottom plate,
	Reinstall	L.F.		6.94		6.94	double top plate and one row of fire
	Minimum Charge	Job		260		260	blocking.
9' High							
	Demolish	L.F.		6.65		6.65	Cost includes material and labor to
	Install	L.F.	10.65	8.70		19.35	install 2" x 4" x 9' wall system
	Demolish and Install	L.F.	10.65	15.35		26	including studs, treated bottom plate,
	Reinstall	L.F.		6.94		6.94	double top plate and one row of fire
	Minimum Charge	Job		260		260	blocking.
10' High							
	Demolish	L.F.		7.40		7.40	Cost includes material and labor to
	Install	L.F.	11.40	8.70		20.10	install 2" x 4" x 10' wall system
	Demolish and Install	L.F.	11.40	16.10		27.50	including studs, treated bottom plate,
	Reinstall	L.F.		6.94		6.94	double top plate and one row of fire
	Minimum Charge	Job		260		260	blocking.

Rough Frame / Structure

Wall Framing System	Unit	Material	Labor	Equip.	Total	Specification
12' High						
Demolish	L.F.		8.90		8.90	Cost includes material and labor to
Install	L.F.	12.85	10.85		23.70	install 2" x 4" x 12' wall system
Demolish and Install	L.F.	12.85	19.75		32.60	including studs, treated bottom plate,
Reinstall	L.F.		8.68		8.68	double top plate and one row of fire
Minimum Charge	Job		260		260	blocking.
2" x 6"						
8' High						
Demolish	L.F.		5.55		5.55	Cost includes material and labor to
Install	L.F.	13.35	9.65		23	install 2" x 6" x 8' wall system
Demolish and Install	L.F.	13.35	15.20		28.55	including studs, treated bottom plate,
Reinstall	L.F.		7.72		7.72	double top plate and one row of fire
Minimum Charge	Job		260		260	blocking.
9' High						
Demolish	L.F.		6.25		6.25	Cost includes material and labor to
Install	L.F.	14.35	9.65		24	install 2" x 6" x 9' wall system
Demolish and Install	L.F.	14.35	15.90		30.25	including studs, treated bottom plate,
Reinstall	L.F.		7.72		7.72	double top plate and one row of fire
Minimum Charge	Job		260		260	blocking.
10' High						
Demolish	L.F.		6.85		6.85	Cost includes material and labor to
Install	L.F.	15.35	9.65		25	install 2" x 6" x 10' wall system
Demolish and Install	L.F.	15.35	16.50		31.85	including studs, treated bottom plate,
Reinstall	L.F.		7.72		7.72	double top plate and one row of fire
Minimum Charge	Job		260		260	blocking.
12' High						
Demolish	L.F.		8.05		8.05	Cost includes material and labor to
Install	L.F.	17.40	11.85		29.25	install 2" x 6" x 12' wall system
Demolish and Install	L.F.	17.40	19.90		37.30	including studs, treated bottom plate,
Reinstall	L.F.		9.47		9.47	double top plate and one row of fire
Minimum Charge	Job		260		260	blocking.
Fireblock						
2" x 4", 16" O.C. System						
Demolish	L.F.		.74		.74	Cost includes material and labor to
Install	L.F.	.99	1.74		2.73	install 2" x 4" fireblocks in wood frame
Demolish and Install	L.F.	.99	2.48		3.47	walls per L.F. of wall to be blocked.
Reinstall	L.F.		1.39		1.39	
Minimum Charge	Job		260		260	
2" x 4", 24" O.C. System						
Demolish	L.F.		.74		.74	Cost includes material and labor to
Install	L.F.	.99	1.74		2.73	install 2" x 4" fireblocks in wood frame
Demolish and Install	L.F.	.99	2.48		3.47	walls per L.F. of wall to be blocked.
Reinstall	L.F.		1.39		1.39	
Minimum Charge	Job		260		260	
2" x 6", 16" O.C. System						
Demolish	L.F.		1.11		1.11	Cost includes material and labor to
Install	L.F.	1.34	1.74		3.08	install 2" x 6" fireblocks in wood frame
Demolish and Install	L.F.	1.34	2.85		4.19	walls per L.F. of wall to be blocked.
Reinstall	L.F.		1.39		1.39	
Minimum Charge	Job		260		260	
2" x 6", 24" O.C. System						
Demolish	L.F.		1.11		1.11	Cost includes material and labor to
Install	L.F.	1.34	1.74		3.08	install 2" x 6" fireblocks in wood frame
Demolish and Install	L.F.	1.34	2.85		4.19	walls per L.F. of wall to be blocked.
Reinstall	L.F.		1.39		1.39	
Minimum Charge	Job		260		260	

For customer support on your Contractor's Pricing Guide: Residential Repair & Remodeling, call 800.448.8182.

Rough Frame / Structure

Wall Framing System

Wall Framing System	Unit	Material	Labor	Equip.	Total	Specification
Bracing						
1″ x 3″						
Demolish	L.F.		.42		.42	Cost includes material and labor to
Install	L.F.	.52	3.26		3.78	install 1″ x 3″ let-in bracing.
Demolish and Install	L.F.	.52	3.68		4.20	
Minimum Charge	Job		260		260	
1″ x 4″						
Demolish	L.F.		.42		.42	Cost includes material and labor to
Install	L.F.	.78	3.26		4.04	install 1″ x 4″ let-in bracing.
Demolish and Install	L.F.	.78	3.68		4.46	
Minimum Charge	Job		260		260	
1″ x 6″						
Demolish	L.F.		.42		.42	Cost includes material and labor to
Install	L.F.	1.11	3.47		4.58	install 1″ x 6″ let-in bracing.
Demolish and Install	L.F.	1.11	3.89		5	
Minimum Charge	Job		260		260	
2″ x 3″						
Demolish	L.F.		.55		.55	Cost includes material and labor to
Install	L.F.	.50	3.47		3.97	install 2″ x 3″ let-in bracing.
Demolish and Install	L.F.	.50	4.02		4.52	
Reinstall	L.F.		2.78		2.78	
Minimum Charge	Job		260		260	
2″ x 4″						
Demolish	L.F.		.55		.55	Cost includes material and labor to
Install	L.F.	.99	3.47		4.46	install 2″ x 4″ let-in bracing.
Demolish and Install	L.F.	.99	4.02		5.01	
Reinstall	L.F.		2.78		2.78	
Minimum Charge	Job		260		260	
2″ x 6″						
Demolish	L.F.		.55		.55	Cost includes material and labor to
Install	L.F.	1.34	3.72		5.06	install 2″ x 6″ let-in bracing.
Demolish and Install	L.F.	1.34	4.27		5.61	
Reinstall	L.F.		2.98		2.98	
Minimum Charge	Job		260		260	
2″ x 8″						
Demolish	L.F.		.55		.55	Cost includes material and labor to
Install	L.F.	1.46	3.72		5.18	install 2″ x 8″ let-in bracing.
Demolish and Install	L.F.	1.46	4.27		5.73	
Reinstall	L.F.		2.98		2.98	
Minimum Charge	Job		260		260	
Earthquake Strapping						
Install	Ea.	2.76	3.26		6.02	Includes labor and material to replace
Minimum Charge	Job		260		260	an earthquake strap.
Hurricane Clips						
Install	Ea.	1.72	3.59		5.31	Includes labor and material to install a
Minimum Charge	Job		260		260	hurricane clip.
Stud						
2″ x 4″						
Demolish	L.F.		.44		.44	Cost includes material and labor to
Install	L.F.	.97	.95		1.92	install 2″ x 4″ wall stud per L.F. of stud.
Demolish and Install	L.F.	.97	1.39		2.36	
Reinstall	L.F.		.76		.76	
Minimum Charge	Job		260		260	

Rough Frame / Structure

Wall Framing System

		Unit	Material	Labor	Equip.	Total	Specification
2" x 6"							
	Demolish	L.F.		.55		.55	Cost includes material and labor to
	Install	L.F.	1.34	1.04		2.38	install 2" x 6" stud per L.F. of stud.
	Demolish and Install	L.F.	1.34	1.59		2.93	
	Reinstall	L.F.		.83		.83	
	Minimum Charge	Job		260		260	

Plates

		Unit	Material	Labor	Equip.	Total	Specification
2" x 4"							
	Demolish	L.F.		.40		.40	Cost includes material and labor to
	Install	L.F.	.99	1.30		2.29	install 2" x 4" plate per L.F. of plate.
	Demolish and Install	L.F.	.99	1.70		2.69	
	Reinstall	L.F.		1.04		1.04	
	Minimum Charge	Job		260		260	
2" x 6"							
	Demolish	L.F.		.42		.42	Cost includes material and labor to
	Install	L.F.	1.34	1.39		2.73	install 2" x 6" plate per L.F. of plate.
	Demolish and Install	L.F.	1.34	1.81		3.15	
	Reinstall	L.F.		1.11		1.11	
	Minimum Charge	Job		260		260	

Headers

		Unit	Material	Labor	Equip.	Total	Specification
2" x 6"							
	Demolish	L.F.		.39		.39	Includes material and labor to install
	Install	L.F.	1.34	2.89		4.23	header over wall openings and
	Demolish and Install	L.F.	1.34	3.28		4.62	around floor, ceiling and roof
	Clean	L.F.	.01	.50		.51	openings or flush beam.
	Paint	L.F.	.28	1.73		2.01	
	Minimum Charge	Job		260		260	
2" x 8"							
	Demolish	L.F.		.42		.42	Includes material and labor to install
	Install	L.F.	1.46	3.06		4.52	header over wall openings and
	Demolish and Install	L.F.	1.46	3.48		4.94	around floor, ceiling and roof
	Clean	L.F.	.02	.51		.53	openings or flush beam.
	Paint	L.F.	.36	2.33		2.69	
	Minimum Charge	Job		260		260	
2" x 10"							
	Demolish	L.F.		.42		.42	Includes material and labor to install
	Install	L.F.	3.88	3.26		7.14	header over wall openings and
	Demolish and Install	L.F.	3.88	3.68		7.56	around floor, ceiling and roof
	Clean	L.F.	.02	.51		.53	openings or flush beam.
	Paint	L.F.	.46	2.88		3.34	
	Minimum Charge	Job		260		260	
2" x 12"							
	Demolish	L.F.		.44		.44	Includes material and labor to install
	Install	L.F.	4.09	3.47		7.56	header over wall openings and
	Demolish and Install	L.F.	4.09	3.91		8	around floor, ceiling and roof
	Clean	L.F.	.03	.51		.54	openings or flush beam.
	Paint	L.F.	.55	3.45		4	
	Minimum Charge	Job		260		260	
4" x 8"							
	Demolish	L.F.		.79		.79	Includes material and labor to install
	Install	L.F.	12.25	4.01		16.26	header over wall openings and
	Demolish and Install	L.F.	12.25	4.80		17.05	around floor, ceiling and roof
	Clean	L.F.	.02	.51		.53	openings or flush beam.
	Paint	L.F.	.36	2.33		2.69	
	Minimum Charge	Job		260		260	

For customer support on your Contractor's Pricing Guide: Residential Repair & Remodeling, call 800.448.8182.

Rough Frame / Structure

Wall Framing System

	Unit	Material	Labor	Equip.	Total	Specification
4" x 10"						
Demolish	L.F.		.85		.85	Includes material and labor to install
Install	L.F.	8.70	4.34		13.04	header over wall openings and
Demolish and Install	L.F.	8.70	5.19		13.89	around floor, ceiling and roof
Clean	L.F.	.02	.51		.53	openings or flush beam.
Paint	L.F.	.46	2.88		3.34	
Minimum Charge	Job		260		260	
4" x 12"						
Demolish	L.F.		.89		.89	Includes material and labor to install
Install	L.F.	11.85	5.50		17.35	header over wall openings and
Demolish and Install	L.F.	11.85	6.39		18.24	around floor, ceiling and roof
Clean	L.F.	.03	.51		.54	openings or flush beam.
Paint	L.F.	.55	3.45		4	
Minimum Charge	Job		260		260	
6" x 8"						
Demolish	L.F.		.79		.79	Includes material and labor to install
Install	L.F.	12.55	2.89		15.44	header over wall openings and
Demolish and Install	L.F.	12.55	3.68		16.23	around floor, ceiling and roof
Clean	L.F.	.02	.51		.53	openings or flush beam.
Paint	L.F.	.36	2.33		2.69	
Minimum Charge	Job		260		260	
6" x 10"						
Demolish	L.F.		.85		.85	Includes material and labor to install
Install	L.F.	16.05	6.30		22.35	header over wall openings and
Demolish and Install	L.F.	16.05	7.15		23.20	around floor, ceiling and roof
Clean	L.F.	.02	.51		.53	openings or flush beam.
Paint	L.F.	.46	2.88		3.34	
Minimum Charge	Job		260		260	
6" x 12"						
Demolish	L.F.		.89		.89	Includes material and labor to install
Install	L.F.	23	7.45		30.45	header over wall openings and
Demolish and Install	L.F.	23	8.34		31.34	around floor, ceiling and roof
Clean	L.F.	.03	.51		.54	openings or flush beam.
Paint	L.F.	.55	1.20		1.75	
Minimum Charge	Job		260		260	

Rough-in Opening

	Unit	Material	Labor	Equip.	Total	Specification
Door w / 2" x 4" Lumber						
3' Wide						
Demolish	Ea.		13.90		13.90	Includes material and labor to install
Install	Ea.	40.50	16.30		56.80	header, double studs each side,
Demolish and Install	Ea.	40.50	30.20		70.70	cripples, blocking and nails, up to 3'
Reinstall	Ea.		13.02		13.02	opening in 2" x 4" stud wall 8' high.
Minimum Charge	Job		260		260	
4' Wide						
Demolish	Ea.		13.90		13.90	Includes material and labor to install
Install	Ea.	43	16.30		59.30	header, double studs each side,
Demolish and Install	Ea.	43	30.20		73.20	cripples, blocking and nails, up to 4'
Reinstall	Ea.		13.02		13.02	opening in 2" x 4" stud wall 8' high.
Minimum Charge	Job		260		260	
5' Wide						
Demolish	Ea.		13.90		13.90	Includes material and labor to install
Install	Ea.	47	16.30		63.30	header, double studs each side,
Demolish and Install	Ea.	47	30.20		77.20	cripples, blocking and nails, up to 5'
Reinstall	Ea.		13.02		13.02	opening in 2" x 4" stud wall 8' high.
Minimum Charge	Job		260		260	

For customer support on your Contractor's Pricing Guide: Residential Repair & Remodeling, call 800.448.8182.

Rough Frame / Structure

Rough-in Opening		Unit	Material	Labor	Equip.	Total	Specification
6' Wide							
	Demolish	Ea.		13.90		13.90	Includes material and labor to install
	Install	Ea.	49.50	16.30		65.80	header, double studs each side,
	Demolish and Install	Ea.	49.50	30.20		79.70	cripples, blocking and nails, up to 6'
	Reinstall	Ea.		13.02		13.02	opening in 2" x 4" stud wall 8' high.
	Minimum Charge	Job		260		260	
8' Wide							
	Demolish	Ea.		14.80		14.80	Includes material and labor to install
	Install	Ea.	95	17.35		112.35	header, double studs each side,
	Demolish and Install	Ea.	95	32.15		127.15	cripples, blocking and nails, up to 8'
	Reinstall	Ea.		13.89		13.89	opening in 2" x 4" stud wall 8' high.
	Minimum Charge	Job		260		260	
10' Wide							
	Demolish	Ea.		14.80		14.80	Includes material and labor to install
	Install	Ea.	114	17.35		131.35	header, double studs each side,
	Demolish and Install	Ea.	114	32.15		146.15	cripples, blocking and nails, up to 10'
	Reinstall	Ea.		13.89		13.89	opening in 2" x 4" stud wall 8' high.
	Minimum Charge	Job		260		260	
12' Wide							
	Demolish	Ea.		14.80		14.80	Includes material and labor to install
	Install	Ea.	220	17.35		237.35	header, double studs each side,
	Demolish and Install	Ea.	220	32.15		252.15	cripples, blocking and nails, up to 12'
	Reinstall	Ea.		13.89		13.89	opening in 2" x 4" stud wall 8' high.
	Minimum Charge	Job		260		260	

Door w / 2" x 6" Lumber

		Unit	Material	Labor	Equip.	Total	Specification
3' Wide							
	Demolish	Ea.		13.90		13.90	Includes material and labor to install
	Install	Ea.	51.50	16.30		67.80	header, double studs each side,
	Demolish and Install	Ea.	51.50	30.20		81.70	cripples, blocking and nails, up to 3'
	Reinstall	Ea.		13.02		13.02	opening in 2" x 6" stud wall 8' high.
	Minimum Charge	Job		260		260	
4' Wide							
	Demolish	Ea.		13.90		13.90	Includes material and labor to install
	Install	Ea.	54.50	16.30		70.80	header, double studs each side,
	Demolish and Install	Ea.	54.50	30.20		84.70	cripples, blocking and nails, up to 4'
	Reinstall	Ea.		13.02		13.02	opening in 2" x 6" stud wall 8' high.
	Minimum Charge	Job		260		260	
5' Wide							
	Demolish	Ea.		13.90		13.90	Includes material and labor to install
	Install	Ea.	58.50	16.30		74.80	header, double studs each side,
	Demolish and Install	Ea.	58.50	30.20		88.70	cripples, blocking and nails up to 5'
	Reinstall	Ea.		13.02		13.02	opening in 2" x 6" stud wall 8' high.
	Minimum Charge	Job		260		260	
6' Wide							
	Demolish	Ea.		13.90		13.90	Includes material and labor to install
	Install	Ea.	61	16.30		77.30	header, double studs each side,
	Demolish and Install	Ea.	61	30.20		91.20	cripples, blocking and nails up to 6'
	Reinstall	Ea.		13.02		13.02	opening in 2" x 6" stud wall 8' high.
	Minimum Charge	Job		260		260	
8' Wide							
	Demolish	Ea.		14.80		14.80	Includes material and labor to install
	Install	Ea.	106	17.35		123.35	header, double studs each side,
	Demolish and Install	Ea.	106	32.15		138.15	cripples, blocking and nails up to 8'
	Reinstall	Ea.		13.89		13.89	opening in 2" x 6" stud wall 8' high.
	Minimum Charge	Job		260		260	
10' Wide							
	Demolish	Ea.		14.80		14.80	Includes material and labor to install
	Install	Ea.	127	17.35		144.35	header, double studs each side,
	Demolish and Install	Ea.	127	32.15		159.15	cripples, blocking and nails up to 10'
	Reinstall	Ea.		13.89		13.89	opening in 2" x 6" stud wall 8' high.
	Minimum Charge	Job		260		260	

For customer support on your Contractor's Pricing Guide: Residential Repair & Remodeling, call 800.448.8182.

Rough Frame / Structure

Rough-in Opening

Rough-in Opening		Unit	Material	Labor	Equip.	Total	Specification
12' Wide							
	Demolish	Ea.		14.80		14.80	Includes material and labor to install
	Install	Ea.	231	17.35		248.35	header, double studs each side,
	Demolish and Install	Ea.	231	32.15		263.15	cripples, blocking and nails, up to 12'
	Reinstall	Ea.		13.89		13.89	opening in 2" x 6" stud wall 8' high.
	Minimum Charge	Job		260		260	

Rough-in Opening

Rough-in Opening		Unit	Material	Labor	Equip.	Total	Specification
Window w / 2" x 4" Lumber							
2' Wide							
	Demolish	Ea.		18.50		18.50	Includes material and labor to install
	Install	Ea.	44	21.50		65.50	header, double studs each side of
	Demolish and Install	Ea.	44	40		84	opening, cripples, blocking, nails and
	Reinstall	Ea.		17.36		17.36	sub-sills for opening up to 2' in a 2" x
	Minimum Charge	Job		260		260	4" stud wall 8' high.
3' Wide							
	Demolish	Ea.		18.50		18.50	Includes material and labor to install
	Install	Ea.	50.50	21.50		72	header, double studs each side of
	Demolish and Install	Ea.	50.50	40		90.50	opening, cripples, blocking, nails and
	Reinstall	Ea.		17.36		17.36	sub-sills for opening up to 3' in a 2" x
	Minimum Charge	Job		260		260	4" stud wall 8' high.
4' Wide							
	Demolish	Ea.		18.50		18.50	Includes material and labor to install
	Install	Ea.	55	21.50		76.50	header, double studs each side of
	Demolish and Install	Ea.	55	40		95	opening, cripples, blocking, nails and
	Reinstall	Ea.		17.36		17.36	sub-sills for opening up to 4' in a 2" x
	Minimum Charge	Job		260		260	4" stud wall 8' high.
5' Wide							
	Demolish	Ea.		18.50		18.50	Includes material and labor to install
	Install	Ea.	59	21.50		80.50	header, double studs each side of
	Demolish and Install	Ea.	59	40		99	opening, cripples, blocking, nails and
	Reinstall	Ea.		17.36		17.36	sub-sills for opening up to 5' in a 2" x
	Minimum Charge	Job		260		260	4" stud wall 8' high.
6' Wide							
	Demolish	Ea.		18.50		18.50	Includes material and labor to install
	Install	Ea.	65	21.50		86.50	header, double studs each side of
	Demolish and Install	Ea.	65	40		105	opening, cripples, blocking, nails and
	Reinstall	Ea.		17.36		17.36	sub-sills for opening up to 6' in a 2" x
	Minimum Charge	Job		260		260	4" stud wall 8' high.
7' Wide							
	Demolish	Ea.		18.50		18.50	Includes material and labor to install
	Install	Ea.	104	21.50		125.50	header, double studs each side of
	Demolish and Install	Ea.	104	40		144	opening, cripples, blocking, nails and
	Reinstall	Ea.		17.36		17.36	sub-sills for opening up to 7' in a 2" x
	Minimum Charge	Job		260		260	4" stud wall 8' high.
8' Wide							
	Demolish	Ea.		20		20	Includes material and labor to install
	Install	Ea.	117	23.50		140.50	header, double studs each side of
	Demolish and Install	Ea.	117	43.50		160.50	opening, cripples, blocking, nails and
	Reinstall	Ea.		18.94		18.94	sub-sills for opening up to 8' in a 2" x
	Minimum Charge	Job		260		260	4" stud wall 8' high.

Rough Frame / Structure

Rough-in Opening		Unit	Material	Labor	Equip.	Total	Specification
10' Wide							Includes material and labor to install
	Demolish	Ea.		20		20	header, double studs each side of
	Install	Ea.	139	23.50		162.50	opening, cripples, blocking, nails and
	Demolish and Install	Ea.	139	43.50		182.50	sub-sills for opening up to 10' in a 2"
	Reinstall	Ea.		18.94		18.94	x 4" stud wall 8' high.
	Minimum Charge	Job		260		260	
12' Wide							Includes material and labor to install
	Demolish	Ea.		20		20	header, double studs each side of
	Install	Ea.	250	23.50		273.50	opening, cripples, blocking, nails and
	Demolish and Install	Ea.	250	43.50		293.50	sub-sills for opening up to 12' in a 2"
	Reinstall	Ea.		18.94		18.94	x 4" stud wall 8' high.
	Minimum Charge	Job		260		260	

Window w / 2" x 6" Lumber

		Unit	Material	Labor	Equip.	Total	Specification
2' Wide							Includes material and labor to install
	Demolish	Ea.		18.50		18.50	header, double studs each side of
	Install	Ea.	58.50	21.50		80	opening, cripples, blocking, nails and
	Demolish and Install	Ea.	58.50	40		98.50	sub-sills for opening up to 2' in a 2" x
	Reinstall	Ea.		17.36		17.36	6" stud wall 8' high.
	Minimum Charge	Job		260		260	
3' Wide							Includes material and labor to install
	Demolish	Ea.		18.50		18.50	header, double studs each side of
	Install	Ea.	62.50	21.50		84	opening, cripples, blocking, nails and
	Demolish and Install	Ea.	62.50	40		102.50	sub-sills for opening up to 3' in a 2" x
	Reinstall	Ea.		17.36		17.36	6" stud wall 8' high.
	Minimum Charge	Job		260		260	
4' Wide							Includes material and labor to install
	Demolish	Ea.		18.50		18.50	header, double studs each side of
	Install	Ea.	70	21.50		91.50	opening, cripples, blocking, nails and
	Demolish and Install	Ea.	70	40		110	sub-sills for opening up to 4' in a 2" x
	Reinstall	Ea.		17.36		17.36	6" stud wall 8' high.
	Minimum Charge	Job		260		260	
5' Wide							Includes material and labor to install
	Demolish	Ea.		18.50		18.50	header, double studs each side of
	Install	Ea.	74.50	21.50		96	opening, cripples, blocking, nails and
	Demolish and Install	Ea.	74.50	40		114.50	sub-sills for opening up to 5' in a 2" x
	Reinstall	Ea.		17.36		17.36	6" stud wall 8' high.
	Minimum Charge	Job		260		260	
6' Wide							Includes material and labor to install
	Demolish	Ea.		18.50		18.50	header, double studs each side of
	Install	Ea.	81	21.50		102.50	opening, cripples, blocking, nails and
	Demolish and Install	Ea.	81	40		121	sub-sills for opening up to 6' in a 2" x
	Reinstall	Ea.		17.36		17.36	6" stud wall 8' high.
	Minimum Charge	Job		260		260	
7' Wide							Includes material and labor to install
	Demolish	Ea.		18.50		18.50	header, double studs each side of
	Install	Ea.	123	21.50		144.50	opening, cripples, blocking, nails and
	Demolish and Install	Ea.	123	40		163	sub-sills for opening up to 7' in a 2" x
	Reinstall	Ea.		17.36		17.36	6" stud wall 8' high.
	Minimum Charge	Job		260		260	

Rough Frame / Structure

Rough-in Opening

Rough-in Opening	Unit	Material	Labor	Equip.	Total	Specification
8' Wide						
Demolish	Ea.		20		20	Includes material and labor to install
Install	Ea.	134	23.50		157.50	header, double studs each side of
Demolish and Install	Ea.	134	43.50		177.50	opening, cripples, blocking, nails and
Reinstall	Ea.		18.94		18.94	sub-sills for opening up to 8' in a 2" x
Minimum Charge	Job		260		260	6" stud wall 8' high.
10' Wide						
Demolish	Ea.		20		20	Includes material and labor to install
Install	Ea.	158	23.50		181.50	header, double studs each side of
Demolish and Install	Ea.	158	43.50		201.50	opening, cripples, blocking, nails and
Reinstall	Ea.		18.94		18.94	sub-sills for opening up to 10' in a 2"
Minimum Charge	Job		260		260	x 6" stud wall 8' high.
12' Wide						
Demolish	Ea.		20		20	Includes material and labor to install
Install	Ea.	272	23.50		295.50	header, double studs each side of
Demolish and Install	Ea.	272	43.50		315.50	opening, cripples, blocking, nails and
Reinstall	Ea.		18.94		18.94	sub-sills for opening up to 12' in a 2"
Minimum Charge	Job		260		260	x 6" stud wall 8' high.

Glue-Laminated Beams

Glue-Laminated Beams	Unit	Material	Labor	Equip.	Total	Specification
3-1/2" x 6"						
Demolish	L.F.		3.29		3.29	Includes material and labor to install
Install	L.F.	9.25	4.04	4.26	17.55	glue-laminated wood beam.
Demolish and Install	L.F.	9.25	7.33	4.26	20.84	
Reinstall	L.F.		3.23	3.41	6.64	
Clean	L.F.	.11	.50		.61	
Paint	L.F.	.72	1.27		1.99	
Minimum Charge	Job		260		260	
3-1/2" x 9"						
Demolish	L.F.		3.29		3.29	Includes material and labor to install
Install	L.F.	12.40	4.04	4.26	20.70	glue-laminated wood beam.
Demolish and Install	L.F.	12.40	7.33	4.26	23.99	
Reinstall	L.F.		3.23	3.41	6.64	
Clean	L.F.	.14	.52		.66	
Paint	L.F.	.92	1.65		2.57	
Minimum Charge	Job		260		260	
3-1/2" x 12"						
Demolish	L.F.		4.95		4.95	Includes material and labor to install
Install	L.F.	16.55	4.04	4.26	24.85	glue-laminated wood beam.
Demolish and Install	L.F.	16.55	8.99	4.26	29.80	
Reinstall	L.F.		3.23	3.41	6.64	
Clean	L.F.	.17	.53		.70	
Paint	L.F.	1.14	2.06		3.20	
Minimum Charge	Job		260		260	
3-1/2" x 15"						
Demolish	L.F.		4.95		4.95	Includes material and labor to install
Install	L.F.	20.50	4.18	4.40	29.08	glue-laminated wood beam.
Demolish and Install	L.F.	20.50	9.13	4.40	34.03	
Reinstall	L.F.		3.34	3.52	6.86	
Clean	L.F.	.20	.55		.75	
Paint	L.F.	1.34	2.40		3.74	
Minimum Charge	Job		260		260	

Rough Frame / Structure

Glue-Laminated Beams

Glue-Laminated Beams		Unit	Material	Labor	Equip.	Total	Specification
3-1/2″ x 18″							
	Demolish	L.F.		6.20		6.20	Includes material and labor to install
	Install	L.F.	28	4.18	4.40	36.58	glue-laminated wood beam.
	Demolish and Install	L.F.	28	10.38	4.40	42.78	
	Reinstall	L.F.		3.34	3.52	6.86	
	Clean	L.F.	.23	.57		.80	
	Paint	L.F.	1.57	2.78		4.35	
	Minimum Charge	Job		260		260	
5-1/8″ x 6″							
	Demolish	L.F.		4.95		4.95	Includes material and labor to install
	Install	L.F.	13.55	4.04	4.26	21.85	glue-laminated wood beam.
	Demolish and Install	L.F.	13.55	8.99	4.26	26.80	
	Reinstall	L.F.		3.23	3.41	6.64	
	Clean	L.F.	.11	.50		.61	
	Paint	L.F.	.77	1.39		2.16	
	Minimum Charge	Job		260		260	
5-1/8″ x 9″							
	Demolish	L.F.		6.20		6.20	Includes material and labor to install
	Install	L.F.	20.50	4.04	4.26	28.80	glue-laminated wood beam.
	Demolish and Install	L.F.	20.50	10.24	4.26	35	
	Reinstall	L.F.		3.23	3.41	6.64	
	Clean	L.F.	.14	.52		.66	
	Paint	L.F.	.99	1.78		2.77	
	Minimum Charge	Job		260		260	
5-1/8″ x 12″							
	Demolish	L.F.		7.35		7.35	Includes material and labor to install
	Install	L.F.	27	4.04	4.26	35.30	glue-laminated wood beam.
	Demolish and Install	L.F.	27	11.39	4.26	42.65	
	Reinstall	L.F.		3.23	3.41	6.64	
	Clean	L.F.	.18	.53		.71	
	Paint	L.F.	1.22	2.16		3.38	
	Minimum Charge	Job		260		260	
5-1/8″ x 18″							
	Demolish	L.F.		12.35		12.35	Includes material and labor to install
	Install	L.F.	40.50	4.18	4.40	49.08	glue-laminated wood beam.
	Demolish and Install	L.F.	40.50	16.53	4.40	61.43	
	Reinstall	L.F.		3.34	3.52	6.86	
	Clean	L.F.	.24	.57		.81	
	Paint	L.F.	1.41	2.92		4.33	
	Minimum Charge	Job		260		260	
6-3/4″ x 12″							
	Demolish	L.F.		7.35		7.35	Includes material and labor to install
	Install	L.F.	36	4.18	4.40	44.58	glue-laminated wood beam.
	Demolish and Install	L.F.	36	11.53	4.40	51.93	
	Reinstall	L.F.		3.34	3.52	6.86	
	Clean	L.F.	.20	.55		.75	
	Paint	L.F.	1.34	2.40		3.74	
	Minimum Charge	Job		260		260	
6-3/4″ x 15″							
	Demolish	L.F.		12.35		12.35	Includes material and labor to install
	Install	L.F.	44.50	4.18	4.40	53.08	glue-laminated wood beam.
	Demolish and Install	L.F.	44.50	16.53	4.40	65.43	
	Reinstall	L.F.		3.34	3.52	6.86	
	Clean	L.F.	.23	.57		.80	
	Paint	L.F.	1.57	2.78		4.35	
	Minimum Charge	Job		260		260	

Rough Frame / Structure

Glue-Laminated Beams

Glue-Laminated Beams	Unit	Material	Labor	Equip.	Total	Specification
6-3/4" x 18"						
Demolish	L.F.		12.35		12.35	Includes material and labor to install
Install	L.F.	53.50	4.32	4.56	62.38	glue-laminated wood beam.
Demolish and Install	L.F.	53.50	16.67	4.56	74.73	
Reinstall	L.F.		3.46	3.65	7.11	
Clean	L.F.	.26	.59		.85	
Paint	L.F.	1.78	3.17		4.95	
Minimum Charge	Job		260		260	
Hardware						
5-1/4" Glue-Lam Seat						
Install	Ea.	54	2.89		56.89	Includes labor and material to install
Minimum Charge	Job		260		260	beam hangers.
6-3/4" Glue-Lam Seat						
Install	Ea.	77.50	2.89		80.39	Includes labor and material to install
Minimum Charge	Job		260		260	beam hangers.
Earthquake Strapping						
Install	Ea.	2.76	3.26		6.02	Includes labor and material to replace
Minimum Charge	Job		260		260	an earthquake strap.
Hurricane Clips						
Install	Ea.	1.72	3.59		5.31	Includes labor and material to install a
Minimum Charge	Job		260		260	hurricane clip.

Metal Stud Framing

Metal Stud Framing	Unit	Material	Labor	Equip.	Total	Specification
16" O.C. System						
4", 16 Ga.						
Demolish	S.F.		1.62		1.62	Includes material and labor to install
Install	S.F.	3.05	1.60		4.65	load bearing cold rolled metal stud
Demolish and Install	S.F.	3.05	3.22		6.27	walls, to 10' high, including studs, top
Reinstall	S.F.		1.28		1.28	and bottom track and screws.
Minimum Charge	Job		260		260	
6", 16 Ga.						
Demolish	S.F.		1.62		1.62	Includes material and labor to install
Install	S.F.	3.86	1.63		5.49	load bearing cold rolled metal stud
Demolish and Install	S.F.	3.86	3.25		7.11	walls, to 10' high, including studs, top
Reinstall	S.F.		1.30		1.30	and bottom track and screws.
Minimum Charge	Job		260		260	
4", 25 Ga.						
Demolish	S.F.		1.12		1.12	Includes material and labor to install
Install	S.F.	1.01	1.10		2.11	cold rolled metal stud walls, to 10'
Demolish and Install	S.F.	1.01	2.22		3.23	high, including studs, top and bottom
Reinstall	S.F.		.88		.88	track and screws.
Minimum Charge	Job		260		260	
6", 25 Ga.						
Demolish	S.F.		1.12		1.12	Includes material and labor to install
Install	S.F.	1.08	1.11		2.19	cold rolled metal stud walls, to 10'
Demolish and Install	S.F.	1.08	2.23		3.31	high, including studs, top and bottom
Reinstall	S.F.		.89		.89	track and screws.
Minimum Charge	Job		260		260	
24" O.C. System						
4", 25 Ga.						
Demolish	S.F.		.95		.95	Includes material and labor to install
Install	S.F.	.73	.70		1.43	cold rolled metal stud walls, to 10'
Demolish and Install	S.F.	.73	1.65		2.38	high, including studs, top and bottom
Reinstall	S.F.		.56		.56	track and screws.
Minimum Charge	Job		260		260	

Rough Frame / Structure

Metal Stud Framing

		Unit	Material	Labor	Equip.	Total	Specification
6", 25 Ga.							
	Demolish	S.F.		.95		.95	Includes material and labor to install
	Install	S.F.	.79	.72		1.51	cold rolled metal stud walls, to 10'
	Demolish and Install	S.F.	.79	1.67		2.46	high, including studs, top and bottom
	Reinstall	S.F.		.57		.57	track and screws.
	Minimum Charge	Job		260		260	
4", 16 Ga.							
	Demolish	L.F.		10.90		10.90	Includes material and labor to install
	Install	S.F.	2.21	1.16		3.37	load bearing cold rolled metal stud
	Demolish and Install	S.F.	2.21	12.06		14.27	walls, to 10' high, including studs, top
	Reinstall	S.F.		.93		.93	and bottom track and screws.
	Minimum Charge	Job		260		260	
6", 16 Ga.							
	Demolish	L.F.		10.90		10.90	Includes material and labor to install
	Install	S.F.	2.82	1.18		4	load bearing cold rolled metal stud
	Demolish and Install	S.F.	2.82	12.08		14.90	walls, to 10' high, including studs, top
	Reinstall	S.F.		.95		.95	and bottom track and screws.
	Minimum Charge	Job		260		260	

Metal Joist

		Unit	Material	Labor	Equip.	Total	Specification
Bar Joist							
18K9							
	Demolish	L.F.		1.31	1.34	2.65	Includes material, labor and
	Install	L.F.	10.50	2.65	1.79	14.94	equipment to install open web joist.
	Demolish and Install	L.F.	10.50	3.96	3.13	17.59	
	Reinstall	L.F.		2.12	1.43	3.55	
	Minimum Charge	Job		2400	2550	4950	
16K6							
	Demolish	L.F.		1.31	1.34	2.65	Includes material, labor and
	Install	L.F.	11.55	1.76	.14	13.45	equipment to install open web joist.
	Demolish and Install	L.F.	11.55	3.07	1.48	16.10	
	Reinstall	L.F.		1.41	.11	1.52	
	Minimum Charge	Job		2400	2550	4950	

Exterior Sheathing

		Unit	Material	Labor	Equip.	Total	Specification
CDX Plywood							
5/16"							
	Demolish	S.F.		.76		.76	Cost includes material and labor to
	Install	S.F.	1.12	.65		1.77	install 5/16" CDX plywood sheathing.
	Demolish and Install	S.F.	1.12	1.41		2.53	
	Minimum Charge	Job		260		260	
3/8"							
	Demolish	S.F.		.77		.77	Cost includes material and labor to
	Install	S.F.	.97	.87		1.84	install 3/8" CDX plywood sheathing.
	Demolish and Install	S.F.	.97	1.64		2.61	
	Minimum Charge	Job		260		260	
1/2"							
	Demolish	S.F.		.79		.79	Cost includes material and labor to
	Install	S.F.	1.09	.93		2.02	install 1/2" CDX plywood sheathing.
	Demolish and Install	S.F.	1.09	1.72		2.81	
	Minimum Charge	Job		260		260	

For customer support on your Contractor's Pricing Guide: Residential Repair & Remodeling, call 800.448.8182.

Rough Frame / Structure

Exterior Sheathing

Exterior Sheathing	Unit	Material	Labor	Equip.	Total	Specification
5/8″						
Demolish	S.F.		.81		.81	Cost includes material and labor to
Install	S.F.	1.45	.99		2.44	install 5/8″ CDX plywood sheathing.
Demolish and Install	S.F.	1.45	1.80		3.25	
Minimum Charge	Job		260		260	
3/4″						
Demolish	S.F.		.83		.83	Cost includes material and labor to
Install	S.F.	1.45	1.07		2.52	install 3/4″ CDX plywood sheathing.
Demolish and Install	S.F.	1.45	1.90		3.35	
Minimum Charge	Job		260		260	
OSB						
1/2″						
Demolish	S.F.		.79		.79	Cost includes material and labor to
Install	S.F.	.68	.74		1.42	install 4′ x 8′ x 1/2″ OSB sheathing.
Demolish and Install	S.F.	.68	1.53		2.21	
Minimum Charge	Job		260		260	
5/8″						
Demolish	S.F.		.81		.81	Cost includes material and labor to
Install	S.F.	.68	.60		1.28	install 4′ x 8′ x 5/8″ OSB sheathing.
Demolish and Install	S.F.	.68	1.41		2.09	
Minimum Charge	Job		260		260	
Vapor Barrier						
Black Paper						
Install	S.F.	.06	.14		.20	Includes material and labor to install
Minimum Charge	Job		260		260	#15 felt building paper.

Plywood Sheathing

Plywood Sheathing	Unit	Material	Labor	Equip.	Total	Specification
Finish Plywood						
5/16″						
Demolish	S.F.		.76		.76	Includes material and labor to install
Install	S.F.	1.12	.33		1.45	exterior 5/16″ AC plywood on walls.
Demolish and Install	S.F.	1.12	1.09		2.21	
Minimum Charge	Job		260		260	
3/8″						
Demolish	S.F.		.77		.77	Includes material and labor to install
Install	S.F.	1.12	.43		1.55	exterior 3/8″ AC plywood on walls.
Demolish and Install	S.F.	1.12	1.20		2.32	
Minimum Charge	Job		260		260	
1/2″						
Demolish	S.F.		.79		.79	Includes material and labor to install
Install	S.F.	2.05	.46		2.51	exterior 1/2″ AC plywood on walls.
Demolish and Install	S.F.	2.05	1.25		3.30	
Minimum Charge	Job		260		260	
5/8″						
Demolish	S.F.		.81		.81	Includes material and labor to install
Install	S.F.	1.83	.46		2.29	exterior 5/8″ AC plywood on walls.
Demolish and Install	S.F.	1.83	1.27		3.10	
Reinstall	S.F.		.46		.46	
Minimum Charge	Job		260		260	
3/4″						
Demolish	S.F.		.83		.83	Includes material and labor to install
Install	S.F.	1.89	.46		2.35	exterior 3/4″ AC plywood on walls.
Demolish and Install	S.F.	1.89	1.29		3.18	
Reinstall	S.F.		.46		.46	
Minimum Charge	Job		260		260	

Rough Frame / Structure

Stairs

	Unit	Material	Labor	Equip.	Total	Specification
Job-built						
Treads and Risers						
Demolish	Ea.		55.50		55.50	Includes material and labor to install
Install	Ea.	1600	520		2120	three 2″ x 12″ stringers, treads and
Demolish and Install	Ea.	1600	575.50		2175.50	risers of 3/4″ CDX plywood, installed
Clean	Flight	3.17	34		37.17	in a straight or ″L″ shaped run
Paint	Ea.	.23	86.50		86.73	including carpet.
Minimum Charge	Job		260		260	
Landing						
Demolish	S.F.		.44		.44	Includes material and labor to install
Install	S.F.	65.50	3.62		69.12	landing framing, 3/4″ CDX plywood
Demolish and Install	S.F.	65.50	4.06		69.56	surface and carpet.
Clean	S.F.		.50		.50	
Paint	S.F.	.88	1.07		1.95	
Minimum Charge	Job		260		260	

Ceiling Framing

	Unit	Material	Labor	Equip.	Total	Specification
16″ O.C. System						
2″ x 6″ Joists						
Demolish	S.F.		.69		.69	Cost includes material and labor to
Install	S.F.	1.01	.86		1.87	install 2″ x 6″ joists including end and
Demolish and Install	S.F.	1.01	1.55		2.56	header joist installed 16″ O.C. Does
Reinstall	S.F.		.69		.69	not include beams, ledger strips,
Minimum Charge	Job		260		260	blocking or bridging.
2″ x 8″ Joists						
Demolish	S.F.		.71		.71	Cost includes material and labor to
Install	S.F.	1.11	1.05		2.16	install 2″ x 8″ joists including end and
Demolish and Install	S.F.	1.11	1.76		2.87	header joist installed 16″ O.C. Does
Reinstall	S.F.		.84		.84	not include beams, ledger strips,
Minimum Charge	Job		260		260	blocking or bridging.
2″ x 10″ Joists						
Demolish	S.F.		.73		.73	Cost includes material and labor to
Install	S.F.	2.95	1.25		4.20	install 2″ x 10″ joists including end
Demolish and Install	S.F.	2.95	1.98		4.93	and header joist installed 16″ O.C.
Reinstall	S.F.		1		1	Does not include beams, ledger strips,
Minimum Charge	Job		260		260	blocking or bridging.
2″ x 12″ Joists						
Demolish	S.F.		.76		.76	Cost includes material and labor to
Install	S.F.	3.11	1.52		4.63	install 2″ x 12″ joists including end
Demolish and Install	S.F.	3.11	2.28		5.39	and header joist installed 16″ O.C.
Reinstall	S.F.		1.22		1.22	Does not include beams, ledger strips,
Minimum Charge	Job		260		260	blocking or bridging.
Block / Bridge						
Demolish	Ea.		1.39		1.39	Includes material and labor to install
Install	Ea.	5.20	2.93		8.13	set of cross bridging or per block of
Demolish and Install	Ea.	5.20	4.32		9.52	solid bridging for 2″ x 10″ joists cut to
Minimum Charge	Job		260		260	size on site.

For customer support on your Contractor's Pricing Guide: Residential Repair & Remodeling, call 800.448.8182.

Rough Frame / Structure

Ceiling Framing

	Unit	Material	Labor	Equip.	Total	Specification
24″ O.C. System						
2″ x 6″ Joists						
Demolish	S.F.		.46		.46	Cost includes material and labor to
Install	S.F.	.68	.57		1.25	install 2″ x 6″ joists including end and
Demolish and Install	S.F.	.68	1.03		1.71	header joist installed 24″ O.C. Does
Reinstall	S.F.		.46		.46	not include beams, ledger strips,
Minimum Charge	Job		260		260	blocking or bridging.
2″ x 8″ Joists						
Demolish	S.F.		.47		.47	Cost includes material and labor to
Install	S.F.	.75	.70		1.45	install 2″ x 8″ joists including end and
Demolish and Install	S.F.	.75	1.17		1.92	header joist installed 24″ O.C. Does
Reinstall	S.F.		.56		.56	not include beams, ledger strips,
Minimum Charge	Job		260		260	blocking or bridging.
2″ x 10″ Joists						
Demolish	S.F.		.49		.49	Cost includes material and labor to
Install	S.F.	1.98	.83		2.81	install 2″ x 10″ joists including end
Demolish and Install	S.F.	1.98	1.32		3.30	and header joist installed 24″ O.C.
Reinstall	S.F.		.66		.66	Does not include beams, ledger strips,
Minimum Charge	Job		260		260	blocking or bridging.
2″ x 12″ Joists						
Demolish	S.F.		.50		.50	Cost includes material and labor to
Install	S.F.	2.09	1.02		3.11	install 2″ x 12″ joists including end
Demolish and Install	S.F.	2.09	1.52		3.61	and header joist installed 24″ O.C.
Reinstall	S.F.		.81		.81	Does not include beams, ledger strips,
Minimum Charge	Job		260		260	blocking or bridging.
Block / Bridge						
Demolish	Ea.		1.39		1.39	Includes material and labor to install
Install	Ea.	5.20	2.93		8.13	set of cross bridging or per block of
Demolish and Install	Ea.	5.20	4.32		9.52	solid bridging for 2″ x 10″ joists cut to
Minimum Charge	Job		260		260	size on site.
Ledger Strips						
1″ x 2″						
Demolish	L.F.		.37		.37	Cost includes material and labor to
Install	L.F.	.18	1.51		1.69	install 1″ x 2″ ledger strip nailed to the
Demolish and Install	L.F.	.18	1.88		2.06	face of studs, beams or joist.
Minimum Charge	Job		260		260	
1″ x 3″						
Demolish	L.F.		.37		.37	Includes material and labor to install
Install	L.F.	.78	1.74		2.52	up to 1″ x 4″ ledger strip nailed to the
Demolish and Install	L.F.	.78	2.11		2.89	face of studs, beams or joist.
Minimum Charge	Job		260		260	
1″ x 4″						
Demolish	L.F.		.37		.37	Includes material and labor to install
Install	L.F.	.78	1.74		2.52	up to 1″ x 4″ ledger strip nailed to the
Demolish and Install	L.F.	.78	2.11		2.89	face of studs, beams or joist.
Minimum Charge	Job		260		260	
2″ x 2″						
Demolish	L.F.		.40		.40	Cost includes material and labor to
Install	L.F.	.50	1.58		2.08	install 2″ x 2″ ledger strip nailed to the
Demolish and Install	L.F.	.50	1.98		2.48	face of studs, beams or joist.
Minimum Charge	Job		260		260	
2″ x 4″						
Demolish	L.F.		.44		.44	Cost includes material and labor to
Install	L.F.	.99	2.08		3.07	install 2″ x 4″ ledger strip nailed to the
Demolish and Install	L.F.	.99	2.52		3.51	face of studs, beams or joist.
Reinstall	L.F.		1.67		1.67	
Minimum Charge	Job		260		260	

Rough Frame / Structure

Ceiling Framing

Ceiling Framing	Unit	Material	Labor	Equip.	Total	Specification
Ledger Boards						
2" x 4"						
Demolish	L.F.		.44		.44	Cost includes material and labor to
Install	L.F.	4.98	2.89		7.87	install 2" x 4" ledger board fastened to
Demolish and Install	L.F.	4.98	3.33		8.31	a wall, joists or studs.
Reinstall	L.F.		2.31		2.31	
Minimum Charge	Job		260		260	
4" x 6"						
Demolish	L.F.		.74		.74	Cost includes material and labor to
Install	L.F.	10.60	3.72		14.32	install 4" x 6" ledger board fastened to
Demolish and Install	L.F.	10.60	4.46		15.06	a wall, joists or studs.
Reinstall	L.F.		2.98		2.98	
Minimum Charge	Job		260		260	
4" x 8"						
Demolish	L.F.		.99		.99	Cost includes material and labor to
Install	L.F.	17.20	4.34		21.54	install 4" x 8" ledger board fastened to
Demolish and Install	L.F.	17.20	5.33		22.53	a wall, joists or studs.
Reinstall	L.F.		3.47		3.47	
Minimum Charge	Job		260		260	
Earthquake Strapping						
Install	Ea.	2.76	3.26		6.02	Includes labor and material to replace
Minimum Charge	Job		260		260	an earthquake strap.
Hurricane Clips						
Install	Ea.	1.72	3.59		5.31	Includes labor and material to install a
Minimum Charge	Job		260		260	hurricane clip.

Roof Framing

Roof Framing	Unit	Material	Labor	Equip.	Total	Specification
16" O.C. System						
2" x 4" Rafters						
Demolish	S.F.		1.01		1.01	Cost includes material and labor to
Install	S.F.	.75	.93		1.68	install 2" x 4", 16" O.C. rafter framing
Demolish and Install	S.F.	.75	1.94		2.69	for flat, shed or gable roofs, 25' span,
Reinstall	S.F.		.74		.74	up to 5/12 slope, per S.F. of roof.
Minimum Charge	Job		260		260	
2" x 6" Rafters						
Demolish	S.F.		1.06		1.06	Cost includes material and labor to
Install	S.F.	1	.98		1.98	install 2" x 6", 16" O.C. rafter framing
Demolish and Install	S.F.	1	2.04		3.04	for flat, shed or gable roofs, 25' span,
Reinstall	S.F.		.78		.78	up to 5/12 slope, per S.F. of roof.
Minimum Charge	Job		260		260	
2" x 8" Rafters						
Demolish	S.F.		1.08		1.08	Cost includes material and labor to
Install	S.F.	1.10	1.04		2.14	install 2" x 8", 16" O.C. rafter framing
Demolish and Install	S.F.	1.10	2.12		3.22	for flat, shed or gable roofs, 25' span,
Reinstall	S.F.		.83		.83	up to 5/12 slope, per S.F. of roof.
Minimum Charge	Job		260		260	
2" x 10" Rafters						
Demolish	S.F.		1.08		1.08	Cost includes material and labor to
Install	S.F.	2.92	1.58		4.50	install 2" x 10", 16" O.C. rafter
Demolish and Install	S.F.	2.92	2.66		5.58	framing for flat, shed or gable roofs,
Reinstall	S.F.		1.26		1.26	25' span, up to 5/12 slope, per S.F.
Minimum Charge	Job		260		260	of roof.

For customer support on your Contractor's Pricing Guide: Residential Repair & Remodeling, call 800.448.8182.

Rough Frame / Structure

Roof Framing

Roof Framing		Unit	Material	Labor	Equip.	Total	Specification
2" x 12" Rafters							Cost includes material and labor to install 2" x 12", 16" O.C. rafter framing for flat, shed or gable roofs, 25' span, up to 5/12 slope, per S.F. of roof.
	Demolish	S.F.		1.10		1.10	
	Install	S.F.	3.07	1.72		4.79	
	Demolish and Install	S.F.	3.07	2.82		5.89	
	Reinstall	S.F.		1.37		1.37	
	Minimum Charge	Job		260		260	
24" O.C. System							
2" x 4" Rafters							Cost includes material and labor to install 2" x 4", 24" O.C. rafter framing for flat, shed or gable roofs, 25' span, up to 5/12 slope, per S.F. of roof.
	Demolish	S.F.		.76		.76	
	Install	S.F.	.50	.70		1.20	
	Demolish and Install	S.F.	.50	1.46		1.96	
	Reinstall	S.F.		.56		.56	
	Minimum Charge	Job		260		260	
2" x 6" Rafters							Cost includes material and labor to install 2" x 6", 24" O.C. rafter framing for flat, shed or gable roofs, 25' span, up to 5/12 slope, per S.F. of roof.
	Demolish	S.F.		.80		.80	
	Install	S.F.	.67	.73		1.40	
	Demolish and Install	S.F.	.67	1.53		2.20	
	Reinstall	S.F.		.59		.59	
	Minimum Charge	Job		260		260	
2" x 8" Rafters							Cost includes material and labor to install 2" x 8", 24" O.C. rafter framing for flat, shed or gable roofs, 25' span, up to 5/12 slope, per S.F. of roof.
	Demolish	S.F.		.81		.81	
	Install	S.F.	.73	.78		1.51	
	Demolish and Install	S.F.	.73	1.59		2.32	
	Reinstall	S.F.		.63		.63	
	Minimum Charge	Job		260		260	
2" x 10" Rafters							Cost includes material and labor to install 2" x 10", 24" O.C. rafter framing for flat, shed or gable roofs, 25' span, up to 5/12 slope, per S.F. of roof.
	Demolish	S.F.		.81		.81	
	Install	S.F.	1.95	1.19		3.14	
	Demolish and Install	S.F.	1.95	2		3.95	
	Reinstall	S.F.		.95		.95	
	Minimum Charge	Job		260		260	
2" x 12" Rafters							Cost includes material and labor to install 2" x 12", 24" O.C. rafter framing for flat, shed or gable roofs, 25' span, up to 5/12 slope, per S.F. of roof.
	Demolish	S.F.		.82		.82	
	Install	S.F.	2.05	1.29		3.34	
	Demolish and Install	S.F.	2.05	2.11		4.16	
	Reinstall	S.F.		1.03		1.03	
	Minimum Charge	Job		260		260	
Rafter							
2" x 4"							Cost includes material and labor to install 2" x 4" rafters for flat, shed or gable roofs, up to 5/12 slope, 25' span, per L.F.
	Demolish	L.F.		1.03		1.03	
	Install	L.F.	.99	1.30		2.29	
	Demolish and Install	L.F.	.99	2.33		3.32	
	Reinstall	L.F.		1.04		1.04	
	Minimum Charge	Job		260		260	
2" x 6"							Cost includes material and labor to install 2" x 6" rafters for flat, shed or gable roofs, up to 5/12 slope, 25' span, per L.F.
	Demolish	L.F.		1.04		1.04	
	Install	L.F.	1.34	1.30		2.64	
	Demolish and Install	L.F.	1.34	2.34		3.68	
	Reinstall	L.F.		1.04		1.04	
	Minimum Charge	Job		260		260	
2" x 8"							Cost includes material and labor to install 2" x 8" rafters for flat, shed or gable roofs, up to 5/12 slope, 25' span, per L.F.
	Demolish	L.F.		1.06		1.06	
	Install	L.F.	1.46	1.39		2.85	
	Demolish and Install	L.F.	1.46	2.45		3.91	
	Reinstall	L.F.		1.11		1.11	
	Minimum Charge	Job		260		260	

Rough Frame / Structure

Roof Framing

Roof Framing		Unit	Material	Labor	Equip.	Total	Specification
2″ x 10″							
	Demolish	L.F.		1.08		1.08	Cost includes material and labor to
	Install	L.F.	3.88	2.10		5.98	install 2″ x 10″ rafters for flat, shed or
	Demolish and Install	L.F.	3.88	3.18		7.06	gable roofs, up to 5/12 slope, 25′
	Reinstall	L.F.		1.68		1.68	span, per L.F.
	Minimum Charge	Job		260		260	
2″ x 12″							
	Demolish	L.F.		1.09		1.09	Cost includes material and labor to
	Install	L.F.	4.09	2.29		6.38	install 2″ x 12″ rafters for flat, shed or
	Demolish and Install	L.F.	4.09	3.38		7.47	gable roofs, up to 5/12 slope, 25′
	Reinstall	L.F.		1.83		1.83	span, per L.F.
	Minimum Charge	Job		260		260	
2″x 4″ Valley / Jack							
	Demolish	L.F.		1.03		1.03	Includes material and labor to install
	Install	L.F.	1.34	2.19		3.53	up to 2″ x 6″ valley/jack rafters for
	Demolish and Install	L.F.	1.34	3.22		4.56	flat, shed or gable roofs, up to 5/12
	Reinstall	L.F.		1.75		1.75	slope, 25′ span, per L.F.
	Minimum Charge	Job		260		260	
2″x 6″ Valley / Jack							
	Demolish	L.F.		1.04		1.04	Includes material and labor to install
	Install	L.F.	1.34	2.19		3.53	up to 2″ x 6″ valley/jack rafters for
	Demolish and Install	L.F.	1.34	3.23		4.57	flat, shed or gable roofs, up to 5/12
	Reinstall	L.F.		1.75		1.75	slope, 25′ span, per L.F.
	Minimum Charge	Job		260		260	

Ridgeboard

Ridgeboard		Unit	Material	Labor	Equip.	Total	Specification
2″ x 4″							
	Demolish	L.F.		.99		.99	Cost includes material and labor to
	Install	L.F.	.99	1.89		2.88	install 2″ x 4″ ridgeboard for flat, shed
	Demolish and Install	L.F.	.99	2.88		3.87	or gable roofs, up to 5/12 slope, 25′
	Reinstall	L.F.		1.52		1.52	span, per L.F.
	Minimum Charge	Job		260		260	
2″ x 6″							
	Demolish	L.F.		1.01		1.01	Cost includes material and labor to
	Install	L.F.	1.34	2.08		3.42	install 2″ x 6″ ridgeboard for flat, shed
	Demolish and Install	L.F.	1.34	3.09		4.43	or gable roofs, up to 5/12 slope, 25′
	Reinstall	L.F.		1.67		1.67	span, per L.F.
	Minimum Charge	Job		260		260	
2″ x 8″							
	Demolish	L.F.		1.04		1.04	Cost includes material and labor to
	Install	L.F.	1.46	2.31		3.77	install 2″ x 8″ ridgeboard for flat, shed
	Demolish and Install	L.F.	1.46	3.35		4.81	or gable roofs, up to 5/12 slope, 25′
	Reinstall	L.F.		1.85		1.85	span, per L.F.
	Minimum Charge	Job		260		260	
2″ x 10″							
	Demolish	L.F.		1.08		1.08	Cost includes material and labor to
	Install	L.F.	3.88	2.60		6.48	install 2″ x 10″ ridgeboard for flat,
	Demolish and Install	L.F.	3.88	3.68		7.56	shed or gable roofs, up to 5/12
	Reinstall	L.F.		2.08		2.08	slope, 25′ span, per L.F.
	Minimum Charge	Job		260		260	
2″ x 12″							
	Demolish	L.F.		1.11		1.11	Cost includes material and labor to
	Install	L.F.	4.09	1.49		5.58	install 2″ x 12″ ridgeboard for flat,
	Demolish and Install	L.F.	4.09	2.60		6.69	shed or gable roofs, up to 5/12
	Reinstall	L.F.		1.19		1.19	slope, 25′ span, per L.F.
	Minimum Charge	Job		260		260	

For customer support on your Contractor's Pricing Guide: Residential Repair & Remodeling, call 800.448.8182.

Rough Frame / Structure

Roof Framing

Roof Framing	Unit	Material	Labor	Equip.	Total	Specification
Collar Beam						
1″ x 6″						
Demolish	L.F.		.78		.78	Cost includes material and labor to
Install	L.F.	1.11	.65		1.76	install 1″ x 6″ collar beam or tie for
Demolish and Install	L.F.	1.11	1.43		2.54	roof framing.
Reinstall	L.F.		.52		.52	
Minimum Charge	Job		260		260	
2″ x 6″						
Demolish	L.F.		.92		.92	Cost includes material and labor to
Install	L.F.	1.34	.65		1.99	install 2″ x 6″ collar beam or tie for
Demolish and Install	L.F.	1.34	1.57		2.91	roof framing.
Reinstall	L.F.		.52		.52	
Minimum Charge	Job		260		260	
Purlins						
2″ x 6″						
Demolish	L.F.		.89		.89	Cost includes material and labor to
Install	L.F.	1.34	.58		1.92	install 2″ x 6″ purlins below roof
Demolish and Install	L.F.	1.34	1.47		2.81	rafters.
Reinstall	L.F.		.46		.46	
Minimum Charge	Job		260		260	
2″ x 8″						
Demolish	L.F.		.90		.90	Cost includes material and labor to
Install	L.F.	1.46	.59		2.05	install 2″ x 8″ purlins below roof
Demolish and Install	L.F.	1.46	1.49		2.95	rafters.
Reinstall	L.F.		.47		.47	
Minimum Charge	Job		260		260	
2″ x 10″						
Demolish	L.F.		.91		.91	Cost includes material and labor to
Install	L.F.	3.88	.59		4.47	install 2″ x 10″ purlins below roof
Demolish and Install	L.F.	3.88	1.50		5.38	rafters.
Reinstall	L.F.		.47		.47	
Minimum Charge	Job		260		260	
2″ x 12″						
Demolish	L.F.		.92		.92	Cost includes material and labor to
Install	L.F.	4.09	.60		4.69	install 2″ x 12″ purlins below roof
Demolish and Install	L.F.	4.09	1.52		5.61	rafters.
Reinstall	L.F.		.48		.48	
Minimum Charge	Job		260		260	
4″ x 6″						
Demolish	L.F.		.92		.92	Cost includes material and labor to
Install	L.F.	5.60	.60		6.20	install 4″ x 6″ purlins below roof
Demolish and Install	L.F.	5.60	1.52		7.12	rafters.
Reinstall	L.F.		.48		.48	
Minimum Charge	Job		260		260	
4″ x 8″						
Demolish	L.F.		.93		.93	Cost includes material and labor to
Install	L.F.	12.25	.61		12.86	install 4″ x 8″ purlins below roof
Demolish and Install	L.F.	12.25	1.54		13.79	rafters.
Reinstall	L.F.		.49		.49	
Minimum Charge	Job		260		260	
Ledger Board						
2″ x 6″						
Demolish	L.F.		.44		.44	Cost includes material and labor to
Install	L.F.	1.34	2.35		3.69	install 2″ x 6″ ledger board nailed to
Demolish and Install	L.F.	1.34	2.79		4.13	the face of a rafter.
Reinstall	L.F.		1.88		1.88	
Minimum Charge	Job		260		260	

For customer support on your Contractor's Pricing Guide: Residential Repair & Remodeling, call 800.448.8182.

Rough Frame / Structure

Roof Framing

Roof Framing		Unit	Material	Labor	Equip.	Total	Specification
4" x 6"							
	Demolish	L.F.		.74		.74	Cost includes material and labor to
	Install	L.F.	5.60	3.45		9.05	install 4" x 6" ledger board nailed to
	Demolish and Install	L.F.	5.60	4.19		9.79	the face of a rafter.
	Reinstall	L.F.		2.76		2.76	
	Minimum Charge	Job		260		260	
4" x 8"							
	Demolish	L.F.		.99		.99	Cost includes material and labor to
	Install	L.F.	12.25	3.98		16.23	install 4" x 8" ledger board nailed to
	Demolish and Install	L.F.	12.25	4.97		17.22	the face of a rafter.
	Reinstall	L.F.		3.18		3.18	
	Minimum Charge	Job		260		260	
Outriggers							
2" x 4"							
	Demolish	L.F.		.31		.31	Cost includes material and labor to
	Install	L.F.	.99	2.08		3.07	install 2" x 4" outrigger rafters for flat,
	Demolish and Install	L.F.	.99	2.39		3.38	shed or gabled roofs, up to 5/12
	Reinstall	L.F.		1.67		1.67	slope, per L.F.
	Minimum Charge	Job		260		260	
2" x 6"							
	Demolish	L.F.		.46		.46	Cost includes material and labor to
	Install	L.F.	1.34	2.08		3.42	install 2" x 6" outrigger rafters for flat,
	Demolish and Install	L.F.	1.34	2.54		3.88	shed or gabled roofs, up to 5/12
	Reinstall	L.F.		1.67		1.67	slope, per L.F.
	Minimum Charge	Job		260		260	
Lookout Rafter							
2" x 4"							
	Demolish	L.F.		1.03		1.03	Cost includes material and labor to
	Install	L.F.	.99	.87		1.86	install 2" x 4" lookout rafters for flat,
	Demolish and Install	L.F.	.99	1.90		2.89	shed or gabled roofs, up to 5/12
	Reinstall	L.F.		.69		.69	slope, per L.F.
	Minimum Charge	Job		260		260	
2" x 6"							
	Demolish	L.F.		1.04		1.04	Cost includes material and labor to
	Install	L.F.	1.34	.88		2.22	install 2" x 6" lookout rafters for flat,
	Demolish and Install	L.F.	1.34	1.92		3.26	shed or gabled roofs, up to 5/12
	Reinstall	L.F.		.71		.71	slope, per L.F.
	Minimum Charge	Job		260		260	
Fly Rafter							
2" x 4"							
	Demolish	L.F.		1.03		1.03	Cost includes material and labor to
	Install	L.F.	.99	.87		1.86	install 2" x 4" fly rafters for flat, shed
	Demolish and Install	L.F.	.99	1.90		2.89	or gabled roofs, up to 5/12 slope,
	Reinstall	L.F.		.69		.69	per L.F.
	Minimum Charge	Job		260		260	
2" x 6"							
	Demolish	L.F.		1.04		1.04	Cost includes material and labor to
	Install	L.F.	1.30	.88		2.18	install 2" x 6" fly rafters for flat, shed
	Demolish and Install	L.F.	1.30	1.92		3.22	or gabled roofs, up to 5/12 slope,
	Reinstall	L.F.		.71		.71	per L.F.
	Minimum Charge	Job		260		260	

For customer support on your Contractor's Pricing Guide: Residential Repair & Remodeling, call 800.448.8182.

Rough Frame / Structure

Residential Trusses

	Unit	Material	Labor	Equip.	Total	Specification
W or Fink Truss						
24' Span						
Demolish	Ea.		41	43	84	Includes material, labor and
Install	Ea.	118	40	42.50	200.50	equipment to install wood gang-nailed
Demolish and Install	Ea.	118	81	85.50	284.50	residential truss, up to 24' span.
Reinstall	Ea.		40.12	42.57	82.69	
Minimum Charge	Job		260		260	
28' Span						
Demolish	Ea.		43.50	46	89.50	Includes material, labor and
Install	Ea.	138	45.50	48	231.50	equipment to install wood gang-nailed
Demolish and Install	Ea.	138	89	94	321	residential truss, up to 28' span.
Reinstall	Ea.		45.42	48.20	93.62	
Minimum Charge	Job		260		260	
32' Span						
Demolish	Ea.		48	50.50	98.50	Includes material, labor and
Install	Ea.	187	48	51	286	equipment to install wood gang-nailed
Demolish and Install	Ea.	187	96	101.50	384.50	residential truss, up to 32' span.
Reinstall	Ea.		48.14	51.09	99.23	
Minimum Charge	Job		260		260	
36' Span						
Demolish	Ea.		51.50	54.50	106	Includes material, labor and
Install	Ea.	206	52.50	55.50	314	equipment to install wood gang-nailed
Demolish and Install	Ea.	206	104	110	420	residential truss, up to 36' span.
Reinstall	Ea.		52.33	55.53	107.86	
Minimum Charge	Job		260		260	
Gable End						
24' Span						
Demolish	Ea.		41	43	84	Includes material, labor and
Install	Ea.	117	43	45.50	205.50	equipment to install wood gang-nailed
Demolish and Install	Ea.	117	84	88.50	289.50	residential gable end truss, up to 24'
Reinstall	Ea.		42.99	45.61	88.60	span.
Clean	Ea.	1.32	53.50		54.82	
Paint	Ea.	8.90	57.50		66.40	
Minimum Charge	Job		260		260	
28' Span						
Demolish	Ea.		43.50	46	89.50	Includes material, labor and
Install	Ea.	149	51	54.50	254.50	equipment to install wood gang-nailed
Demolish and Install	Ea.	149	94.50	100.50	344	residential gable end truss, up to 30'
Reinstall	Ea.		51.22	54.35	105.57	span.
Clean	Ea.	1.38	72.50		73.88	
Paint	Ea.	9.30	78.50		87.80	
Minimum Charge	Job		260		260	
32' Span						
Demolish	Ea.		48	50.50	98.50	Includes material, labor and
Install	Ea.	162	52.50	55.50	270	equipment to install wood gang-nailed
Demolish and Install	Ea.	162	100.50	106	368.50	residential gable end truss, up to 32'
Reinstall	Ea.		52.33	55.53	107.86	span.
Clean	Ea.	1.44	95		96.44	
Paint	Ea.	9.75	103		112.75	
Minimum Charge	Job		260		260	
36' Span						
Demolish	Ea.		51.50	54.50	106	Includes material, labor and
Install	Ea.	250	60	64	374	equipment to install wood gang-nailed
Demolish and Install	Ea.	250	111.50	118.50	480	residential gable end truss, up to 36'
Reinstall	Ea.		60.18	63.86	124.04	span.
Clean	Ea.	1.51	120		121.51	
Paint	Ea.	10.20	130		140.20	
Minimum Charge	Job		260		260	

For customer support on your Contractor's Pricing Guide: Residential Repair & Remodeling, call 800.448.8182.

Rough Frame / Structure

Residential Trusses

	Unit	Material	Labor	Equip.	Total	Specification
Truss Hardware						
5-1/4″ Glue-Lam Seat						
Install	Ea.	54	2.89		56.89	Includes labor and material to install
Minimum Charge	Job		260		260	beam hangers.
6-3/4″ Glue-Lam Seat						
Install	Ea.	77.50	2.89		80.39	Includes labor and material to install
Minimum Charge	Job		260		260	beam hangers.
2″ x 4″ Joist Hanger						
Install	Ea.	1.02	2.98		4	Includes labor and material to install
Minimum Charge	Job		260		260	joist and beam hangers.
2″ x 12″ Joist Hanger						
Install	Ea.	5.10	3.16		8.26	Cost includes labor and material to
Minimum Charge	Job		260		260	install 2″ x 12″ joist hanger.

Commercial Trusses

	Unit	Material	Labor	Equip.	Total	Specification
Engineered Lumber, Truss / Joist						
9-1/2″						
Demolish	S.F.		.73		.73	Includes material and labor to install
Install	S.F.	3.74	.52		4.26	engineered lumber truss / joists per
Demolish and Install	S.F.	3.74	1.25		4.99	S.F. of floor area based on joists 16″
Reinstall	S.F.		.42		.42	O.C. Beams, supports and bridging
Minimum Charge	Job		260		260	are not included.
11-7/8″						
Demolish	S.F.		.76		.76	Includes material and labor to install
Install	S.F.	1.65	.53		2.18	engineered lumber truss / joists per
Demolish and Install	S.F.	1.65	1.29		2.94	S.F. of floor area based on joists 16″
Reinstall	S.F.		.43		.43	O.C. Beams, supports and bridging
Minimum Charge	Job		260		260	are not included.
14″						
Demolish	S.F.		.78		.78	Includes material and labor to install
Install	S.F.	8.95	.57		9.52	engineered lumber truss / joists per
Demolish and Install	S.F.	8.95	1.35		10.30	S.F. of floor area based on joists 16″
Reinstall	S.F.		.46		.46	O.C. Beams, supports and bridging
Minimum Charge	Job		260		260	are not included.
16″						
Demolish	S.F.		.81		.81	Includes material and labor to install
Install	S.F.	9.35	.60		9.95	engineered lumber truss / joists per
Demolish and Install	S.F.	9.35	1.41		10.76	S.F. of floor area based on joists 16″
Reinstall	S.F.		.48		.48	O.C. Beams, supports and bridging
Minimum Charge	Job		260		260	are not included.
Bowstring Truss						
100′ Clear Span						
Demolish	SF Flr.		.74	.87	1.61	Includes material, labor and
Install	SF Flr.	14.25	.61	.73	15.59	equipment to install bow string truss
Demolish and Install	SF Flr.	14.25	1.35	1.60	17.20	system up to 100′ clear span per S.F.
Minimum Charge	Job		260		260	of floor area.
120′ Clear Span						
Demolish	SF Flr.		.83	.97	1.80	Includes material, labor and
Install	SF Flr.	15.20	.67	.81	16.68	equipment to install bow string truss
Demolish and Install	SF Flr.	15.20	1.50	1.78	18.48	system up to 120′ clear span per S.F.
Minimum Charge	Job		260		260	of floor area.

Rough Frame / Structure

Roof Sheathing		Unit	Material	Labor	Equip.	Total	Specification
Plywood							
3/8"							
	Demolish	S.F.		.58		.58	Cost includes material and labor to
	Install	S.F.	.97	.68		1.65	install 3/8" CDX plywood roof
	Demolish and Install	S.F.	.97	1.26		2.23	sheathing.
	Reinstall	S.F.		.68		.68	
	Minimum Charge	Job		260		260	
1/2"							
	Demolish	S.F.		.58		.58	Cost includes material and labor to
	Install	S.F.	1.09	.74		1.83	install 1/2" CDX plywood roof
	Demolish and Install	S.F.	1.09	1.32		2.41	sheathing.
	Minimum Charge	Job		260		260	
5/8"							
	Demolish	S.F.		.58		.58	Cost includes material and labor to
	Install	S.F.	1.45	.80		2.25	install 5/8" CDX plywood roof
	Demolish and Install	S.F.	1.45	1.38		2.83	sheathing.
	Minimum Charge	Job		260		260	
3/4"							
	Demolish	S.F.		.58		.58	Cost includes material and labor to
	Install	S.F.	1.45	.87		2.32	install 3/4" CDX plywood roof
	Demolish and Install	S.F.	1.45	1.45		2.90	sheathing.
	Minimum Charge	Job		260		260	
OSB							
1/2"							
	Demolish	S.F.		.58		.58	Cost includes material and labor to
	Install	S.F.	.68	.74		1.42	install 4' x 8' x 1/2" OSB sheathing.
	Demolish and Install	S.F.	.68	1.32		2	
	Minimum Charge	Job		260		260	
5/8"							
	Demolish	S.F.		.58		.58	Cost includes material and labor to
	Install	S.F.	.68	.60		1.28	install 4' x 8' x 5/8" OSB sheathing.
	Demolish and Install	S.F.	.68	1.18		1.86	
	Minimum Charge	Job		260		260	
Plank							
	Demolish	S.F.		.58		.58	Cost includes material and labor to
	Install	S.F.	1.35	1.44		2.79	install 1" x 6" or 1" x 8" utility T&G
	Demolish and Install	S.F.	1.35	2.02		3.37	board sheathing laid diagonal.
	Reinstall	S.F.		1.44		1.44	
	Minimum Charge	Job		260		260	
Skip-type							
1" x 4"							
	Demolish	S.F.		.89		.89	Includes material and labor to install
	Install	S.F.	.97	.43		1.40	sheathing board material 1" x 4", 7"
	Demolish and Install	S.F.	.97	1.32		2.29	O.C.
	Minimum Charge	Job		260		260	
1" x 6"							
	Demolish	S.F.		.89		.89	Includes material and labor to install
	Install	S.F.	1.09	.36		1.45	sheathing board material 1" x 6", 7"
	Demolish and Install	S.F.	1.09	1.25		2.34	or 9" O.C.
	Minimum Charge	Job		260		260	

Rough Frame / Structure

Wood Deck

Wood Deck		Unit	Material	Labor	Equip.	Total	Specification
Stairs							
	Demolish	Ea.		515		515	Includes material and labor to install
	Install	Ea.	200	925		1125	wood steps. 5 - 7 risers included.
	Demolish and Install	Ea.	200	1440		1640	
	Reinstall	Ea.		739.20		739.20	
	Clean	Flight	3.17	34		37.17	
	Paint	Flight	37.50	57.50		95	
	Minimum Charge	Job		260		260	
Bench Seating							
	Demolish	L.F.		7.40		7.40	Includes material and labor to install
	Install	L.F.	3.56	26		29.56	pressure treated bench seating.
	Demolish and Install	L.F.	3.56	33.40		36.96	
	Reinstall	L.F.		20.83		20.83	
	Clean	L.F.	.13	2		2.13	
	Paint	L.F.	1.28	3.45		4.73	
	Minimum Charge	Job		260		260	
Privacy Wall							
	Demolish	SF Flr.		1.78		1.78	Includes material, labor and
	Install	SF Flr.	1.83	.97	.40	3.20	equipment to install privacy wall.
	Demolish and Install	SF Flr.	1.83	2.75	.40	4.98	
	Reinstall	SF Flr.		.78	.32	1.10	
	Clean	SF Flr.	.07	.41		.48	
	Paint	S.F.	.15	.33		.48	
	Minimum Charge	Job		260		260	
Treated Column or Post							
4" x 4"							
	Demolish	L.F.		1.11		1.11	Includes material and labor to install
	Install	L.F.	1.63	2.67		4.30	4" x 4" treated column or post.
	Demolish and Install	L.F.	1.63	3.78		5.41	
	Reinstall	L.F.		2.14		2.14	
	Clean	L.F.	.11	.36		.47	
	Paint	L.F.	.48	1.08		1.56	
	Minimum Charge	Job		260		260	
4" x 6"							
	Demolish	L.F.		1.61		1.61	Includes material and labor to install
	Install	L.F.	4.17	3.79		7.96	4" x 6" treated column or post.
	Demolish and Install	L.F.	4.17	5.40		9.57	
	Reinstall	L.F.		3.03		3.03	
	Clean	L.F.	.14	.50		.64	
	Paint	L.F.	.48	1.08		1.56	
	Minimum Charge	Job		260		260	
Cedar Columns or Posts							
4" x 4"							
	Demolish	L.F.		1.11		1.11	Includes material and labor to install
	Install	L.F.	6.65	2.67		9.32	4" x 4" cedar columns or posts.
	Demolish and Install	L.F.	6.65	3.78		10.43	
	Reinstall	L.F.		2.14		2.14	
	Clean	L.F.	.11	.36		.47	
	Paint	L.F.	.48	1.08		1.56	
	Minimum Charge	Job		260		260	
4" x 6"							
	Demolish	L.F.		1.61		1.61	Includes material and labor to install
	Install	L.F.	10.85	3.79		14.64	4" x 6" cedar columns or posts.
	Demolish and Install	L.F.	10.85	5.40		16.25	
	Reinstall	L.F.		3.03		3.03	
	Clean	L.F.	.14	.50		.64	
	Paint	L.F.	.48	1.08		1.56	
	Minimum Charge	Job		260		260	

For customer support on your Contractor's Pricing Guide: Residential Repair & Remodeling, call 800.448.8182.

Rough Frame / Structure

Wood Deck	Unit	Material	Labor	Equip.	Total	Specification
Redwood Columns or Posts						
4″ x 4″						
Demolish	L.F.		1.11		1.11	Includes material and labor to install
Install	L.F.	16.25	2.67		18.92	4″ x 4″ redwood columns or posts.
Demolish and Install	L.F.	16.25	3.78		20.03	
Reinstall	L.F.		2.14		2.14	
Clean	L.F.	.11	.36		.47	
Paint	L.F.	.48	1.08		1.56	
Minimum Charge	Job		260		260	
4″ x 6″						
Demolish	L.F.		1.61		1.61	Includes material and labor to install
Install	L.F.	19.95	3.79		23.74	4″ x 6″ redwood columns or posts.
Demolish and Install	L.F.	19.95	5.40		25.35	
Reinstall	L.F.		3.03		3.03	
Clean	L.F.	.14	.50		.64	
Paint	L.F.	.48	1.08		1.56	
Minimum Charge	Job		260		260	
Treated Girder - Single						
4″ x 8″						
Demolish	L.F.		3.29		3.29	Includes material and labor to install a
Install	L.F.	9.25	1.98		11.23	treated lumber girder consisting of one
Demolish and Install	L.F.	9.25	5.27		14.52	4″ x 8″ for porch or deck construction.
Reinstall	L.F.		1.59		1.59	
Clean	L.F.	.13	.43		.56	
Minimum Charge	Job		260		260	
Treated Girder - Double						
2″ x 8″						
Demolish	L.F.		3.29		3.29	Includes material and labor to install a
Install	L.F.	3.26	1.81		5.07	treated lumber girder consisting of two
Demolish and Install	L.F.	3.26	5.10		8.36	2″ x 8″s for porch or deck
Reinstall	L.F.		1.45		1.45	construction.
Clean	L.F.	.13	.43		.56	
Minimum Charge	Job		260		260	
2″ x 10″						
Demolish	L.F.		4.12		4.12	Includes material and labor to install a
Install	L.F.	4.53	1.89		6.42	treated lumber girder consisting of two
Demolish and Install	L.F.	4.53	6.01		10.54	2″ x 10″s for porch or deck
Reinstall	L.F.		1.52		1.52	construction.
Clean	L.F.	.14	.50		.64	
Minimum Charge	Job		260		260	
2″ x 12″						
Demolish	L.F.		4.95		4.95	Includes material and labor to install a
Install	L.F.	6.55	1.98		8.53	treated lumber girder consisting of two
Demolish and Install	L.F.	6.55	6.93		13.48	2″ x 12″s for porch or deck
Reinstall	L.F.		1.59		1.59	construction.
Clean	L.F.	.17	.57		.74	
Minimum Charge	Job		260		260	
Treated Girder - Triple						
2″ x 8″						
Demolish	L.F.		4.95		4.95	Includes material and labor to install a
Install	L.F.	4.88	1.98		6.86	treated lumber girder consisting of
Demolish and Install	L.F.	4.88	6.93		11.81	three 2″ x 8″s for porch or deck
Reinstall	L.F.		1.59		1.59	construction.
Clean	L.F.	.14	.50		.64	
Minimum Charge	Job		260		260	

Rough Frame / Structure

Wood Deck

Wood Deck	Unit	Material	Labor	Equip.	Total	Specification
2″ x 10″						
Demolish	L.F.		6.20		6.20	Includes material and labor to install a
Install	L.F.	6.75	2.08		8.83	treated lumber girder consisting of
Demolish and Install	L.F.	6.75	8.28		15.03	three 2″ x 10″s for porch or deck
Reinstall	L.F.		1.67		1.67	construction.
Clean	L.F.	.17	.57		.74	
Minimum Charge	Job		260		260	
2″ x 12″						
Demolish	L.F.		7.35		7.35	Includes material and labor to install a
Install	L.F.	9.85	2.19		12.04	treated lumber girder consisting of
Demolish and Install	L.F.	9.85	9.54		19.39	three 2″ x 12″s for porch or deck
Reinstall	L.F.		1.75		1.75	construction.
Clean	L.F.	.19	.64		.83	
Minimum Charge	Job		260		260	

Treated Ledger - Bolted

	Unit	Material	Labor	Equip.	Total	Specification
2″ x 8″						
Demolish	L.F.		.55		.55	Includes material and labor to install a
Install	L.F.	1.83	2.67		4.50	treated lumber 2″ x 8″ ledger with
Demolish and Install	L.F.	1.83	3.22		5.05	carriage bolts 4′ on center for porch
Reinstall	L.F.		2.14		2.14	or deck construction.
Clean	L.F.	.02	.29		.31	
Minimum Charge	Job		260		260	
2″ x 10″						
Demolish	L.F.		.55		.55	Includes material and labor to install a
Install	L.F.	2.43	2.71		5.14	treated lumber 2″ x 10″ ledger with
Demolish and Install	L.F.	2.43	3.26		5.69	carriage bolts 4′ on center for porch
Reinstall	L.F.		2.16		2.16	or deck construction.
Clean	L.F.	.03	.33		.36	
Minimum Charge	Job		260		260	
2″ x 12″						
Demolish	L.F.		.55		.55	Includes material and labor to install a
Install	L.F.	3.42	2.74		6.16	treated lumber 2″ x 12″ ledger with
Demolish and Install	L.F.	3.42	3.29		6.71	carriage bolts 4′ on center for porch
Reinstall	L.F.		2.19		2.19	or deck construction.
Clean	L.F.	.03	.33		.36	
Minimum Charge	Job		260		260	

Treated Joists

	Unit	Material	Labor	Equip.	Total	Specification
2″ x 8″						
Demolish	L.F.		.94		.94	Includes material and labor to install
Install	L.F.	1.57	.95		2.52	treated lumber 2″ x 8″ deck or floor
Demolish and Install	L.F.	1.57	1.89		3.46	joists for porch or deck construction.
Reinstall	L.F.		.76		.76	
Clean	L.F.	.11	.36		.47	
Minimum Charge	Job		260		260	
2″ x 10″						
Demolish	L.F.		.98		.98	Includes material and labor to install
Install	L.F.	2.20	1.16		3.36	treated lumber 2″ x 10″ deck or floor
Demolish and Install	L.F.	2.20	2.14		4.34	joists for porch or deck construction.
Reinstall	L.F.		.93		.93	
Clean	L.F.	.13	.43		.56	
Minimum Charge	Job		260		260	
2″ x 12″						
Demolish	L.F.		1.01		1.01	Includes material and labor to install
Install	L.F.	2.64	1.19		3.83	treated lumber 2″ x 12″ deck or floor
Demolish and Install	L.F.	2.64	2.20		4.84	joists for porch or deck construction.
Reinstall	L.F.		.95		.95	
Clean	L.F.	.14	.50		.64	
Minimum Charge	Job		260		260	

For customer support on your Contractor's Pricing Guide: Residential Repair & Remodeling, call 800.448.8182.

Rough Frame / Structure

Wood Deck	Unit	Material	Labor	Equip.	Total	Specification
Treated Railings and Trim						
2″ x 2″						
Demolish	L.F.		.37		.37	Includes material and labor to install
Install	L.F.	.67	1.74		2.41	treated lumber 2″ x 2″ railings and
Demolish and Install	L.F.	.67	2.11		2.78	trim for porch or deck construction.
Reinstall	L.F.		1.39		1.39	
Clean	L.F.	.01	.27		.28	
Paint	L.F.	.48	1.08		1.56	
Minimum Charge	Job		260		260	
2″ x 4″						
Demolish	L.F.		.37		.37	Includes material and labor to install
Install	L.F.	.96	1.74		2.70	treated lumber 2″ x 4″ railings and
Demolish and Install	L.F.	.96	2.11		3.07	trim for porch or deck construction.
Reinstall	L.F.		1.39		1.39	
Clean	L.F.	.02	.29		.31	
Paint	L.F.	.48	1.08		1.56	
Minimum Charge	Job		260		260	
2″ x 6″						
Demolish	L.F.		.37		.37	Includes material and labor to install
Install	L.F.	1.33	1.74		3.07	treated lumber 2″ x 6″ railings and
Demolish and Install	L.F.	1.33	2.11		3.44	trim for porch or deck construction.
Reinstall	L.F.		1.39		1.39	
Clean	L.F.	.02	.33		.35	
Paint	L.F.	.48	1.08		1.56	
Minimum Charge	Job		260		260	
Cedar Railings and Trim						
1″ x 4″						
Demolish	L.F.		.37		.37	Includes material and labor to install
Install	L.F.	1.79	1.74		3.53	cedar 1″ x 4″ railings and trim for
Demolish and Install	L.F.	1.79	2.11		3.90	porch or deck construction.
Reinstall	L.F.		1.39		1.39	
Clean	L.F.	.01	.27		.28	
Paint	L.F.	.48	1.08		1.56	
Minimum Charge	Job		260		260	
2″ x 4″						
Demolish	L.F.		.37		.37	Includes material and labor to install
Install	L.F.	11.05	1.74		12.79	cedar 2″ x 4″ railings and trim for
Demolish and Install	L.F.	11.05	2.11		13.16	porch or deck construction.
Reinstall	L.F.		1.39		1.39	
Clean	L.F.	.02	.29		.31	
Paint	L.F.	.48	1.08		1.56	
Minimum Charge	Job		260		260	
2″ x 6″						
Demolish	L.F.		.37		.37	Includes material and labor to install
Install	L.F.	8.25	1.74		9.99	cedar 2″ x 6″ railings and trim for
Demolish and Install	L.F.	8.25	2.11		10.36	porch or deck construction.
Reinstall	L.F.		1.39		1.39	
Clean	L.F.	.02	.33		.35	
Paint	L.F.	.48	1.08		1.56	
Minimum Charge	Job		260		260	
Redwood Railings and Trim						
1″ x 4″						
Demolish	L.F.		.37		.37	Includes material and labor to install
Install	L.F.	4.37	1.74		6.11	redwood 1″ x 4″ railings and trim for
Demolish and Install	L.F.	4.37	2.11		6.48	porch or deck construction.
Reinstall	L.F.		1.39		1.39	
Clean	L.F.	.01	.27		.28	
Paint	L.F.	.48	1.08		1.56	
Minimum Charge	Job		260		260	

For customer support on your Contractor's Pricing Guide: Residential Repair & Remodeling, call 800.448.8182.

Rough Frame / Structure

Wood Deck

Wood Deck		Unit	Material	Labor	Equip.	Total	Specification
2″ x 6″							
	Demolish	L.F.		.37		.37	Includes material and labor to install
	Install	L.F.	27	1.74		28.74	redwood 2″ x 6″ railings and trim for
	Demolish and Install	L.F.	27	2.11		29.11	porch or deck construction.
	Reinstall	L.F.		1.39		1.39	
	Clean	L.F.	.02	.33		.35	
	Paint	L.F.	.48	1.08		1.56	
	Minimum Charge	Job		260		260	

Treated Decking

		Unit	Material	Labor	Equip.	Total	Specification
1″ x 4″							
	Demolish	L.F.		.54		.54	Includes material and labor to install
	Install	S.F.	2.49	1.89		4.38	treated lumber 1″ x 4″ decking for
	Demolish and Install	S.F.	2.49	2.43		4.92	porch or deck construction.
	Reinstall	S.F.		1.52		1.52	
	Clean	S.F.	.07	.41		.48	
	Paint	S.F.	.43	.86		1.29	
	Minimum Charge	Job		260		260	
2″ x 4″							
	Demolish	L.F.		.54		.54	Includes material and labor to install
	Install	S.F.	3.18	1.74		4.92	treated lumber 2″ x 4″ decking for
	Demolish and Install	S.F.	3.18	2.28		5.46	porch or deck construction.
	Reinstall	S.F.		1.39		1.39	
	Clean	S.F.	.07	.41		.48	
	Paint	S.F.	.43	.86		1.29	
	Minimum Charge	Job		260		260	
2″ x 6″							
	Demolish	L.F.		.54		.54	Includes material and labor to install
	Install	S.F.	2.84	1.63		4.47	treated lumber 2″ x 6″ decking for
	Demolish and Install	S.F.	2.84	2.17		5.01	porch or deck construction.
	Reinstall	S.F.		1.30		1.30	
	Clean	S.F.	.07	.41		.48	
	Paint	S.F.	.43	.86		1.29	
	Minimum Charge	Job		260		260	
5/4″ x 6″							
	Demolish	L.F.		.54		.54	Includes material and labor to install
	Install	S.F.	2.99	1.63		4.62	treated lumber 5/4″ x 6″ decking for
	Demolish and Install	S.F.	2.99	2.17		5.16	porch or deck construction.
	Reinstall	S.F.		1.30		1.30	
	Clean	S.F.	.07	.41		.48	
	Paint	S.F.	.43	.86		1.29	
	Minimum Charge	Job		260		260	

Wood / Plastic Composite Decking

		Unit	Material	Labor	Equip.	Total	Specification
5/4″ x 6″							
	Demolish	L.F.		.54		.54	Includes material and labor to install
	Install	L.F.	2.57	1.63		4.20	wood/plastic composite 5/4″ x 6″
	Demolish and Install	L.F.	2.57	2.17		4.74	decking for porch or deck
	Reinstall	L.F.		1.30		1.30	construction.
	Clean	S.F.	.07	.41		.48	
	Paint	S.F.	.43	.86		1.29	
	Minimum Charge	Job		260		260	

Square Edge Fir Decking

		Unit	Material	Labor	Equip.	Total	Specification
1″ x 4″							
	Demolish	L.F.		.54		.54	Includes material and labor to install
	Install	L.F.	2.52	1.89		4.41	square edge fir 1″ x 4″ decking for
	Demolish and Install	L.F.	2.52	2.43		4.95	porch or deck construction.
	Reinstall	L.F.		1.52		1.52	
	Clean	S.F.	.07	.41		.48	
	Paint	S.F.	.43	.86		1.29	
	Minimum Charge	Job		260		260	

For customer support on your Contractor's Pricing Guide: Residential Repair & Remodeling, call 800.448.8182.

Rough Frame / Structure

Wood Deck	Unit	Material	Labor	Equip.	Total	Specification
Tongue and Groove Fir Decking						
1″ x 4″						
Demolish	L.F.		.54		.54	Includes material and labor to install
Install	L.F.	1.87	2.31		4.18	treated lumber 1″ x 4″ decking for
Demolish and Install	L.F.	1.87	2.85		4.72	porch or deck construction.
Reinstall	L.F.		1.85		1.85	
Clean	S.F.	.07	.41		.48	
Paint	S.F.	.43	.86		1.29	
Minimum Charge	Job		260		260	
Mahogany Decking						
1″ x 4″						
Demolish	L.F.		.54		.54	Includes material and labor to install
Install	L.F.	3.16	1.89		5.05	treated lumber 1″ x 4″ decking for
Demolish and Install	L.F.	3.16	2.43		5.59	porch or deck construction.
Reinstall	L.F.		1.52		1.52	
Clean	S.F.	.07	.41		.48	
Paint	S.F.	.43	.86		1.29	
Minimum Charge	Job		260		260	
PVC Decking						
5/4″ x 6″						
Demolish	L.F.		.54		.54	Includes material and labor to install
Install	L.F.	3.62	1.89		5.51	PVC 5/4″ x 6″ decking for porch or
Demolish and Install	L.F.	3.62	2.43		6.05	deck construction.
Reinstall	L.F.		1.52		1.52	
Clean	S.F.	.07	.41		.48	
Minimum Charge	Job		260		260	
PVC Handrails and Balusters						
Straight						
Demolish	L.F.		3.70		3.70	Includes material and labor to install
Install	L.F.	32	5.45		37.45	straight PVC handrails and balusters
Demolish and Install	L.F.	32	9.15		41.15	for porch or deck construction.
Reinstall	L.F.		4.34		4.34	
Clean	L.F.	.64	3.12		3.76	
Minimum Charge	Job		260		260	
Angled						
Demolish	L.F.		3.70		3.70	Includes material and labor to install
Install	L.F.	26	7.25		33.25	angled PVC handrails and balusters
Demolish and Install	L.F.	26	10.95		36.95	for porch or deck construction.
Reinstall	L.F.		5.79		5.79	
Clean	L.F.	.64	3.12		3.76	
Minimum Charge	Job		260		260	
PVC Post Sleeve						
for 4″ x 4″ post						
Demolish	L.F.		1.11		1.11	Includes material and labor to install
Install	L.F.	9.70	5.45		15.15	PVC post sleeves over 4″ x 4″ porch
Demolish and Install	L.F.	9.70	6.56		16.26	or deck posts.
Reinstall	L.F.		4.34		4.34	
Clean	L.F.	.11	.36		.47	
Minimum Charge	Job		260		260	
Post Cap						
Demolish	Ea.		5.55		5.55	Includes material and labor to install
Install	Ea.	37	10.85		47.85	PVC post caps for porch or deck
Demolish and Install	Ea.	37	16.40		53.40	construction.
Reinstall	Ea.		8.68		8.68	
Clean	L.F.	.11	.36		.47	
Minimum Charge	Job		260		260	

For customer support on your Contractor's Pricing Guide: Residential Repair & Remodeling, call 800.448.8182.

Rough Frame / Structure

Porch

	Unit	Material	Labor	Equip.	Total	Specification
Stairs						
Demolish	Ea.		515		515	Includes material and labor to install
Install	Ea.	200	925		1125	wood steps. 5 - 7 risers included.
Demolish and Install	Ea.	200	1440		1640	
Reinstall	Ea.		739.20		739.20	
Clean	Flight	3.17	34		37.17	
Paint	Flight	37.50	57.50		95	
Minimum Charge	Job		260		260	
Bench Seating						
Demolish	L.F.		7.40		7.40	Includes material and labor to install
Install	L.F.	3.56	26		29.56	pressure treated bench seating.
Demolish and Install	L.F.	3.56	33.40		36.96	
Reinstall	L.F.		20.83		20.83	
Clean	L.F.	.13	2		2.13	
Paint	L.F.	1.28	3.45		4.73	
Minimum Charge	Job		260		260	
Privacy Wall						
Demolish	SF Flr.		1.78		1.78	Includes material, labor and
Install	SF Flr.	1.83	.97	.40	3.20	equipment to install privacy wall.
Demolish and Install	SF Flr.	1.83	2.75	.40	4.98	
Reinstall	SF Flr.		.78	.32	1.10	
Clean	SF Flr.	.07	.41		.48	
Paint	S.F.	.15	.33		.48	
Minimum Charge	Job		260		260	
Wood Board Ceiling						
Demolish	S.F.		.74		.74	Includes material and labor to install
Install	S.F.	8.60	2.31		10.91	wood board on ceiling.
Demolish and Install	S.F.	8.60	3.05		11.65	
Clean	S.F.	.02	.23		.25	
Paint	S.F.	.55	.54		1.09	
Minimum Charge	Job		260		260	
Re-screen						
Install	S.F.	4.74	2.08		6.82	Includes labor and material to
Minimum Charge	Job		260		260	re-screen wood frame.

Exterior Trim

Exterior Trim

Exterior Trim	Unit	Material	Labor	Equip.	Total	Specification
Stock Lumber						
1" x 2"						
Demolish	L.F.		.37		.37	Cost includes material and labor to
Install	L.F.	.20	1.58		1.78	install 1" x 2" pine trim.
Demolish and Install	L.F.	.20	1.95		2.15	
Clean	L.F.	.01	.27		.28	
Paint	L.F.	.07	.67		.74	
Minimum Charge	Job		260		260	
1" x 3"						
Demolish	L.F.		.37		.37	Cost includes material and labor to
Install	L.F.	.36	1.80		2.16	install 1" x 3" pine trim.
Demolish and Install	L.F.	.36	2.17		2.53	
Clean	L.F.	.01	.27		.28	
Paint	L.F.	.07	.67		.74	
Minimum Charge	Job		260		260	
1" x 4"						
Demolish	L.F.		.37		.37	Cost includes material and labor to
Install	L.F.	.83	2.08		2.91	install 1" x 4" pine trim.
Demolish and Install	L.F.	.83	2.45		3.28	
Clean	L.F.	.01	.27		.28	
Paint	L.F.	.07	.67		.74	
Minimum Charge	Job		260		260	
1" x 6"						
Demolish	L.F.		.37		.37	Cost includes material and labor to
Install	L.F.	1.14	2.08		3.22	install 1" x 6" pine trim.
Demolish and Install	L.F.	1.14	2.45		3.59	
Clean	L.F.	.01	.27		.28	
Paint	L.F.	.24	.67		.91	
Minimum Charge	Job		260		260	
1" x 8"						
Demolish	L.F.		.40		.40	Cost includes material and labor to
Install	L.F.	.90	2.60		3.50	install 1" x 8" pine trim.
Demolish and Install	L.F.	.90	3		3.90	
Reinstall	L.F.		2.08		2.08	
Clean	L.F.	.02	.29		.31	
Paint	L.F.	.24	.67		.91	
Minimum Charge	Job		260		260	

Exterior Trim

Exterior Trim	Unit	Material	Labor	Equip.	Total	Specification
Stock Lumber						
1" x 10"						
Demolish	L.F.		.62		.62	Cost includes material and labor to
Install	L.F.	1.28	2.89		4.17	install 1" x 10" pine trim.
Demolish and Install	L.F.	1.28	3.51		4.79	
Minimum Charge	Job		260		260	
1" x 12"						
Demolish	L.F.		.62		.62	Cost includes material and labor to
Install	L.F.	2.67	2.89		5.56	install 1" x 12" pine trim.
Demolish and Install	L.F.	2.67	3.51		6.18	
Minimum Charge	Job		260		260	
PVC						
1" x 4"						
Demolish	L.F.		.37		.37	Includes material and labor to install
Install	L.F.	3.09	2.08		5.17	1" x 4" PVC exterior trim.
Demolish and Install	L.F.	3.09	2.45		5.54	
Reinstall	L.F.		1.67		1.67	
Clean	L.F.	.01	.27		.28	
Paint	L.F.	.07	.67		.74	
Minimum Charge	Job		260		260	
1" x 6"						
Demolish	L.F.		.37		.37	Includes material and labor to install
Install	L.F.	4.43	2.08		6.51	1" x 6" PVC exterior trim.
Demolish and Install	L.F.	4.43	2.45		6.88	
Reinstall	L.F.		1.67		1.67	
Clean	L.F.	.01	.27		.28	
Paint	L.F.	.24	.67		.91	
Minimum Charge	Job		260		260	
1" x 8"						
Demolish	L.F.		.40		.40	Includes material and labor to install
Install	L.F.	5.65	2.31		7.96	1" x 8" PVC exterior trim.
Demolish and Install	L.F.	5.65	2.71		8.36	
Reinstall	L.F.		1.85		1.85	
Clean	L.F.	.02	.29		.31	
Paint	L.F.	.24	.67		.91	
Minimum Charge	Job		260		260	
1" x 10"						
Demolish	L.F.		.42		.42	Includes material and labor to install
Install	L.F.	7.30	2.31		9.61	1" x 10" PVC exterior trim.
Demolish and Install	L.F.	7.30	2.73		10.03	
Reinstall	L.F.		1.85		1.85	
Clean	L.F.	.03	.33		.36	
Paint	L.F.	.24	.67		.91	
Minimum Charge	Job		260		260	
1" x 12"						
Demolish	L.F.		.44		.44	Includes material and labor to install
Install	L.F.	9.20	2.60		11.80	1" x 12" PVC exterior trim.
Demolish and Install	L.F.	9.20	3.04		12.24	
Reinstall	L.F.		2.08		2.08	
Clean	L.F.	.03	.33		.36	
Paint	L.F.	.24	.67		.91	
Minimum Charge	Job		260		260	

For customer support on your Contractor's Pricing Guide: Residential Repair & Remodeling, call 800.448.8182.

Exterior Trim

Fascia		Unit	Material	Labor	Equip.	Total	Specification
Aluminum							
	Demolish	L.F.		.77		.77	Includes material and labor to install
	Install	S.F.	2.10	2.48		4.58	aluminum fascia.
	Demolish and Install	S.F.	2.10	3.25		5.35	
	Reinstall	S.F.		1.98		1.98	
	Clean	L.F.		.50		.50	
	Minimum Charge	Job		260		260	
Vinyl							
	Demolish	L.F.		.77		.77	Cost includes material and labor to
	Install	L.F.	4.71	2.98		7.69	install 6" vinyl fascia with 12" vinyl
	Demolish and Install	L.F.	4.71	3.75		8.46	soffit and J-channel.
	Reinstall	L.F.		2.38		2.38	
	Clean	L.F.		.50		.50	
	Minimum Charge	Job		260		260	
Plywood							
	Demolish	L.F.		.77		.77	Cost includes material and labor to
	Install	L.F.	1.97	1.16		3.13	install 12" wide plywood fascia.
	Demolish and Install	L.F.	1.97	1.93		3.90	
	Reinstall	L.F.		.93		.93	
	Clean	L.F.	.03	.33		.36	
	Paint	L.F.	.48	1.08		1.56	
	Minimum Charge	Job		260		260	
Cedar							
	Demolish	L.F.		.77		.77	Cost includes material and labor to
	Install	L.F.	7.25	1.16		8.41	install 12" wide cedar fascia.
	Demolish and Install	L.F.	7.25	1.93		9.18	
	Reinstall	L.F.		.93		.93	
	Clean	L.F.	.03	.33		.36	
	Paint	L.F.	.48	1.08		1.56	
	Minimum Charge	Job		260		260	
Redwood							
1" x 6"							
	Demolish	L.F.		.59		.59	Cost includes material and labor to
	Install	L.F.	6.40	2.08		8.48	install 1" x 6" redwood fascia board.
	Demolish and Install	L.F.	6.40	2.67		9.07	
	Reinstall	L.F.		1.67		1.67	
	Minimum Charge	Job		260		260	
1" x 8"							
	Demolish	L.F.		.67		.67	Cost includes material and labor to
	Install	L.F.	7.10	2.26		9.36	install 1" x 8" redwood fascia board.
	Demolish and Install	L.F.	7.10	2.93		10.03	
	Reinstall	L.F.		1.81		1.81	
	Minimum Charge	Job		260		260	
2" x 6"							
	Demolish	L.F.		.59		.59	Cost includes material and labor to
	Install	L.F.	1.34	4.17		5.51	install 2" x 6" exterior trim board.
	Demolish and Install	L.F.	1.34	4.76		6.10	
	Reinstall	L.F.		3.33		3.33	
	Minimum Charge	Job		260		260	
2" x 8"							
	Demolish	L.F.		.67		.67	Cost includes material and labor to
	Install	L.F.	1.46	4.63		6.09	install 2" x 8" exterior trim board.
	Demolish and Install	L.F.	1.46	5.30		6.76	
	Reinstall	L.F.		3.70		3.70	
	Minimum Charge	Job		260		260	

For customer support on your Contractor's Pricing Guide: Residential Repair & Remodeling, call 800.448.8182.

Exterior Trim

Fascia

Fascia		Unit	Material	Labor	Equip.	Total	Specification
Hem-fir Std & Better							
2″ x 6″							
	Demolish	L.F.		.59		.59	Cost includes material and labor to
	Install	L.F.	1.34	4.17		5.51	install 2″ x 6″ exterior trim board.
	Demolish and Install	L.F.	1.34	4.76		6.10	
	Reinstall	L.F.		3.33		3.33	
	Clean	L.F.	.02	.33		.35	
	Paint	L.F.	.48	1.08		1.56	
	Minimum Charge	Job		260		260	
2″ x 8″							
	Demolish	L.F.		.67		.67	Cost includes material and labor to
	Install	L.F.	1.46	4.63		6.09	install 2″ x 8″ exterior trim board.
	Demolish and Install	L.F.	1.46	5.30		6.76	
	Reinstall	L.F.		3.70		3.70	
	Clean	L.F.	.02	.33		.35	
	Paint	L.F.	.48	1.08		1.56	
	Minimum Charge	Job		260		260	
2″ x 10″							
	Demolish	L.F.		.77		.77	Cost includes material and labor to
	Install	L.F.	3.88	5.80		9.68	install 2″ x 10″ exterior trim board.
	Demolish and Install	L.F.	3.88	6.57		10.45	
	Reinstall	L.F.		4.63		4.63	
	Clean	L.F.	.03	.36		.39	
	Paint	L.F.	.48	1.08		1.56	
	Minimum Charge	Job		260		260	
Barge Rafter							
2″ x 6″							
	Demolish	L.F.		1.18		1.18	Cost includes material and labor to
	Install	L.F.	1.34	.58		1.92	install 2″ x 6″ barge rafters.
	Demolish and Install	L.F.	1.34	1.76		3.10	
	Reinstall	L.F.		.46		.46	
	Clean	L.F.	.02	.33		.35	
	Paint	L.F.	.48	1.08		1.56	
	Minimum Charge	Job		260		260	
2″ x 8″							
	Demolish	L.F.		1.20		1.20	Cost includes material and labor to
	Install	L.F.	1.46	.61		2.07	install 2″ x 8″ barge rafters.
	Demolish and Install	L.F.	1.46	1.81		3.27	
	Reinstall	L.F.		.49		.49	
	Clean	L.F.	.02	.33		.35	
	Paint	L.F.	.48	1.08		1.56	
	Minimum Charge	Job		260		260	
2″ x 10″							
	Demolish	L.F.		1.22		1.22	Cost includes material and labor to
	Install	L.F.	3.88	.63		4.51	install 2″ x 10″ barge rafters.
	Demolish and Install	L.F.	3.88	1.85		5.73	
	Reinstall	L.F.		.51		.51	
	Clean	L.F.	.03	.36		.39	
	Paint	L.F.	.48	1.08		1.56	
	Minimum Charge	Job		260		260	

Exterior Trim

Soffit

Soffit		Unit	Material	Labor	Equip.	Total	Specification
Aluminum							
12" w / Fascia							
	Demolish	L.F.		.77		.77	Cost includes material and labor to
	Install	L.F.	4.43	2.98		7.41	install 12" aluminum soffit with 6"
	Demolish and Install	L.F.	4.43	3.75		8.18	fascia and J-channel.
	Reinstall	L.F.		2.38		2.38	
	Clean	L.F.		.50		.50	
	Paint	L.F.	.65	1.08		1.73	
	Minimum Charge	Job		260		260	
18" w / Fascia							
	Demolish	L.F.		.77		.77	Cost includes material and labor to
	Install	L.F.	5.35	3.47		8.82	install 18" aluminum soffit with 6"
	Demolish and Install	L.F.	5.35	4.24		9.59	fascia and J-channel.
	Reinstall	L.F.		2.78		2.78	
	Clean	L.F.		.50		.50	
	Paint	L.F.	.65	1.08		1.73	
	Minimum Charge	Job		260		260	
24" w / Fascia							
	Demolish	L.F.		.77		.77	Cost includes material and labor to
	Install	L.F.	6.40	3.86		10.26	install 24" aluminum soffit with 6"
	Demolish and Install	L.F.	6.40	4.63		11.03	fascia and J-channel.
	Reinstall	L.F.		3.09		3.09	
	Clean	L.F.		.73		.73	
	Paint	L.F.	.65	1.08		1.73	
	Minimum Charge	Job		260		260	
Vinyl							
	Demolish	L.F.		.77		.77	Cost includes material and labor to
	Install	L.F.	4.71	2.98		7.69	install 6" vinyl fascia with 12" vinyl
	Demolish and Install	L.F.	4.71	3.75		8.46	soffit and J-channel.
	Reinstall	L.F.		2.38		2.38	
	Clean	L.F.		.50		.50	
	Paint	L.F.	.65	1.08		1.73	
	Minimum Charge	Job		260		260	
Plywood							
	Demolish	L.F.		.85		.85	Cost includes material and labor to
	Install	L.F.	1.89	1.49		3.38	install 12" wide plywood soffit.
	Demolish and Install	L.F.	1.89	2.34		4.23	
	Reinstall	L.F.		1.19		1.19	
	Clean	L.F.	.03	.33		.36	
	Paint	L.F.	.48	1.08		1.56	
	Minimum Charge	Job		260		260	
Redwood							
	Demolish	L.F.		.85		.85	Cost includes material and labor to
	Install	L.F.	8.15	1.49		9.64	install 12" wide redwood soffit.
	Demolish and Install	L.F.	8.15	2.34		10.49	
	Reinstall	L.F.		1.19		1.19	
	Clean	L.F.	.03	.33		.36	
	Paint	L.F.	.48	1.08		1.56	
	Minimum Charge	Job		260		260	
Cedar							
	Demolish	L.F.		.85		.85	Cost includes material and labor to
	Install	L.F.	7.20	1.49		8.69	install 12" wide cedar soffit.
	Demolish and Install	L.F.	7.20	2.34		9.54	
	Reinstall	L.F.		1.19		1.19	
	Clean	L.F.	.03	.33		.36	
	Paint	L.F.	.48	1.08		1.56	
	Minimum Charge	Job		260		260	

For customer support on your Contractor's Pricing Guide: Residential Repair & Remodeling, call 800.448.8182.

Exterior Trim

Soffit

	Unit	Material	Labor	Equip.	Total	Specification
Douglas Fir						
Demolish	L.F.		.85		.85	Cost includes material and labor to install 12" wide Douglas fir soffit.
Install	L.F.	2.60	1.49		4.09	
Demolish and Install	L.F.	2.60	2.34		4.94	
Reinstall	L.F.		1.19		1.19	
Clean	L.F.	.03	.33		.36	
Paint	L.F.	.48	1.08		1.56	
Minimum Charge	Job		260		260	

For customer support on your Contractor's Pricing Guide: Residential Repair & Remodeling, call 800.448.8182.

Roofing

Roofing and Flashing

Tile Roofing

Metal Flashing	Unit	Material	Labor	Equip.	Total	Specification
Galvanized						
6"						
Demolish	L.F.		.30		.30	Includes material and labor to install flat galvanized steel flashing, 6" wide.
Install	L.F.	.96	1.22		2.18	
Demolish and Install	L.F.	.96	1.52		2.48	
Paint	L.F.	.35	.54		.89	
Minimum Charge	Job		244		244	
14"						
Demolish	L.F.		.30		.30	Includes material and labor to install flat galvanized steel flashing, 14" wide.
Install	L.F.	2.22	1.22		3.44	
Demolish and Install	L.F.	2.22	1.52		3.74	
Paint	L.F.	.35	.54		.89	
Minimum Charge	Job		244		244	
20"						
Demolish	L.F.		.30		.30	Includes material and labor to install flat galvanized steel flashing, 20" wide.
Install	L.F.	3.29	1.22		4.51	
Demolish and Install	L.F.	3.29	1.52		4.81	
Paint	L.F.	.35	.54		.89	
Minimum Charge	Job		244		244	
Aluminum						
6"						
Demolish	L.F.		.30		.30	Includes material and labor to install flat aluminum flashing, 6" wide.
Install	L.F.	.64	1.22		1.86	
Demolish and Install	L.F.	.64	1.52		2.16	
Paint	L.F.	.35	.54		.89	
Minimum Charge	Job		244		244	
14"						
Demolish	L.F.		.30		.30	Includes material and labor to install flat aluminum flashing, 14" wide.
Install	L.F.	1.12	1.22		2.34	
Demolish and Install	L.F.	1.12	1.52		2.64	
Paint	L.F.	.35	.54		.89	
Minimum Charge	Job		244		244	
20"						
Demolish	L.F.		.30		.30	Includes material and labor to install flat aluminum flashing, 20" wide.
Install	L.F.	1.53	1.22		2.75	
Demolish and Install	L.F.	1.53	1.52		3.05	
Paint	L.F.	.35	.54		.89	
Minimum Charge	Job		244		244	

For customer support on your Contractor's Pricing Guide: Residential Repair & Remodeling, call 800.448.8182.

Roofing

Metal Flashing

Metal Flashing	Unit	Material	Labor	Equip.	Total	Specification
Copper						
6"						
Demolish	L.F.		.30		.30	Includes material and labor to install
Install	L.F.	12	1.63		13.63	flat copper flashing, 6" wide.
Demolish and Install	L.F.	12	1.93		13.93	
Minimum Charge	Job		284		284	
14"						
Demolish	L.F.		.30		.30	Includes material and labor to install
Install	L.F.	28	1.63		29.63	flat copper flashing, 14" wide.
Demolish and Install	L.F.	28	1.93		29.93	
Minimum Charge	Job		284		284	
18"						
Demolish	L.F.		.30		.30	Includes material and labor to install
Install	L.F.	40	1.63		41.63	flat copper flashing, 18" wide.
Demolish and Install	L.F.	40	1.93		41.93	
Minimum Charge	Job		284		284	
Aluminum Valley						
Demolish	L.F.		.30		.30	Includes material and labor to install
Install	L.F.	6.15	3.06		9.21	aluminum valley flashing, .024" thick.
Demolish and Install	L.F.	6.15	3.36		9.51	
Paint	L.F.	.35	.54		.89	
Minimum Charge	Job		260		260	
Gravel Stop						
Demolish	L.F.		.47		.47	Includes material and labor to install
Install	L.F.	6.15	3.92		10.07	aluminum gravel stop, .050" thick, 4"
Demolish and Install	L.F.	6.15	4.39		10.54	high, mill finish.
Reinstall	L.F.		3.13		3.13	
Paint	L.F.	.55	.54		1.09	
Minimum Charge	Job		260		260	
Drip Edge						
Demolish	L.F.		.44		.44	Includes material and labor to install
Install	L.F.	.97	1.30		2.27	aluminum drip edge, .016" thick, 5"
Demolish and Install	L.F.	.97	1.74		2.71	wide, mill finish.
Reinstall	L.F.		1.04		1.04	
Paint	L.F.	.55	.54		1.09	
Minimum Charge	Job		260		260	
End Wall						
Demolish	L.F.		1.37		1.37	Includes material and labor to install
Install	L.F.	2.13	3.06		5.19	aluminum end wall flashing, .024"
Demolish and Install	L.F.	2.13	4.43		6.56	thick.
Minimum Charge	Job		260		260	
Side Wall						
Demolish	L.F.		.27		.27	Includes material and labor to install
Install	L.F.	6.25	3.06		9.31	aluminum side wall flashing, .024"
Demolish and Install	L.F.	6.25	3.33		9.58	thick.
Minimum Charge	Job		260		260	
Coping and Wall Cap						
Demolish	L.F.		1.85		1.85	Includes material and labor to install
Install	L.F.	6.20	3.36		9.56	aluminum flashing, mill finish, 0.40
Demolish and Install	L.F.	6.20	5.21		11.41	thick, 12" wide.
Clean	L.F.	.08	1.25		1.33	
Minimum Charge	Job		244		244	

For customer support on your Contractor's Pricing Guide: Residential Repair & Remodeling, call 800.448.8182.

Roofing

Roof Shingles

	Unit	Material	Labor	Equip.	Total	Specification
Underlayment						
15#						
Demolish	Sq.		14.80		14.80	Includes material and labor to install
Install	Sq.	5.20	7.60		12.80	#15 felt underlayment.
Demolish and Install	Sq.	5.20	22.40		27.60	
Minimum Charge	Job		244		244	
30#						
Demolish	Sq.		14.80		14.80	Includes material and labor to install
Install	Sq.	6.65	8.40		15.05	#30 felt underlayment.
Demolish and Install	Sq.	6.65	23.20		29.85	
Minimum Charge	Job		244		244	
Self Adhering						
Demolish	Sq.		37		37	Includes material and labor to install
Install	Sq.	90	22		112	self adhering ice barrier roofing
Demolish and Install	Sq.	90	59		149	underlayment.
Minimum Charge	Job		244		244	

Rolled Roofing

	Unit	Material	Labor	Equip.	Total	Specification
90 lb						
Demolish	Sq.		37		37	Cost includes material and labor to
Install	Sq.	52.50	40.50		93	install 90 lb mineral surface rolled
Demolish and Install	Sq.	52.50	77.50		130	roofing.
Minimum Charge	Job		244		244	

Composition Shingles

	Unit	Material	Labor	Equip.	Total	Specification
Fiberglass (3-tab)						
25 Year						
Demolish	Sq.		40.50		40.50	Cost includes material and labor to
Install	Sq.	103	54.50		157.50	install 25 year fiberglass shingles.
Demolish and Install	Sq.	103	95		198	
Minimum Charge	Job		244		244	
30 Year						
Demolish	Sq.		40.50		40.50	Cost includes material and labor to
Install	Sq.	134	61		195	install 30 year fiberglass shingles.
Demolish and Install	Sq.	134	101.50		235.50	
Minimum Charge	Job		244		244	
Emergency Repairs						
Install	Ea.	228	244		472	Includes labor and material for
Minimum Charge	Job		244		244	emergency repairs.
Architectural						
Laminated, 25 Year						
Demolish	Sq.		40.50		40.50	Cost includes material and labor to
Install	Sq.	161	67.50		228.50	install 25 year fiberglass roof shingles.
Demolish and Install	Sq.	161	108		269	
Minimum Charge	Job		244		244	
Laminated, 30 Year						
Demolish	Sq.		40.50		40.50	Cost includes material and labor to
Install	Sq.	173	87		260	install 30 year laminated roof
Demolish and Install	Sq.	173	127.50		300.50	shingles.
Minimum Charge	Job		244		244	

For customer support on your Contractor's Pricing Guide: Residential Repair & Remodeling, call 800.448.8182.

Roofing

Composition Shingles		Unit	Material	Labor	Equip.	Total	Specification
Laminated, 40 Year							
	Demolish	Sq.		40.50		40.50	Cost includes material and labor to
	Install	Sq.	450	101		551	install 40 year laminated roof
	Demolish and Install	Sq.	450	141.50		591.50	shingles.
	Minimum Charge	Job		244		244	
Laminated Shake, 40 Year							
	Demolish	Sq.		40.50		40.50	Cost includes material and labor to
	Install	Sq.	450	101		551	install 40 year laminated roof
	Demolish and Install	Sq.	450	141.50		591.50	shingles.
	Minimum Charge	Job		244		244	
Emergency Repairs							
	Install	Ea.	228	244		472	Includes labor and material for
	Minimum Charge	Job		244		244	emergency repairs.
Roof Jack							
	Demolish	Ea.		8.05		8.05	Includes material and labor to install
	Install	Ea.	33	51.50		84.50	residential roof jack, w/bird screen,
	Demolish and Install	Ea.	33	59.55		92.55	backdraft damper, 3" & 4" dia. round
	Paint	Ea.	.19	5.40		5.59	duct.
	Minimum Charge	Job		310		310	
Ridge Vent							
	Demolish	L.F.		1.43		1.43	Includes material and labor to install a
	Install	L.F.	4.41	3.66		8.07	mill finish aluminum ridge vent strip.
	Demolish and Install	L.F.	4.41	5.09		9.50	
	Minimum Charge	Job		260		260	
Shingle Molding							
1" x 2"							
	Demolish	L.F.		.28		.28	Cost includes material and labor to
	Install	L.F.	.20	1.58		1.78	install 1" x 2" pine trim.
	Demolish and Install	L.F.	.20	1.86		2.06	
	Paint	L.F.	.07	.67		.74	
	Minimum Charge	Job		260		260	
1" x 3"							
	Demolish	L.F.		.28		.28	Cost includes material and labor to
	Install	L.F.	.36	1.80		2.16	install 1" x 3" pine trim.
	Demolish and Install	L.F.	.36	2.08		2.44	
	Paint	L.F.	.07	.67		.74	
	Minimum Charge	Job		260		260	
1" x 4"							
	Demolish	L.F.		.28		.28	Cost includes material and labor to
	Install	L.F.	.83	2.08		2.91	install 1" x 4" pine trim.
	Demolish and Install	L.F.	.83	2.36		3.19	
	Paint	L.F.	.07	.67		.74	
	Minimum Charge	Job		260		260	
Add for Steep Pitch							
	Install	Sq.		30.50		30.50	Includes labor and material added for
							2nd story or steep roofs.
Add for Additional Story							
	Install	Sq.		30.50		30.50	Includes labor and material added for
							2nd story or steep roofs.

74

For customer support on your Contractor's Pricing Guide: Residential Repair & Remodeling, call 800.448.8182.

Roofing

Wood Shingles

Wood Shingles	Unit	Material	Labor	Equip.	Total	Specification
Cedar Wood Shingles						
16" Red Label (#2)						
Demolish	Sq.		98.50		98.50	Includes material and labor to install
Install	Sq.	232	208		440	wood shingles/shakes.
Demolish and Install	Sq.	232	306.50		538.50	
Minimum Charge	Job		260		260	
16" Blue Label (#1)						
Demolish	Sq.		98.50		98.50	Includes material and labor to install
Install	Sq.	645	160		805	wood shingles/shakes.
Demolish and Install	Sq.	645	258.50		903.50	
Minimum Charge	Job		260		260	
18" Red Label (#2)						
Demolish	Sq.		98.50		98.50	Includes material and labor to install
Install	Sq.	420	134		554	wood shingles/shakes.
Demolish and Install	Sq.	420	232.50		652.50	
Minimum Charge	Job		260		260	
18" Blue Label (#1)						
Demolish	Sq.		98.50		98.50	Includes material and labor to install
Install	Sq.	610	146		756	wood shingles/shakes.
Demolish and Install	Sq.	610	244.50		854.50	
Minimum Charge	Job		260		260	
Replace Shingle						
Install	Ea.	10	13		23	Includes labor and material to replace
Minimum Charge	Job		260		260	individual wood shingle/shake. Cost is per shingle.
Cedar Shake						
18" Medium Handsplit						
Demolish	Sq.		98.50		98.50	Includes material and labor to install
Install	Sq.	510	200		710	wood shingles/shakes.
Demolish and Install	Sq.	510	298.50		808.50	
Minimum Charge	Job		260		260	
18" Heavy Handsplit						
Demolish	Sq.		98.50		98.50	Includes material and labor to install
Install	Sq.	510	223		733	wood shingles/shakes.
Demolish and Install	Sq.	510	321.50		831.50	
Minimum Charge	Job		260		260	
24" Heavy Handsplit						
Demolish	Sq.		98.50		98.50	Includes material and labor to install
Install	Sq.	815	178		993	wood shingles/shakes.
Demolish and Install	Sq.	815	276.50		1091.50	
Minimum Charge	Job		260		260	
24" Medium Handsplit						
Demolish	Sq.		98.50		98.50	Includes material and labor to install
Install	Sq.	815	160		975	wood shingles/shakes.
Demolish and Install	Sq.	815	258.50		1073.50	
Minimum Charge	Job		260		260	
24" Straight Split						
Demolish	Sq.		98.50		98.50	Includes material and labor to install
Install	Sq.	455	237		692	wood shingles/shakes.
Demolish and Install	Sq.	455	335.50		790.50	
Minimum Charge	Job		260		260	
24" Tapersplit						
Demolish	Sq.		98.50		98.50	Includes material and labor to install
Install	Sq.	345	237		582	wood shingles/shakes.
Demolish and Install	Sq.	345	335.50		680.50	
Minimum Charge	Job		260		260	

For customer support on your Contractor's Pricing Guide: Residential Repair & Remodeling, call 800.448.8182.

Roofing

Wood Shingles

Wood Shingles		Unit	Material	Labor	Equip.	Total	Specification
18″ Straight Split							
	Demolish	Sq.		98.50		98.50	Includes material and labor to install
	Install	Sq.	585	237		822	wood shingles/shakes.
	Demolish and Install	Sq.	585	335.50		920.50	
	Minimum Charge	Job		260		260	
Replace Shingle							
	Install	Ea.	10	13		23	Includes labor and material to replace
	Minimum Charge	Job		260		260	individual wood shingle/shake. Cost is per shingle.

Slate Tile Roofing

Slate Tile Roofing		Unit	Material	Labor	Equip.	Total	Specification
Unfading Green							
	Demolish	Sq.		89		89	Includes material and labor to install
	Install	Sq.	980	279		1259	clear Vermont slate tile.
	Demolish and Install	Sq.	980	368		1348	
	Reinstall	Sq.		222.90		222.90	
	Minimum Charge	Job		244		244	
Unfading Purple							
	Demolish	Sq.		89		89	Includes material and labor to install
	Install	Sq.	1025	279		1304	clear Vermont slate tile.
	Demolish and Install	Sq.	1025	368		1393	
	Reinstall	Sq.		222.90		222.90	
	Minimum Charge	Job		244		244	
Replace Slate Tiles							
	Install	Ea.	10.35	25.50		35.85	Includes labor and material to replace
	Minimum Charge	Job		244		244	individual slate tiles. Cost is per tile.
Variegated Purple							
	Demolish	Sq.		89		89	Includes material and labor to install
	Install	Sq.	1025	279		1304	clear Vermont slate tile.
	Demolish and Install	Sq.	1025	368		1393	
	Reinstall	Sq.		222.90		222.90	
	Minimum Charge	Job		244		244	
Unfading Grey / Black							
	Demolish	Sq.		89		89	Includes material and labor to install
	Install	Sq.	925	279		1204	clear Vermont slate tile.
	Demolish and Install	Sq.	925	368		1293	
	Reinstall	Sq.		222.90		222.90	
	Minimum Charge	Job		244		244	
Unfading Red							
	Demolish	Sq.		89		89	Includes material and labor to install
	Install	Sq.	1600	279		1879	clear Vermont slate tile.
	Demolish and Install	Sq.	1600	368		1968	
	Reinstall	Sq.		222.90		222.90	
	Minimum Charge	Job		244		244	
Weathering Black							
	Demolish	Sq.		89		89	Includes material and labor to install
	Install	Sq.	725	279		1004	clear Pennsylvania slate tile.
	Demolish and Install	Sq.	725	368		1093	
	Reinstall	Sq.		222.90		222.90	
	Minimum Charge	Job		244		244	
Weathering Green							
	Demolish	Sq.		89		89	Includes material and labor to install
	Install	Sq.	925	279		1204	clear Vermont slate tile.
	Demolish and Install	Sq.	925	368		1293	
	Reinstall	Sq.		222.90		222.90	
	Minimum Charge	Job		244		244	

Roofing

Flat Clay Tile Roofing		Unit	Material	Labor	Equip.	Total	Specification
Glazed							
	Demolish	Sq.		98.50		98.50	Includes material and labor to install
	Install	Sq.	720	244		964	tile roofing. Cost based on terra cotta
	Demolish and Install	Sq.	720	342.50		1062.50	red flat clay tile and includes ridge
	Reinstall	Sq.		195.04		195.04	and rake tiles.
	Minimum Charge	Job		244		244	
Terra Cotta Red							
	Demolish	Sq.		98.50		98.50	Includes material and labor to install
	Install	Sq.	720	244		964	tile roofing. Cost based on terra cotta
	Demolish and Install	Sq.	720	342.50		1062.50	red flat clay tile and includes ridge
	Reinstall	Sq.		195.04		195.04	and rake tiles.
	Minimum Charge	Job		244		244	

Mission Tile Roofing		Unit	Material	Labor	Equip.	Total	Specification
Glazed Red							
	Demolish	Sq.		98.50		98.50	Includes material and labor to install
	Install	Sq.	620	266		886	tile roofing. Cost based on glazed red
	Demolish and Install	Sq.	620	364.50		984.50	tile and includes birdstop, ridge and
	Reinstall	Sq.		212.77		212.77	rake tiles.
	Minimum Charge	Job		244		244	
Unglazed Red							
	Demolish	Sq.		98.50		98.50	Includes material and labor to install
	Install	Sq.	480	266		746	tile roofing. Cost based on unglazed
	Demolish and Install	Sq.	480	364.50		844.50	red tile and includes birdstop, ridge
	Reinstall	Sq.		212.77		212.77	and rake tiles.
	Minimum Charge	Job		244		244	
Glazed Blue							
	Demolish	Sq.		98.50		98.50	Includes material and labor to install
	Install	Sq.	405	266		671	tile roofing. Cost based on glazed
	Demolish and Install	Sq.	405	364.50		769.50	blue tile and includes birdstop, ridge
	Reinstall	Sq.		212.77		212.77	and rake tiles.
	Minimum Charge	Job		244		244	
Glazed White							
	Demolish	Sq.		98.50		98.50	Includes material and labor to install
	Install	Sq.	415	266		681	tile roofing. Cost based on glazed
	Demolish and Install	Sq.	415	364.50		779.50	white tile and includes birdstop, ridge
	Reinstall	Sq.		212.77		212.77	and rake tiles.
	Minimum Charge	Job		244		244	
Unglazed White							
	Demolish	Sq.		98.50		98.50	Includes material and labor to install
	Install	Sq.	292	266		558	tile roofing. Cost based on unglazed
	Demolish and Install	Sq.	292	364.50		656.50	white tile and includes birdstop, ridge
	Reinstall	Sq.		212.77		212.77	and rake tiles.
	Minimum Charge	Job		244		244	
Color Blend							
	Demolish	Sq.		98.50		98.50	Includes material and labor to install
	Install	Sq.	296	266		562	tile roofing. Cost based on color blend
	Demolish and Install	Sq.	296	364.50		660.50	tile and includes birdstop, ridge and
	Reinstall	Sq.		212.77		212.77	rake tiles.
	Minimum Charge	Job		244		244	

For customer support on your Contractor's Pricing Guide: Residential Repair & Remodeling, call 800.448.8182.

Roofing

Mission Tile Roofing

Mission Tile Roofing		Unit	Material	Labor	Equip.	Total	Specification
Accessories / Extras							
Add for Cap Furring Strips	Install	Sq.	99	305		404	Includes labor and material to install vertically placed wood furring strips under cap tiles.
Add for Tile Adhesive	Install	Sq.	21.50	17.80		39.30	Includes material and labor for adhesive material placed between tiles.
Add for Wire Attachment	Install	Sq.	15	212		227	Includes labor and material to install single wire anchors for attaching tile.
Add for Braided Runners	Install	Sq.	37	203		240	Includes labor and material to install braided wire runners for secure attachment in high wind areas.
Add for Stainless Steel Nails	Install	Sq.	23.50			23.50	Includes material only for stainless steel nails.
Add for Brass Nails	Install	Sq.	42			42	Includes material only for brass nails.
Add for Hurricane Clips	Install	Sq.	23.50	287		310.50	Includes material and labor for hurricane or wind clips for high wind areas.
Add for Copper Nails	Install	Sq.	42			42	Includes material only for copper nails.
Add for Vertical Furring Strips	Install	Sq.	19.75	56.50		76.25	Includes labor and material to install vertically placed wood furring strips under tiles.
Add for Horizontal Furring Strips	Install	Sq.	13.65	25.50		39.15	Includes labor and material to install horizontally placed wood furring strips under tiles.

Spanish Tile Roofing

Spanish Tile Roofing		Unit	Material	Labor	Equip.	Total	Specification
Glazed Red							
	Demolish	Sq.		98.50		98.50	Includes material and labor to install tile roofing. Cost based on glazed red tile and includes birdstop, ridge and rake tiles.
	Install	Sq.	610	203		813	
	Demolish and Install	Sq.	610	301.50		911.50	
	Reinstall	Sq.		162.53		162.53	
	Minimum Charge	Job		244		244	
Glazed White							
	Demolish	Sq.		98.50		98.50	Includes material and labor to install tile roofing. Cost based on glazed white tile and includes birdstop, ridge and rake tiles.
	Install	Sq.	345	271		616	
	Demolish and Install	Sq.	345	369.50		714.50	
	Reinstall	Sq.		216.71		216.71	
	Minimum Charge	Job		244		244	

For customer support on your Contractor's Pricing Guide: Residential Repair & Remodeling, call 800.448.8182.

Roofing

Spanish Tile Roofing		Unit	Material	Labor	Equip.	Total	Specification
Glazed Blue							
	Demolish	Sq.		98.50		98.50	Includes material and labor to install
	Install	Sq.	545	203		748	tile roofing. Cost based on glazed
	Demolish and Install	Sq.	545	301.50		846.50	blue tile and includes birdstop, ridge
	Reinstall	Sq.		162.53		162.53	and rake tiles.
	Minimum Charge	Job		244		244	
Unglazed Red							
	Demolish	Sq.		98.50		98.50	Includes material and labor to install
	Install	Sq.	655	203		858	tile roofing. Cost based on unglazed
	Demolish and Install	Sq.	655	301.50		956.50	red tile and includes birdstop, ridge
	Reinstall	Sq.		162.53		162.53	and rake tiles.
	Minimum Charge	Job		244		244	
Unglazed White							
	Demolish	Sq.		98.50		98.50	Includes material and labor to install
	Install	Sq.	570	203		773	tile roofing. Cost based on unglazed
	Demolish and Install	Sq.	570	301.50		871.50	white tile and includes birdstop, ridge
	Reinstall	Sq.		162.53		162.53	and rake tiles.
	Minimum Charge	Job		244		244	
Color Blend							
	Demolish	Sq.		98.50		98.50	Includes material and labor to install
	Install	Sq.	700	203		903	tile roofing. Cost based on color blend
	Demolish and Install	Sq.	700	301.50		1001.50	tile and includes birdstop, ridge and
	Reinstall	Sq.		162.53		162.53	rake tiles.
	Minimum Charge	Job		244		244	
Accessories / Extras							
Add for Vertical Furring Strips							
	Install	Sq.	19.75	56.50		76.25	Includes labor and material to install vertically placed wood furring strips under tiles.
Add for Wire Attachment							
	Install	Sq.	7.45	163		170.45	Includes labor and material to install single wire ties.
Add for Braided Runners							
	Install	Sq.	28.50	139		167.50	Includes labor and material to install braided wire runners for secure attachment in high wind areas.
Add for Stainless Steel Nails							
	Install	Sq.	23			23	Includes material only for stainless steel nails.
Add for Copper Nails							
	Install	Sq.	36			36	Includes material only for copper nails.
Add for Brass Nails							
	Install	Sq.	32			32	Includes material only for brass nails.
Add for Hurricane Clips							
	Install	Sq.	11.70	168		179.70	Includes material and labor to install hurricane or wind clips for high wind areas.
Add for Mortar Set Tiles							
	Install	Sq.	52.50	244		296.50	Includes material and labor to install setting tiles by using mortar material placed between tiles.
Add for Tile Adhesive							
	Install	Sq.	13.95	15.85		29.80	Includes material and labor for adhesive placed between tiles.

For customer support on your Contractor's Pricing Guide: Residential Repair & Remodeling, call 800.448.8182.

Roofing

Spanish Tile Roofing		Unit	Material	Labor	Equip.	Total	Specification
Add for Cap Furring Strips	Install	Sq.	99	305		404	Includes labor and material to install vertically placed wood furring strips under cap tiles.
Add for Horizontal Furring Strips	Install	Sq.	13.65	25.50		39.15	Includes labor and material to install horizontally placed wood furring strips under tiles.

Concrete Tile Roofing		Unit	Material	Labor	Equip.	Total	Specification
Corrugated Red / Brown	Demolish	Sq.		98.50		98.50	Includes material and labor to install tile roofing. Cost includes birdstop, booster, ridge and rake tiles.
	Install	Sq.	27.50	360		387.50	
	Demolish and Install	Sq.	27.50	458.50		486	
	Reinstall	Sq.		288.95		288.95	
	Minimum Charge	Job		244		244	
Corrugated Gray	Demolish	Sq.		98.50		98.50	Includes material and labor to install tile roofing. Cost includes birdstop, booster, ridge and rake tiles.
	Install	Sq.	117	360		477	
	Demolish and Install	Sq.	117	458.50		575.50	
	Reinstall	Sq.		288.95		288.95	
	Minimum Charge	Job		244		244	
Corrugated Bright Red	Demolish	Sq.		98.50		98.50	Includes material and labor to install tile roofing. Cost includes birdstop, booster, ridge and rake tiles.
	Install	Sq.	128	360		488	
	Demolish and Install	Sq.	128	458.50		586.50	
	Reinstall	Sq.		288.95		288.95	
	Minimum Charge	Job		244		244	
Corrugated Black	Demolish	Sq.		98.50		98.50	Includes material and labor to install tile roofing. Cost includes birdstop, booster, ridge and rake tiles.
	Install	Sq.	128	360		488	
	Demolish and Install	Sq.	128	458.50		586.50	
	Reinstall	Sq.		288.95		288.95	
	Minimum Charge	Job		244		244	
Corrugated Green	Demolish	Sq.		98.50		98.50	Includes material and labor to install tile roofing. Cost includes birdstop, booster, ridge and rake tiles.
	Install	Sq.	128	360		488	
	Demolish and Install	Sq.	128	458.50		586.50	
	Reinstall	Sq.		288.95		288.95	
	Minimum Charge	Job		244		244	
Corrugated Blue	Demolish	Sq.		98.50		98.50	Includes material and labor to install tile roofing. Cost includes birdstop, booster, ridge and rake tiles.
	Install	Sq.	128	360		488	
	Demolish and Install	Sq.	128	458.50		586.50	
	Reinstall	Sq.		288.95		288.95	
	Minimum Charge	Job		244		244	
Flat Natural Gray	Demolish	Sq.		98.50		98.50	Includes material and labor to install tile roofing. Cost includes birdstop, booster, ridge and rake tiles.
	Install	Sq.	177	360		537	
	Demolish and Install	Sq.	177	458.50		635.50	
	Reinstall	Sq.		288.95		288.95	
	Minimum Charge	Job		244		244	

For customer support on your Contractor's Pricing Guide: Residential Repair & Remodeling, call 800.448.8182.

Roofing

Concrete Tile Roofing

Concrete Tile Roofing	Unit	Material	Labor	Equip.	Total	Specification
Flat Red / Brown						
Demolish	Sq.		98.50		98.50	Includes material and labor to install tile roofing. Cost includes birdstop, booster, ridge and rake tiles.
Install	Sq.	177	360		537	
Demolish and Install	Sq.	177	458.50		635.50	
Reinstall	Sq.		288.95		288.95	
Minimum Charge	Job		244		244	
Flat Bright Red						
Demolish	Sq.		98.50		98.50	Includes material and labor to install tile roofing. Cost includes birdstop, booster, ridge and rake tiles.
Install	Sq.	177	360		537	
Demolish and Install	Sq.	177	458.50		635.50	
Reinstall	Sq.		288.95		288.95	
Minimum Charge	Job		244		244	
Flat Green						
Demolish	Sq.		98.50		98.50	Includes material and labor to install tile roofing. Cost includes birdstop, booster, ridge and rake tiles.
Install	Sq.	177	360		537	
Demolish and Install	Sq.	177	458.50		635.50	
Minimum Charge	Job		244		244	
Flat Blue						
Demolish	Sq.		98.50		98.50	Includes material and labor to install tile roofing. Cost includes birdstop, booster, ridge and rake tiles.
Install	Sq.	177	360		537	
Demolish and Install	Sq.	177	458.50		635.50	
Reinstall	Sq.		288.95		288.95	
Minimum Charge	Job		244		244	
Flat Black						
Demolish	Sq.		98.50		98.50	Includes material and labor to install tile roofing. Cost includes birdstop, booster, ridge and rake tiles.
Install	Sq.	177	360		537	
Demolish and Install	Sq.	177	458.50		635.50	
Reinstall	Sq.		288.95		288.95	
Minimum Charge	Job		244		244	

Aluminum Sheet

Aluminum Sheet	Unit	Material	Labor	Equip.	Total	Specification
Corrugated						
.019" Thick Natural						
Demolish	S.F.		1.05		1.05	Includes material and labor to install aluminum sheet metal roofing.
Install	S.F.	1.52	2.37		3.89	
Demolish and Install	S.F.	1.52	3.42		4.94	
Reinstall	S.F.		1.89		1.89	
Minimum Charge	Job		570		570	
.016" Thick Natural						
Demolish	S.F.		1.05		1.05	Includes material and labor to install aluminum sheet metal roofing.
Install	S.F.	3.83	2.37		6.20	
Demolish and Install	S.F.	3.83	3.42		7.25	
Reinstall	S.F.		1.89		1.89	
Minimum Charge	Job		570		570	
.016" Colored Finish						
Demolish	S.F.		1.05		1.05	Includes material and labor to install aluminum sheet metal roofing.
Install	S.F.	1.63	2.37		4	
Demolish and Install	S.F.	1.63	3.42		5.05	
Reinstall	S.F.		1.89		1.89	
Minimum Charge	Job		570		570	
.019" Colored Finish						
Demolish	S.F.		1.05		1.05	Includes material and labor to install aluminum sheet metal roofing.
Install	S.F.	1.50	2.37		3.87	
Demolish and Install	S.F.	1.50	3.42		4.92	
Reinstall	S.F.		1.89		1.89	
Minimum Charge	Job		570		570	

For customer support on your Contractor's Pricing Guide: Residential Repair & Remodeling, call 800.448.8182.

Roofing

Aluminum Sheet	Unit	Material	Labor	Equip.	Total	Specification
Ribbed						
.019" Thick Natural						
Demolish	S.F.		1.05		1.05	Includes material and labor to install
Install	S.F.	1.52	2.37		3.89	aluminum sheet metal roofing.
Demolish and Install	S.F.	1.52	3.42		4.94	
Reinstall	S.F.		1.89		1.89	
Minimum Charge	Job		570		570	
.016" Thick Natural						
Demolish	S.F.		1.05		1.05	Includes material and labor to install
Install	S.F.	3.83	2.37		6.20	aluminum sheet metal roofing.
Demolish and Install	S.F.	3.83	3.42		7.25	
Reinstall	S.F.		1.89		1.89	
Minimum Charge	Job		570		570	
.050" Colored Finish						
Demolish	S.F.		1.05		1.05	Includes material and labor to install
Install	S.F.	5.60	2.37		7.97	aluminum sheet metal roofing.
Demolish and Install	S.F.	5.60	3.42		9.02	
Reinstall	S.F.		1.89		1.89	
Minimum Charge	Job		570		570	
.032" Thick Natural						
Demolish	S.F.		1.05		1.05	Includes material and labor to install
Install	S.F.	3.11	2.37		5.48	aluminum sheet metal roofing.
Demolish and Install	S.F.	3.11	3.42		6.53	
Reinstall	S.F.		1.89		1.89	
Minimum Charge	Job		570		570	
.040" Thick Natural						
Demolish	S.F.		1.05		1.05	Includes material and labor to install
Install	S.F.	4.54	2.37		6.91	aluminum sheet metal roofing.
Demolish and Install	S.F.	4.54	3.42		7.96	
Reinstall	S.F.		1.89		1.89	
Minimum Charge	Job		570		570	
.050" Thick Natural						
Demolish	S.F.		1.05		1.05	Includes material and labor to install
Install	S.F.	4.66	2.37		7.03	aluminum sheet metal roofing.
Demolish and Install	S.F.	4.66	3.42		8.08	
Reinstall	S.F.		1.89		1.89	
Minimum Charge	Job		570		570	
.016" Colored Finish						
Demolish	S.F.		1.05		1.05	Includes material and labor to install
Install	S.F.	1.63	2.37		4	aluminum sheet metal roofing.
Demolish and Install	S.F.	1.63	3.42		5.05	
Reinstall	S.F.		1.89		1.89	
Minimum Charge	Job		570		570	
.019" Colored Finish						
Demolish	S.F.		1.05		1.05	Includes material and labor to install
Install	S.F.	1.50	2.37		3.87	aluminum sheet metal roofing.
Demolish and Install	S.F.	1.50	3.42		4.92	
Reinstall	S.F.		1.89		1.89	
Minimum Charge	Job		570		570	
.032" Colored Finish						
Demolish	S.F.		1.05		1.05	Includes material and labor to install
Install	S.F.	6.35	2.37		8.72	aluminum sheet metal roofing.
Demolish and Install	S.F.	6.35	3.42		9.77	
Reinstall	S.F.		1.89		1.89	
Minimum Charge	Job		570		570	

Roofing

Aluminum Sheet

.040" Colored Finish	Unit	Material	Labor	Equip.	Total	Specification
Demolish	S.F.		1.05		1.05	Includes material and labor to install
Install	S.F.	6.05	2.37		8.42	aluminum sheet metal roofing.
Demolish and Install	S.F.	6.05	3.42		9.47	
Reinstall	S.F.		1.89		1.89	
Minimum Charge	Job		570		570	

Fiberglass Sheet

	Unit	Material	Labor	Equip.	Total	Specification
12 Ounce Corrugated						
Demolish	S.F.		1.87		1.87	Cost includes material and labor to
Install	S.F.	5.50	1.94		7.44	install 12 ounce corrugated fiberglass
Demolish and Install	S.F.	5.50	3.81		9.31	sheet roofing.
Reinstall	S.F.		1.55		1.55	
Minimum Charge	Job		244		244	
8 Ounce Corrugated						
Demolish	S.F.		1.87		1.87	Cost includes material and labor to
Install	S.F.	2.84	1.94		4.78	install 8 ounce corrugated fiberglass
Demolish and Install	S.F.	2.84	3.81		6.65	roofing.
Reinstall	S.F.		1.55		1.55	
Minimum Charge	Job		244		244	

Galvanized Steel

	Unit	Material	Labor	Equip.	Total	Specification
Corrugated						
24 Gauge						
Demolish	S.F.		1.05		1.05	Includes material and labor to install
Install	S.F.	3.48	2.04		5.52	galvanized steel sheet metal roofing.
Demolish and Install	S.F.	3.48	3.09		6.57	Cost based on 24 gauge corrugated
Reinstall	S.F.		1.63		1.63	or ribbed steel roofing.
Minimum Charge	Job		244		244	
26 Gauge						
Demolish	S.F.		1.05		1.05	Includes material and labor to install
Install	S.F.	1.57	1.94		3.51	galvanized steel sheet metal roofing.
Demolish and Install	S.F.	1.57	2.99		4.56	Cost based on 26 gauge corrugated
Reinstall	S.F.		1.55		1.55	or ribbed steel roofing.
Minimum Charge	Job		244		244	
28 Gauge						
Demolish	S.F.		1.05		1.05	Includes material and labor to install
Install	S.F.	1.57	1.84		3.41	galvanized steel sheet metal roofing.
Demolish and Install	S.F.	1.57	2.89		4.46	Cost based on 28 gauge corrugated
Reinstall	S.F.		1.47		1.47	or ribbed steel roofing.
Minimum Charge	Job		244		244	
30 Gauge						
Demolish	S.F.		1.05		1.05	Includes material and labor to install
Install	S.F.	.98	1.76		2.74	galvanized steel sheet metal roofing.
Demolish and Install	S.F.	.98	2.81		3.79	Cost based on 30 gauge corrugated
Reinstall	S.F.		1.41		1.41	or ribbed steel roofing.
Minimum Charge	Job		244		244	

For customer support on your Contractor's Pricing Guide: Residential Repair & Remodeling, call 800.448.8182.

Roofing

Galvanized Steel

Galvanized Steel	Unit	Material	Labor	Equip.	Total	Specification
Ribbed						
24 Gauge						
Demolish	S.F.		1.05		1.05	Includes material and labor to install
Install	S.F.	3.48	2.04		5.52	galvanized steel sheet metal roofing.
Demolish and Install	S.F.	3.48	3.09		6.57	Cost based on 24 gauge corrugated
Reinstall	S.F.		1.63		1.63	or ribbed steel roofing.
Minimum Charge	Job		244		244	
26 Gauge						
Demolish	S.F.		1.05		1.05	Includes material and labor to install
Install	S.F.	1.57	1.94		3.51	galvanized steel sheet metal roofing.
Demolish and Install	S.F.	1.57	2.99		4.56	Cost based on 26 gauge corrugated
Reinstall	S.F.		1.55		1.55	or ribbed steel roofing.
Minimum Charge	Job		244		244	
28 Gauge						
Demolish	S.F.		1.05		1.05	Includes material and labor to install
Install	S.F.	1.57	1.84		3.41	galvanized steel sheet metal roofing.
Demolish and Install	S.F.	1.57	2.89		4.46	Cost based on 28 gauge corrugated
Reinstall	S.F.		1.47		1.47	or ribbed steel roofing.
Minimum Charge	Job		244		244	
30 Gauge						
Demolish	S.F.		1.05		1.05	Includes material and labor to install
Install	S.F.	.98	1.76		2.74	galvanized steel sheet metal roofing.
Demolish and Install	S.F.	.98	2.81		3.79	Cost based on 30 gauge corrugated
Reinstall	S.F.		1.41		1.41	or ribbed steel roofing.
Minimum Charge	Job		244		244	

Standing Seam

Standing Seam	Unit	Material	Labor	Equip.	Total	Specification
Copper						
16 Ounce						
Demolish	Sq.		105		105	Includes material and labor to install
Install	Sq.	3000	435		3435	copper standing seam metal roofing.
Demolish and Install	Sq.	3000	540		3540	Cost based on 16 ounce copper
Reinstall	Sq.		349.54		349.54	roofing.
Minimum Charge	Job		284		284	
18 Ounce						
Demolish	Sq.		105		105	Includes material and labor to install
Install	Sq.	3350	475		3825	copper standing seam metal roofing.
Demolish and Install	Sq.	3350	580		3930	Cost based on 18 ounce copper
Reinstall	Sq.		378.67		378.67	roofing.
Minimum Charge	Job		284		284	
20 Ounce						
Demolish	Sq.		105		105	Includes material and labor to install
Install	Sq.	4200	515		4715	copper standing seam metal roofing.
Demolish and Install	Sq.	4200	620		4820	Cost based on 20 ounce copper
Reinstall	Sq.		413.09		413.09	roofing.
Stainless Steel						
26 Gauge						
Demolish	Sq.		105		105	Cost includes material and labor to
Install	Sq.	1525	495		2020	install 26 gauge stainless steel
Demolish and Install	Sq.	1525	600		2125	standing seam metal roofing.
Reinstall	Sq.		395.13		395.13	
Minimum Charge	Job		284		284	

For customer support on your Contractor's Pricing Guide: Residential Repair & Remodeling, call 800.448.8182.

Roofing

Standing Seam	Unit	Material	Labor	Equip.	Total	Specification
28 Gauge						
Demolish	Sq.		105		105	Includes material and labor to install
Install	Sq.	1875	475		2350	standing seam metal roofing. Cost
Demolish and Install	Sq.	1875	580		2455	based on 28 gauge stainless steel
Reinstall	Sq.		378.67		378.67	roofing.
Minimum Charge	Job		284		284	

Elastomeric Roofing	Unit	Material	Labor	Equip.	Total	Specification
Neoprene						
1/16″						
Demolish	Sq.		36.50		36.50	Includes material and labor to install
Install	Sq.	227	232	61	520	neoprene membrane, fully adhered,
Demolish and Install	Sq.	227	268.50	61	556.50	1/16″ thick.
Minimum Charge	Job		1100	114	1214	
GRM Membrane						
1/16″						
Demolish	Sq.		36.50		36.50	Includes material and labor to install
Install	Sq.	82.50	370	38	490.50	coal tar based elastomeric roofing
Demolish and Install	Sq.	82.50	406.50	38	527	membrane, polyester fiber reinforced.
Minimum Charge	Job		1100	114	1214	
Cant Strips						
3″						
Demolish	L.F.		1.12		1.12	Includes material and labor to install
Install	L.F.	1.18	1.50		2.68	cants 3″ x 3″, treated timber, cut
Demolish and Install	L.F.	1.18	2.62		3.80	diagonally.
Minimum Charge	Job		122		122	
4″						
Demolish	L.F.		1.12		1.12	Includes material and labor to install
Install	L.F.	3.86	1.50		5.36	cants 4″ x 4″, treated timber, cut
Demolish and Install	L.F.	3.86	2.62		6.48	diagonally.
Minimum Charge	Job		122		122	
Polyvinyl Chloride (PVC)						
45 mil (loose laid)						
Demolish	Sq.		36.50		36.50	Cost includes material and labor to
Install	Sq.	72	43.50	4.47	119.97	install 45 mil PVC elastomeric roofing,
Demolish and Install	Sq.	72	80	4.47	156.47	loose-laid with stone.
Minimum Charge	Job		1100	114	1214	
48 mil (loose laid)						
Demolish	Sq.		36.50		36.50	Cost includes material and labor to
Install	Sq.	101	43.50	4.47	148.97	install 48 mil PVC elastomeric roofing,
Demolish and Install	Sq.	101	80	4.47	185.47	loose-laid with stone.
Minimum Charge	Job		1100	114	1214	
60 mil (loose laid)						
Demolish	Sq.		36.50		36.50	Cost includes material and labor to
Install	Sq.	239	43.50	4.47	286.97	install 60 mil PVC elastomeric roofing,
Demolish and Install	Sq.	239	80	4.47	323.47	loose-laid with stone.
Minimum Charge	Job		1100	114	1214	
45 mil (attached)						
Demolish	Sq.		57.50		57.50	Cost includes material and labor to
Install	Sq.	56	85.50	8.75	150.25	install 45 mil PVC elastomeric roofing
Demolish and Install	Sq.	56	143	8.75	207.75	fully attached to roof deck.
Minimum Charge	Job		1100	114	1214	

For customer support on your Contractor's Pricing Guide: Residential Repair & Remodeling, call 800.448.8182.

Roofing

Elastomeric Roofing

Elastomeric Roofing	Unit	Material	Labor	Equip.	Total	Specification
48 mil (attached)						
Demolish	Sq.		57.50		57.50	Cost includes material and labor to install 48 mil PVC elastomeric roofing fully attached to roof deck.
Install	Sq.	83	85.50	8.75	177.25	
Demolish and Install	Sq.	83	143	8.75	234.75	
Minimum Charge	Job		1100	114	1214	
60 mil (attached)						
Demolish	Sq.		57.50		57.50	Cost includes material and labor to install 60 mil PVC elastomeric roofing fully attached to roof deck.
Install	Sq.	220	85.50	8.75	314.25	
Demolish and Install	Sq.	220	143	8.75	371.75	
Minimum Charge	Job		1100	114	1214	

CSPE Type

CSPE Type	Unit	Material	Labor	Equip.	Total	Specification
45 mil (loose laid w / stone)						
Demolish	Sq.		36.50		36.50	Cost includes material and labor to install 45 mil CSPE (chlorosulfonated polyethylene-hypalon) roofing material loose laid on deck and ballasted with stone.
Install	Sq.	310	85.50	8.75	404.25	
Demolish and Install	Sq.	310	122	8.75	440.75	
Minimum Charge	Job		1100	114	1214	
45 mil (attached at seams)						
Demolish	Sq.		43		43	Cost includes material and labor to install 45 mil CSPE (chlorosulfonated polyethylene-hypalon) roofing material attached to deck at seams with batten strips.
Install	Sq.	400	63.50	6.50	470	
Demolish and Install	Sq.	400	106.50	6.50	513	
Minimum Charge	Job		1100	114	1214	
45 mil (fully attached)						
Demolish	Sq.		57.50		57.50	Cost includes material and labor to install 45 mil CSPE (chlorosulfonated polyethylene-hypalon) roofing material fully attached to deck.
Install	Sq.	294	63.50	6.50	364	
Demolish and Install	Sq.	294	121	6.50	421.50	
Minimum Charge	Job		1100	114	1214	

EPDM Type

EPDM Type	Unit	Material	Labor	Equip.	Total	Specification
45 mil (attached at seams)						
Demolish	Sq.		43		43	Cost includes material and labor to install 45 mil EPDM (ethylene propylene diene monomer) roofing material attached to deck at seams.
Install	Sq.	162	63.50	6.50	232	
Demolish and Install	Sq.	162	106.50	6.50	275	
Minimum Charge	Job		1100	114	1214	
45 mil (fully attached)						
Demolish	Sq.		57.50		57.50	Cost includes material and labor to install 45 mil EPDM (ethylene propylene diene monomer) roofing material fully attached to deck.
Install	Sq.	153	85.50	8.75	247.25	
Demolish and Install	Sq.	153	143	8.75	304.75	
Minimum Charge	Job		1100	114	1214	
45 mil (loose laid w / stone)						
Demolish	Sq.		36.50		36.50	Cost includes material and labor to install 45 mil EPDM (ethylene propylene diene monomer) roofing material loose laid on deck and ballasted with stones (approx. 1/2 ton of stone per square).
Install	Sq.	167	43.50	4.47	214.97	
Demolish and Install	Sq.	167	80	4.47	251.47	
Minimum Charge	Job		1100	114	1214	
60 mil (attached at seams)						
Demolish	Sq.		43		43	Cost includes material and labor to install 60 mil EPDM (ethylene propylene diene monomer) roofing material attached to deck at seams.
Install	Sq.	179	63.50	6.50	249	
Demolish and Install	Sq.	179	106.50	6.50	292	
Minimum Charge	Job		1100	114	1214	
60 mil (fully attached)						
Demolish	Sq.		57.50		57.50	Cost includes material and labor to install 60 mil EPDM (ethylene propylene diene monomer) roofing material fully attached to deck.
Install	Sq.	171	85.50	8.75	265.25	
Demolish and Install	Sq.	171	143	8.75	322.75	
Minimum Charge	Job		1100	114	1214	

Roofing

Elastomeric Roofing

	Unit	Material	Labor	Equip.	Total	Specification
60 mil (loose laid w / stone)						Cost includes material and labor to
Demolish	Sq.		36.50		36.50	install 60 mil EPDM (ethylene
Install	Sq.	186	43.50	4.47	233.97	propylene diene monomer) roofing
Demolish and Install	Sq.	186	80	4.47	270.47	material loose laid on deck and
Minimum Charge	Job		1100	114	1214	ballasted with stones (approx. 1/2 ton of stone per square).

Modified Bitumen Roofing

	Unit	Material	Labor	Equip.	Total	Specification
Fully Attached						
120 mil						Cost includes material and labor to
Demolish	Sq.		86.50		86.50	install 120 mil modified bitumen
Install	Sq.	72	10.85	2.83	85.68	roofing, fully attached to deck by
Demolish and Install	Sq.	72	97.35	2.83	172.18	torch.
Minimum Charge	Job		122		122	
150 mil						Cost includes material and labor to
Demolish	Sq.		86.50		86.50	install 150 mil modified bitumen
Install	Sq.	119	26.50	6.95	152.45	roofing, fully attached to deck by
Demolish and Install	Sq.	119	113	6.95	238.95	torch.
Minimum Charge	Job		122		122	
160 mil						Cost includes material and labor to
Demolish	Sq.		86.50		86.50	install 160 mil modified bitumen
Install	Sq.	106	28	7.25	141.25	roofing, fully attached to deck by
Demolish and Install	Sq.	106	114.50	7.25	227.75	torch.
Minimum Charge	Job		122		122	

Built-up Roofing

	Unit	Material	Labor	Equip.	Total	Specification
Emergency Repairs						Minimum charge for work requiring
Install	Ea.	265	2275	595	3135	hot asphalt.
Minimum Charge	Job		1600	420	2020	
Asphalt						
3-ply Asphalt						Cost includes material and labor to
Demolish	Sq.		90		90	install 3-ply asphalt built-up roof
Install	Sq.	105	133	35	273	including one 30 lb and two 15 lb
Demolish and Install	Sq.	105	223	35	363	layers of felt.
Minimum Charge	Job		1600	420	2020	
4-ply Asphalt						Cost includes material and labor to
Demolish	Sq.		107		107	install 4-ply asphalt built-up roof
Install	Sq.	176	145	38	359	including one 30 lb and three 15 lb
Demolish and Install	Sq.	176	252	38	466	layers of felt.
Minimum Charge	Job		1600	420	2020	
5-ply Asphalt						Cost includes material and labor to
Demolish	Sq.		141		141	install 5-ply asphalt built-up roof
Install	Sq.	217	160	42	419	including one 30 lb and four 15 lb
Demolish and Install	Sq.	217	301	42	560	layers of felt.
Minimum Charge	Job		1600	420	2020	

For customer support on your Contractor's Pricing Guide: Residential Repair & Remodeling, call 800.448.8182.

Roofing

Built-up Roofing

Built-up Roofing		Unit	Material	Labor	Equip.	Total	Specification
Components / Accessories							
Flood Coat							
	Install	Sq.	47.50	75		122.50	Includes labor and material to install flood coat.
R / R Gravel and Flood Coat							
	Install	Sq.	55	244		299	Includes labor and material to install flood coat and gravel.
Aluminum UV Coat							
	Install	Sq.	40.50	278	28.50	347	Includes labor and material to install aluminum-based UV coating.
Gravel Coat							
	Demolish	Sq.		45		45	Includes labor and material to install gravel coat.
	Install	Sq.	7.85	40.50		48.35	
	Demolish and Install	Sq.	7.85	85.50		93.35	
Mineral Surface Cap Sheet							
	Install	Sq.	91	8.40		99.40	Cost includes labor and material to install mineral surfaced cap sheet.
Fibrated Aluminum UV Coating							
	Install	Sq.	34.50	75		109.50	Includes labor and material to install fibered aluminum-based UV coating.
Impregnated Walk							
1"							
	Install	S.F.	2.99	2.60		5.59	Cost includes labor and material to install 1" layer of asphalt coating.
1/2"							
	Install	S.F.	2.22	2.05		4.27	Cost includes labor and material to install 1/2" layer of asphalt coating.
3/4"							
	Install	S.F.	2.70	2.17		4.87	Cost includes labor and material to install 3/4" layer of asphalt coating.

Exterior Doors and Windows

Doors   Roof Window   Double Hung Window

Exterior Type Door		Unit	Material	Labor	Equip.	Total	Specification
Entry							
Flush (1-3/4")							
	Demolish	Ea.		48		48	Cost includes material and labor to
	Install	Ea.	540	32.50		572.50	install 3' x 6' 8" pre-hung 1-3/4"
	Demolish and Install	Ea.	540	80.50		620.50	door, including casing and stop,
	Reinstall	Ea.		32.55		32.55	hinges, jamb, aluminum sill,
	Clean	Ea.	3.17	16.65		19.82	weatherstripped.
	Paint	Ea.	21	61.50		82.50	
	Minimum Charge	Job		260		260	
Raised Panel							
	Demolish	Ea.		48		48	Cost includes material and labor to
	Install	Ea.	330	29		359	install 3' x 6' 8" pre-hung 1-3/4"
	Demolish and Install	Ea.	330	77		407	door, jamb, including casing and
	Reinstall	Ea.		28.93		28.93	stop, hinges, aluminum sill,
	Clean	Ea.	3.17	16.65		19.82	weatherstripping.
	Paint	Ea.	21	61.50		82.50	
	Minimum Charge	Job		260		260	
Entry w / Lights							
Good Quality							
	Demolish	Ea.		48		48	Cost includes material and labor to
	Install	Ea.	1300	65		1365	install 3' x 6' 8" pre-hung 1-3/4"
	Demolish and Install	Ea.	1300	113		1413	door, solid wood lower panel, with up
	Reinstall	Ea.		65.10		65.10	to 9-light glass panels, casing, sill,
	Clean	Ea.	3.17	16.65		19.82	weatherstripped.
	Paint	Ea.	20.50	29		49.50	
	Minimum Charge	Job		260		260	
Better Quality							
	Demolish	Ea.		48		48	Cost includes material and labor to
	Install	Ea.	1325	65		1390	install 3' x 6' 8" pre-hung 1-3/4" Ash
	Demolish and Install	Ea.	1325	113		1438	door, solid wood, raised panel door
	Reinstall	Ea.		65.10		65.10	without glass panels.
	Clean	Ea.	3.17	16.65		19.82	
	Paint	Ea.	21	61.50		82.50	
	Minimum Charge	Job		260		260	
Premium Quality							
	Demolish	Ea.		48		48	Cost includes material and labor to
	Install	Ea.	3075	30.50		3105.50	install 3' x 6' 8" pre-hung 1-3/4" Oak
	Demolish and Install	Ea.	3075	78.50		3153.50	door, solid wood, raised panel door
	Reinstall	Ea.		30.64		30.64	without glass panels.
	Clean	Ea.	3.17	16.65		19.82	
	Paint	Ea.	21	61.50		82.50	
	Minimum Charge	Job		260		260	

Exterior Doors and Windows

Exterior Type Door		Unit	Material	Labor	Equip.	Total	Specification
Custom Quality							
	Demolish	Ea.		48		48	Cost includes material and labor to
	Install	Ea.	1800	69.50		1869.50	install 3' x 6' 8" pre-hung 1-3/4"
	Demolish and Install	Ea.	1800	117.50		1917.50	hand-made, solid wood, raised panel
	Reinstall	Ea.		69.44		69.44	door with custom stained glass.
	Clean	Ea.	.22	16.65		16.87	
	Paint	Ea.	6.55	144		150.55	
	Minimum Charge	Job		260		260	
Miami Style							
	Demolish	Ea.		48		48	Cost includes material and labor to
	Install	Ea.	540	30.50		570.50	install 3' x 6' 8" pre-hung 1-3/4"
	Demolish and Install	Ea.	540	78.50		618.50	lauan door of solid wood.
	Reinstall	Ea.		30.64		30.64	
	Clean	Ea.	.22	16.65		16.87	
	Paint	Ea.	21	61.50		82.50	
	Minimum Charge	Job		260		260	
French Style							
1-Light							
	Demolish	Ea.		48		48	Cost includes material and labor to
	Install	Ea.	700	87		787	install 3' x 6' 8" pre-hung 1-3/4"
	Demolish and Install	Ea.	700	135		835	French style door with 1 light.
	Reinstall	Ea.		86.80		86.80	
	Clean	Ea.	1.98	25		26.98	
	Paint	Ea.	30.50	108		138.50	
	Minimum Charge	Job		260		260	
9-Light							
	Demolish	Ea.		48		48	Cost includes material and labor to
	Install	Ea.	1100	37		1137	install 3' x 6' 8" pre-hung 1-3/4"
	Demolish and Install	Ea.	1100	85		1185	French style door with 9 lights.
	Reinstall	Ea.		37.20		37.20	
	Clean	Ea.	.44	33.50		33.94	
	Paint	Ea.	30.50	108		138.50	
	Minimum Charge	Job		260		260	
15-Light							
	Demolish	Ea.		55.50		55.50	Cost includes material and labor to
	Install	Ea.	1925	74.50		1999.50	install 3' x 6' 8" pre-hung 1-3/4"
	Demolish and Install	Ea.	1925	130		2055	French style door with 15 lights.
	Reinstall	Ea.		74.40		74.40	
	Clean	Ea.	.44	33.50		33.94	
	Paint	Ea.	30.50	108		138.50	
	Minimum Charge	Job		260		260	
Jamb							
Wood							
	Demolish	Ea.		36		36	Includes material and labor to install
	Install	Ea.	161	14.10		175.10	flat pine jamb with square cut heads
	Demolish and Install	Ea.	161	50.10		211.10	and rabbeted sides for 6' 8" high and
	Clean	Ea.	.10	10		10.10	3-9/16" door including trim sets for
	Paint	Ea.	20.50	29		49.50	both sides.
	Minimum Charge	Job		260		260	
Metal 6' 8" High							
	Demolish	Ea.		29		29	Cost includes material and labor to
	Install	Ea.	450	32.50		482.50	install 18 gauge hollow metal door
	Demolish and Install	Ea.	450	61.50		511.50	frame to fit a 4-1/2" jamb 6' 8" high
	Reinstall	Ea.		32.55		32.55	and up to 3' 6" opening.
	Clean	Ea.	.66	8.35		9.01	
	Paint	Ea.	3.45	8.65		12.10	
	Minimum Charge	Job		260		260	

For customer support on your Contractor's Pricing Guide: Residential Repair & Remodeling, call 800.448.8182.

Exterior Doors and Windows

Exterior Type Door		Unit	Material	Labor	Equip.	Total	Specification
Metal 7' High							
	Demolish	Ea.		29		29	Cost includes material and labor to
	Install	Ea.	455	34.50		489.50	install 18 gauge hollow metal door
	Demolish and Install	Ea.	455	63.50		518.50	frame to fit a 4-1/2" jamb 7' high and
	Reinstall	Ea.		34.72		34.72	up to 3' 6" opening.
	Clean	Ea.	.66	8.35		9.01	
	Paint	Ea.	3.45	8.65		12.10	
	Minimum Charge	Job		260		260	
Metal 8' High							
	Demolish	Ea.		29		29	Cost includes material and labor to
	Install	Ea.	273	43.50		316.50	install 18 gauge hollow metal door
	Demolish and Install	Ea.	273	72.50		345.50	frame to fit a 4-1/2" jamb 8' high and
	Reinstall	Ea.		43.40		43.40	up to 3' 6" opening.
	Clean	Ea.	.94	10		10.94	
	Paint	Ea.	3.86	9.60		13.46	
	Minimum Charge	Job		260		260	
Metal 9' High							
	Demolish	Ea.		29		29	Cost includes material and labor to
	Install	Ea.	340	52		392	install 18 gauge hollow metal door
	Demolish and Install	Ea.	340	81		421	frame to fit a 4-1/2" jamb 9' high and
	Reinstall	Ea.		52.08		52.08	up to 3' 6" opening.
	Clean	Ea.	.94	10		10.94	
	Paint	Ea.	3.86	9.60		13.46	
	Minimum Charge	Job		260		260	
Casing Trim							
Single Width							
	Demolish	Ea.		9.20		9.20	Cost includes material and labor to
	Install	Opng.	31.50	43.50		75	install 11/16" x 2-1/2" pine ranch
	Demolish and Install	Opng.	31.50	52.70		84.20	style casing for one side of a standard
	Reinstall	Opng.		34.72		34.72	door opening.
	Clean	Opng.	.45	4		4.45	
	Paint	Ea.	10.20	8.65		18.85	
	Minimum Charge	Job		260		260	
Double Width							
	Demolish	Ea.		10.90		10.90	Cost includes material and labor to
	Install	Opng.	37	52		89	install 11/16" x 2-1/2" pine ranch
	Demolish and Install	Opng.	37	62.90		99.90	style door casing for one side of a
	Reinstall	Opng.		41.66		41.66	double door opening.
	Clean	Opng.	.53	8		8.53	
	Paint	Ea.	10.20	9.80		20	
	Minimum Charge	Job		260		260	

Residential Metal Door		Unit	Material	Labor	Equip.	Total	Specification
Pre-hung, Steel-clad							
2' 8" x 6' 8"							
	Demolish	Ea.		48		48	Cost includes material and labor to
	Install	Ea.	1450	61.50	·	1511.50	install 2' 6" x 6' 8" pre-hung 1-3/4"
	Demolish and Install	Ea.	1450	109.50		1559.50	24 gauge steel clad door with 4-1/2"
	Reinstall	Ea.		61.27		61.27	wood frame, trims, hinges, aluminum
	Clean	Ea.	3.17	16.65		19.82	sill, weatherstripped.
	Paint	Ea.	30.50	61.50		92	
	Minimum Charge	Job		260		260	

For customer support on your **Contractor's Pricing Guide: Residential Repair & Remodeling,** call 800.448.8182.

Exterior Doors and Windows

Residential Metal Door	Unit	Material	Labor	Equip.	Total	Specification
3' x 6' 8"						Cost includes material and labor to
Demolish	Ea.		48		48	install 3' x 6' 8" pre-hung 1-3/4" 24
Install	Ea.	1100	69.50		1169.50	gauge steel clad door with 4-1/2"
Demolish and Install	Ea.	1100	117.50		1217.50	wood frame, trims, hinges, aluminum
Reinstall	Ea.		69.44		69.44	sill, weatherstripped.
Clean	Ea.	3.17	16.65		19.82	
Paint	Ea.	30.50	61.50		92	
Minimum Charge	Job		260		260	

Metal Door	Unit	Material	Labor	Equip.	Total	Specification
Steel, Door Only						
2' 8" x 6' 8", 18 Gauge						Cost includes material and labor to
Demolish	Ea.		10.35		10.35	install 18 gauge 2' 8" x 6' 8" steel
Install	Ea.	670	32.50		702.50	door only, 1-3/4" thick, hinges.
Demolish and Install	Ea.	670	42.85		712.85	
Reinstall	Ea.		32.55		32.55	
Clean	Ea.	1.32	12.50		13.82	
Paint	Ea.	12.55	36		48.55	
Minimum Charge	Job		260		260	
3' x 6' 8", 18 Gauge						Cost includes material and labor to
Demolish	Ea.		10.35		10.35	install 18 gauge 3' x 6' 8" steel door
Install	Ea.	490	34.50		524.50	only, 1-3/4" thick, hinges.
Demolish and Install	Ea.	490	44.85		534.85	
Reinstall	Ea.		34.72		34.72	
Clean	Ea.	1.32	12.50		13.82	
Paint	Ea.	12.55	36		48.55	
Minimum Charge	Job		260		260	
2' 8" x 7', 18 Gauge						Cost includes material and labor to
Demolish	Ea.		10.35		10.35	install 18 gauge 2' 8" x 7' steel door
Install	Ea.	505	34.50		539.50	only, 1-3/4" thick, hinges.
Demolish and Install	Ea.	505	44.85		549.85	
Reinstall	Ea.		34.72		34.72	
Clean	Ea.	1.32	12.50		13.82	
Paint	Ea.	12.55	36		48.55	
Minimum Charge	Job		260		260	
3' x 7', 18 Gauge						Cost includes material and labor to
Demolish	Ea.		10.35		10.35	install 18 gauge 3' x 7' steel door
Install	Ea.	580	37		617	only, 1-3/4" thick, hinges.
Demolish and Install	Ea.	580	47.35		627.35	
Reinstall	Ea.		37.20		37.20	
Clean	Ea.	1.32	12.50		13.82	
Paint	Ea.	12.55	36		48.55	
Minimum Charge	Job		260		260	
3' 6" x 7', 18 Gauge						Cost includes material and labor to
Demolish	Ea.		10.35		10.35	install 18 gauge 3' 6" x 7' steel door
Install	Ea.	680	40		720	only, 1-3/4" thick, hinges.
Demolish and Install	Ea.	680	50.35		730.35	
Reinstall	Ea.		40.06		40.06	
Clean	Ea.	1.32	12.50		13.82	
Paint	Ea.	12.55	36		48.55	
Minimum Charge	Job		260		260	

For customer support on your Contractor's Pricing Guide: Residential Repair & Remodeling, call 800.448.8182.

Exterior Doors and Windows

Metal Door

Metal Door	Unit	Material	Labor	Equip.	Total	Specification
Casing Trim						
Single Width						Cost includes material and labor to
Demolish	Ea.		9.20		9.20	install 11/16" x 2-1/2" pine ranch
Install	Opng.	31.50	43.50		75	style casing for one side of a standard
Demolish and Install	Opng.	31.50	52.70		84.20	door opening.
Reinstall	Opng.		34.72		34.72	
Clean	Opng.	.45	4		4.45	
Paint	Ea.	10.20	8.65		18.85	
Minimum Charge	Job		260		260	
Double Width						Cost includes material and labor to
Demolish	Ea.		10.90		10.90	install 11/16" x 2-1/2" pine ranch
Install	Opng.	37	52		89	style door casing for one side of a
Demolish and Install	Opng.	37	62.90		99.90	double door opening.
Reinstall	Opng.		41.66		41.66	
Clean	Opng.	.53	8		8.53	
Paint	Ea.	10.20	9.80		20	
Minimum Charge	Job		260		260	

Solid Core Door

Solid Core Door	Unit	Material	Labor	Equip.	Total	Specification
Exterior Type						
Flush (1-3/4")						Cost includes material and labor to
Demolish	Ea.		48		48	install 3' x 6' 8" pre-hung 1-3/4"
Install	Ea.	1400	65		1465	door, including casing and stop,
Demolish and Install	Ea.	1400	113		1513	hinges, jamb, aluminum sill,
Reinstall	Ea.		65.10		65.10	weatherstripped.
Clean	Ea.	3.17	16.65		19.82	
Paint	Ea.	30.50	61.50		92	
Minimum Charge	Job		260		260	
Raised Panel						Cost includes material and labor to
Demolish	Ea.		48		48	install 3' x 6' 8" pre-hung 1-3/4"
Install	Ea.	330	29		359	door, jamb, including casing and
Demolish and Install	Ea.	330	77		407	stop, hinges, aluminum sill,
Reinstall	Ea.		28.93		28.93	weatherstripping.
Clean	Ea.	3.17	16.65		19.82	
Paint	Ea.	30.50	144		174.50	
Minimum Charge	Job		260		260	
Entry w / Lights						Cost includes material and labor to
Demolish	Ea.		48		48	install 3' x 6' 8" pre-hung 1-3/4"
Install	Ea.	1300	65		1365	door, solid wood lower panel, with up
Demolish and Install	Ea.	1300	113		1413	to 9-light glass panels, casing, sill,
Reinstall	Ea.		65.10		65.10	weatherstripped.
Clean	Ea.	3.17	16.65		19.82	
Paint	Ea.	6.55	144		150.55	
Minimum Charge	Job		260		260	
Custom Entry						Cost includes material and labor to
Demolish	Ea.		48		48	install prehung, fir, single glazed with
Install	Ea.	2225	104		2329	lead caming, 1-3/4" x 6' 8" x 3'
Demolish and Install	Ea.	2225	152		2377	
Reinstall	Ea.		104.16		104.16	
Clean	Ea.	3.17	16.65		19.82	
Paint	Ea.	12.55	36		48.55	
Minimum Charge	Job		260		260	

Exterior Doors and Windows

Solid Core Door		Unit	Material	Labor	Equip.	Total	Specification
Dutch Style							
	Demolish	Ea.		48		48	Includes material and labor to install
	Install	Ea.	2225	87		2312	dutch door, pine, 1-3/4" x 6' 8" x 2'
	Demolish and Install	Ea.	2225	135		2360	8" wide.
	Reinstall	Ea.		86.80		86.80	
	Clean	Ea.	3.17	16.65		19.82	
	Paint	Ea.	6.55	144		150.55	
	Minimum Charge	Job		260		260	
Miami Style							
	Demolish	Ea.		48		48	Cost includes material and labor to
	Install	Ea.	540	30.50		570.50	install 3' x 6' 8" pre-hung 1-3/4"
	Demolish and Install	Ea.	540	78.50		618.50	lauan door of solid wood.
	Reinstall	Ea.		30.64		30.64	
	Clean	Ea.	3.17	16.65		19.82	
	Paint	Ea.	21	61.50		82.50	
	Minimum Charge	Job		260		260	
French Style							
	Demolish	Ea.		55.50		55.50	Cost includes material and labor to
	Install	Ea.	1925	74.50		1999.50	install 3' x 6' 8" pre-hung 1-3/4"
	Demolish and Install	Ea.	1925	130		2055	French style door with 15 lights.
	Reinstall	Ea.		74.40		74.40	
	Clean	Ea.	.22	16.65		16.87	
	Paint	Ea.	17.50	216		233.50	
	Minimum Charge	Job		260		260	
Double							
	Demolish	Ea.		37		37	Includes material and labor to install a
	Install	Ea.	1650	130		1780	pre-hung double entry door including
	Demolish and Install	Ea.	1650	167		1817	frame and exterior trim.
	Clean	Ea.	.12	64.50		64.62	
	Paint	Ea.	61	216		277	
	Minimum Charge	Job		130		130	
Commercial Grade							
3' x 6' 8"							
	Demolish	Ea.		48		48	Includes material and labor to install
	Install	Ea.	985	30.50		1015.50	flush solid core 1-3/4" hardwood
	Demolish and Install	Ea.	985	78.50		1063.50	veneer 3' x 6' 8" door only.
	Reinstall	Ea.		30.64		30.64	
	Clean	Ea.	3.17	16.65		19.82	
	Paint	Ea.	21	36		57	
	Minimum Charge	Job		260		260	
3' 6" x 6' 8"							
	Demolish	Ea.		48		48	Includes material and labor to install
	Install	Ea.	790	30.50		820.50	flush solid core 1-3/4" hardwood
	Demolish and Install	Ea.	790	78.50		868.50	veneer 3' 6" x 6' 8" door only.
	Reinstall	Ea.		30.64		30.64	
	Clean	Ea.	3.17	16.65		19.82	
	Paint	Ea.	21	36		57	
	Minimum Charge	Job		260		260	
3' x 7'							
	Demolish	Ea.		48		48	Includes material and labor to install
	Install	Ea.	920	74.50		994.50	flush solid core 1-3/4" hardwood
	Demolish and Install	Ea.	920	122.50		1042.50	veneer 3' x 7' door only.
	Reinstall	Ea.		74.40		74.40	
	Clean	Ea.	3.17	16.65		19.82	
	Paint	Ea.	21	36		57	
	Minimum Charge	Job		260		260	

For customer support on your Contractor's Pricing Guide: Residential Repair & Remodeling, call 800.448.8182.

Exterior Doors and Windows

Solid Core Door		Unit	Material	Labor	Equip.	Total	Specification
3′ 6″ x 7′							
	Demolish	Ea.		48		48	Includes material and labor to install
	Install	Ea.	910	37		947	flush solid core 1-3/4″ hardwood
	Demolish and Install	Ea.	910	85		995	veneer 3′ 6″ x 7′ door only.
	Reinstall	Ea.		37.20		37.20	
	Clean	Ea.	3.17	16.65		19.82	
	Paint	Ea.	21	36		57	
	Minimum Charge	Job		260		260	
3′ x 8′							
	Demolish	Ea.		48		48	Includes material and labor to install
	Install	Ea.	780	65		845	flush solid core 1-3/4″ hardwood
	Demolish and Install	Ea.	780	113		893	veneer 3′ x 8′ door only.
	Reinstall	Ea.		65.10		65.10	
	Clean	Ea.	3.80	25		28.80	
	Paint	Ea.	21	36		57	
	Minimum Charge	Job		260		260	
3′ 6″ x 8′							
	Demolish	Ea.		48		48	Includes material and labor to install
	Install	Ea.	910	65		975	flush solid core 1-3/4″ hardwood
	Demolish and Install	Ea.	910	113		1023	veneer 3′ 6″ x 8′ door only.
	Reinstall	Ea.		65.10		65.10	
	Clean	Ea.	3.80	25		28.80	
	Paint	Ea.	21	36		57	
	Minimum Charge	Job		260		260	
3′ x 9′							
	Demolish	Ea.		48		48	Includes material and labor to install
	Install	Ea.	815	74.50		889.50	flush solid core 1-3/4″ hardwood
	Demolish and Install	Ea.	815	122.50		937.50	veneer 3′ x 9′ door only.
	Reinstall	Ea.		74.40		74.40	
	Clean	Ea.	3.80	25		28.80	
	Paint	Ea.	21	36		57	
	Minimum Charge	Job		260		260	
Stain							
	Install	Ea.	20	48		68	Includes labor and material to stain
	Minimum Charge	Job		216		216	exterior type door and trim on both sides.
Door Only (SC)							
2′ 4″ x 6′ 8″							
	Demolish	Ea.		10.35		10.35	Cost includes material and labor to
	Install	Ea.	187	29		216	install 2′ 4″ x 6′ 8″ lauan solid core
	Demolish and Install	Ea.	187	39.35		226.35	door slab.
	Reinstall	Ea.		28.93		28.93	
	Clean	Ea.	1.32	12.50		13.82	
	Paint	Ea.	8.70	54		62.70	
	Minimum Charge	Job		260		260	
2′ 8″ x 6′ 8″							
	Demolish	Ea.		10.35		10.35	Cost includes material and labor to
	Install	Ea.	207	29		236	install 2′ 8″ x 6′ 8″ lauan solid core
	Demolish and Install	Ea.	207	39.35		246.35	door slab.
	Reinstall	Ea.		28.93		28.93	
	Clean	Ea.	1.32	12.50		13.82	
	Paint	Ea.	8.70	54		62.70	
	Minimum Charge	Job		260		260	

For customer support on your Contractor's Pricing Guide: Residential Repair & Remodeling, call 800.448.8182.

Exterior Doors and Windows

Solid Core Door

		Unit	Material	Labor	Equip.	Total	Specification
3′ x 6′ 8″							
	Demolish	Ea.		10.35		10.35	Cost includes material and labor to
	Install	Ea.	226	30.50		256.50	install 3′ x 6′ 8″ lauan solid core door
	Demolish and Install	Ea.	226	40.85		266.85	slab.
	Reinstall	Ea.		30.64		30.64	
	Clean	Ea.	1.32	12.50		13.82	
	Paint	Ea.	8.70	54		62.70	
	Minimum Charge	Job		260		260	
3′ 6″ x 6′ 8″							
	Demolish	Ea.		10.35		10.35	Cost includes material and labor to
	Install	Ea.	305	32.50		337.50	install 3′ 6″ x 6′ 8″ lauan solid core
	Demolish and Install	Ea.	305	42.85		347.85	door slab.
	Reinstall	Ea.		32.55		32.55	
	Clean	Ea.	1.32	12.50		13.82	
	Paint	Ea.	8.70	54		62.70	
	Minimum Charge	Job		260		260	
Panel							
	Demolish	Ea.		10.35		10.35	Cost includes material and labor to
	Install	Ea.	275	61.50		336.50	install 3′ x 6′ 8″ six (6) panel interior
	Demolish and Install	Ea.	275	71.85		346.85	door.
	Reinstall	Ea.		61.27		61.27	
	Clean	Ea.	1.58	12.50		14.08	
	Paint	Ea.	8.70	54		62.70	
	Minimum Charge	Job		260		260	

Casing Trim
Single Width

		Unit	Material	Labor	Equip.	Total	Specification
	Demolish	Ea.		9.20		9.20	Cost includes material and labor to
	Install	Opng.	31.50	43.50		75	install 11/16″ x 2-1/2″ pine ranch
	Demolish and Install	Opng.	31.50	52.70		84.20	style casing for one side of a standard
	Reinstall	Opng.		34.72		34.72	door opening.
	Clean	Opng.	.45	4		4.45	
	Paint	Ea.	10.20	8.65		18.85	
	Minimum Charge	Job		260		260	

Double Width

		Unit	Material	Labor	Equip.	Total	Specification
	Demolish	Ea.		10.90		10.90	Cost includes material and labor to
	Install	Opng.	37	52		89	install 11/16″ x 2-1/2″ pine ranch
	Demolish and Install	Opng.	37	62.90		99.90	style door casing for one side of a
	Reinstall	Opng.		41.66		41.66	double door opening.
	Clean	Opng.	.53	8		8.53	
	Paint	Ea.	10.20	9.80		20	
	Minimum Charge	Job		260		260	

Components

		Unit	Material	Labor	Equip.	Total	Specification
Hardware							
Doorknob							
	Demolish	Ea.		20.50		20.50	Includes material and labor to install
	Install	Ea.	63.50	32.50		96	residential passage lockset, keyless
	Demolish and Install	Ea.	63.50	53		116.50	bored type.
	Reinstall	Ea.		26.04		26.04	
	Clean	Ea.	.32	12.50		12.82	
	Minimum Charge	Job		260		260	

For customer support on your Contractor's Pricing Guide: Residential Repair & Remodeling, call 800.448.8182.

Exterior Doors and Windows

Components		Unit	Material	Labor	Equip.	Total	Specification
Doorknob w / Lock							
	Demolish	Ea.		20.50		20.50	Includes material and labor to install
	Install	Ea.	65.50	32.50		98	privacy lockset.
	Demolish and Install	Ea.	65.50	53		118.50	
	Reinstall	Ea.		26.04		26.04	
	Clean	Ea.	.32	12.50		12.82	
	Minimum Charge	Job		260		260	
Lever Handle							
	Demolish	Ea.		20.50		20.50	Includes material and labor to install
	Install	Ea.	273	52		325	residential passage lockset, keyless
	Demolish and Install	Ea.	273	72.50		345.50	bored type with lever handle.
	Reinstall	Ea.		41.66		41.66	
	Clean	Ea.	.32	12.50		12.82	
	Minimum Charge	Job		260		260	
Deadbolt							
	Demolish	Ea.		18.10		18.10	Includes material and labor to install a
	Install	Ea.	136	37		173	deadbolt.
	Demolish and Install	Ea.	136	55.10		191.10	
	Reinstall	Ea.		29.76		29.76	
	Clean	Ea.	.32	12.50		12.82	
	Minimum Charge	Job		260		260	
Grip Handle Entry							
	Demolish	Ea.		16.55		16.55	Includes material and labor to install
	Install	Ea.	164	58		222	residential keyed entry lock set with
	Demolish and Install	Ea.	164	74.55		238.55	grip type handle.
	Reinstall	Ea.		46.29		46.29	
	Clean	Ea.	.32	12.50		12.82	
	Minimum Charge	Job		260		260	
Panic Device							
	Demolish	Ea.		44.50		44.50	Includes material and labor to install
	Install	Ea.	550	87		637	touch bar, low profile, exit only.
	Demolish and Install	Ea.	550	131.50		681.50	
	Reinstall	Ea.		69.44		69.44	
	Clean	Ea.	.15	22		22.15	
	Minimum Charge	Job		260		260	
Closer							
	Demolish	Ea.		10.85		10.85	Includes material and labor to install
	Install	Ea.	415	87		502	pneumatic heavy duty closer for
	Demolish and Install	Ea.	415	97.85		512.85	exterior type doors.
	Reinstall	Ea.		69.44		69.44	
	Clean	Ea.	.15	16.65		16.80	
	Minimum Charge	Job		260		260	
Peep-hole							
	Demolish	Ea.		9.85		9.85	Includes material and labor to install
	Install	Ea.	15.50	16.30		31.80	peep-hole for door.
	Demolish and Install	Ea.	15.50	26.15		41.65	
	Reinstall	Ea.		13.02		13.02	
	Clean	Ea.	.02	4.98		5	
	Minimum Charge	Job		260		260	
Exit Sign							
	Demolish	Ea.		31.50		31.50	Includes material and labor to install a
	Install	Ea.	121	77.50		198.50	wall mounted interior electric exit sign.
	Demolish and Install	Ea.	121	109		230	
	Reinstall	Ea.		62.08		62.08	
	Clean	Ea.	.02	10.10		10.12	
	Minimum Charge	Job		260		260	

For customer support on your Contractor's Pricing Guide: Residential Repair & Remodeling, call 800.448.8182.

Exterior Doors and Windows

Components		Unit	Material	Labor	Equip.	Total	Specification
Kickplate							
	Demolish	Ea.		16.05		16.05	Includes material and labor to install
	Install	Ea.	40	34.50		74.50	aluminum kickplate, 10" x 28".
	Demolish and Install	Ea.	40	50.55		90.55	
	Reinstall	Ea.		27.78		27.78	
	Clean	Ea.	.32	12.50		12.82	
	Minimum Charge	Job		260		260	
Threshold							
	Install	L.F.	3.36	5.20		8.56	Includes labor and material to install
	Minimum Charge	Job		130		130	oak threshold.
Weatherstripping							
	Install	L.F.	2.83	1.30		4.13	Includes labor and material to install
	Minimum Charge	Job		130		130	rubber weatherstripping for door.
Metal Jamb							
6' 8" High							
	Demolish	Ea.		29		29	Cost includes material and labor to
	Install	Ea.	450	32.50		482.50	install 18 gauge hollow metal door
	Demolish and Install	Ea.	450	61.50		511.50	frame to fit a 4-1/2" jamb 6' 8" high
	Reinstall	Ea.		32.55		32.55	and up to 3' 6" opening.
	Clean	Ea.	.66	8.35		9.01	
	Paint	Ea.	3.45	8.65		12.10	
	Minimum Charge	Job		260		260	
7' High							
	Demolish	Ea.		29		29	Cost includes material and labor to
	Install	Ea.	455	34.50		489.50	install 18 gauge hollow metal door
	Demolish and Install	Ea.	455	63.50		518.50	frame to fit a 4-1/2" jamb 7' high and
	Reinstall	Ea.		34.72		34.72	up to 3' 6" opening.
	Clean	Ea.	.66	8.35		9.01	
	Paint	Ea.	3.45	8.65		12.10	
	Minimum Charge	Job		260		260	
8' High							
	Demolish	Ea.		29		29	Cost includes material and labor to
	Install	Ea.	273	43.50		316.50	install 18 gauge hollow metal door
	Demolish and Install	Ea.	273	72.50		345.50	frame to fit a 4-1/2" jamb 8' high and
	Reinstall	Ea.		43.40		43.40	up to 3' 6" opening.
	Clean	Ea.	.94	10		10.94	
	Paint	Ea.	3.86	9.60		13.46	
	Minimum Charge	Job		260		260	
9' High							
	Demolish	Ea.		29		29	Cost includes material and labor to
	Install	Ea.	340	52		392	install 18 gauge hollow metal door
	Demolish and Install	Ea.	340	81		421	frame to fit a 4-1/2" jamb 9' high and
	Reinstall	Ea.		52.08		52.08	up to 3' 6" opening.
	Clean	Ea.	.94	10		10.94	
	Paint	Ea.	3.86	9.60		13.46	
	Minimum Charge	Job		260		260	
Wood Jamb							
	Demolish	Ea.		36		36	Includes material and labor to install
	Install	Ea.	161	14.10		175.10	flat pine jamb with square cut heads
	Demolish and Install	Ea.	161	50.10		211.10	and rabbeted sides for 6' 8" high and
	Clean	Ea.	.10	10		10.10	3-9/16" door including trim sets for
	Paint	Ea.	20.50	29		49.50	both sides.
	Minimum Charge	Job		260		260	
Stain							
	Install	Ea.	13.90	36		49.90	Includes labor and material to stain
	Minimum Charge	Job		216		216	single door and trim on both sides.

For customer support on your Contractor's Pricing Guide: Residential Repair & Remodeling, call 800.448.8182.

Exterior Doors and Windows

Components		Unit	Material	Labor	Equip.	Total	Specification
Shave & Refit							
	Install	Ea.		43.50		43.50	Includes labor to shave and rework
	Minimum Charge	Job		130		130	door to fit opening at the job site.

Sliding Patio Door		Unit	Material	Labor	Equip.	Total	Specification
Aluminum							
6' Single Pane							
	Demolish	Ea.		72.50		72.50	Cost includes material and labor to
	Install	Ea.	3575	260		3835	install 6' x 6' 8" sliding anodized
	Demolish and Install	Ea.	3575	332.50		3907.50	aluminum patio door with single pane
	Reinstall	Ea.		208.32		208.32	tempered safety glass, anodized
	Clean	Ea.	1.98	25		26.98	frame, screen, weatherstripped.
	Minimum Charge	Job		260		260	
8' Single Pane							
	Demolish	Ea.		72.50		72.50	Cost includes material and labor to
	Install	Ea.	4675	345		5020	install 8' x 6' 8" sliding aluminum
	Demolish and Install	Ea.	4675	417.50		5092.50	patio door with single pane tempered
	Reinstall	Ea.		277.76		277.76	safety glass, anodized frame, screen,
	Clean	Ea.	1.98	25		26.98	weatherstripped.
	Minimum Charge	Job		260		260	
10' Single Pane							
	Demolish	Ea.		72.50		72.50	Cost includes material and labor to
	Install	Ea.	6875	520		7395	install 10' x 6' 8" sliding aluminum
	Demolish and Install	Ea.	6875	592.50		7467.50	patio door with single pane tempered
	Reinstall	Ea.		416.64		416.64	safety glass, anodized frame, screen,
	Clean	Ea.	1.98	25		26.98	weatherstripped.
	Minimum Charge	Job		260		260	
6' Insulated							
	Demolish	Ea.		72.50		72.50	Cost includes material and labor to
	Install	Ea.	4675	260		4935	install 6' x 6' 8" sliding aluminum
	Demolish and Install	Ea.	4675	332.50		5007.50	patio door with double insulated
	Reinstall	Ea.		208.32		208.32	tempered glass, baked-on enamel
	Clean	Ea.	1.98	25		26.98	finish, screen, weatherstripped.
	Minimum Charge	Job		260		260	
8' Insulated							
	Demolish	Ea.		72.50		72.50	Cost includes material and labor to
	Install	Ea.	4975	345		5320	install 8' x 6' 8" sliding aluminum
	Demolish and Install	Ea.	4975	417.50		5392.50	patio door with insulated tempered
	Reinstall	Ea.		277.76		277.76	glass, baked-on enamel finish, screen,
	Clean	Ea.	1.98	25		26.98	weatherstripped.
	Minimum Charge	Job		260		260	
10' Insulated							
	Demolish	Ea.		72.50		72.50	Cost includes material and labor to
	Install	Ea.	6875	520		7395	install 10' x 6' 8" sliding aluminum
	Demolish and Install	Ea.	6875	592.50		7467.50	patio door with double insulated
	Reinstall	Ea.		416.64		416.64	tempered glass, baked-on enamel
	Clean	Ea.	1.98	25		26.98	finish, screen, weatherstripped.
	Minimum Charge	Job		260		260	

For customer support on your Contractor's Pricing Guide: Residential Repair & Remodeling, call 800.448.8182.

Exterior Doors and Windows

Sliding Patio Door

Sliding Patio Door	Unit	Material	Labor	Equip.	Total	Specification
Vinyl						
6' Single Pane						
Demolish	Ea.		72.50		72.50	Cost includes material and labor to
Install	Ea.	1875	260		2135	install 6' x 6' 8" sliding vinyl patio
Demolish and Install	Ea.	1875	332.50		2207.50	door with single pane tempered safety
Reinstall	Ea.		208.32		208.32	glass, anodized frame, screen,
Clean	Ea.	1.98	25		26.98	weatherstripped.
Minimum Charge	Job		260		260	
8' Single Pane						
Demolish	Ea.		72.50		72.50	Cost includes material and labor to
Install	Ea.	1975	345		2320	install 8' x 6' 8" sliding vinyl patio
Demolish and Install	Ea.	1975	417.50		2392.50	door with single pane tempered safety
Reinstall	Ea.		277.76		277.76	glass, anodized frame, screen,
Clean	Ea.	1.98	25		26.98	weatherstripped.
Minimum Charge	Job		260		260	
10' Single Pane						
Demolish	Ea.		72.50		72.50	Cost includes material and labor to
Install	Ea.	6875	520		7395	install 10' x 6' 8" sliding vinyl patio
Demolish and Install	Ea.	6875	592.50		7467.50	door with single pane tempered safety
Reinstall	Ea.		416.64		416.64	glass, anodized frame, screen,
Clean	Ea.	1.98	25		26.98	weatherstripped.
Minimum Charge	Job		260		260	
6' Insulated						
Demolish	Ea.		72.50		72.50	Cost includes material and labor to
Install	Ea.	4075	260		4335	install 6' x 6' 8" sliding vinyl patio
Demolish and Install	Ea.	4075	332.50		4407.50	door with double insulated tempered
Reinstall	Ea.		208.32		208.32	glass, anodized frame, screen,
Clean	Ea.	1.98	25		26.98	weatherstripped.
Minimum Charge	Job		260		260	
8' Insulated						
Demolish	Ea.		72.50		72.50	Cost includes material and labor to
Install	Ea.	4675	345		5020	install 8' x 6' 8" sliding vinyl patio
Demolish and Install	Ea.	4675	417.50		5092.50	door with double insulated tempered
Reinstall	Ea.		277.76		277.76	glass, anodized frame, screen,
Clean	Ea.	1.98	25		26.98	weatherstripped.
Minimum Charge	Job		260		260	
10' Insulated						
Demolish	Ea.		72.50		72.50	Cost includes material and labor to
Install	Ea.	6875	260		7135	install 10' x 6' 8" sliding vinyl patio
Demolish and Install	Ea.	6875	332.50		7207.50	door with double insulated tempered
Reinstall	Ea.		208.32		208.32	glass, anodized frame, screen,
Clean	Ea.	1.98	25		26.98	weatherstripped.
Minimum Charge	Job		260		260	
Wood Framed						
6' Premium Insulated						
Demolish	Ea.		72.50		72.50	Cost includes material and labor to
Install	Ea.	9450	260		9710	install 6' x 6' 8" sliding wood patio
Demolish and Install	Ea.	9450	332.50		9782.50	door with pine frame, double insulated
Reinstall	Ea.		208.32		208.32	glass, screen, lock and dead bolt,
Clean	Ea.	1.98	25		26.98	weatherstripped. Add for casing and
Minimum Charge	Job		260		260	finishing.

Exterior Doors and Windows

Sliding Patio Door

Sliding Patio Door	Unit	Material	Labor	Equip.	Total	Specification
8' Premium Insulated						Cost includes material and labor to install 8' x 6' 8" sliding wood patio door with pine frame, double insulated glass, screen, lock and dead bolt, weatherstripped. Add for casing and finishing.
Demolish	Ea.		72.50		72.50	
Install	Ea.	13200	345		13545	
Demolish and Install	Ea.	13200	417.50		13617.50	
Reinstall	Ea.		277.76		277.76	
Clean	Ea.	1.98	25		26.98	
Minimum Charge	Job		260		260	
9' Premium Insulated						Cost includes material and labor to install 9' x 6' 8" sliding wood patio door with pine frame, double insulated glass, screen, lock and dead bolt, weatherstripped. Add for casing and finishing.
Demolish	Ea.		72.50		72.50	
Install	Ea.	13200	345		13545	
Demolish and Install	Ea.	13200	417.50		13617.50	
Reinstall	Ea.		277.76		277.76	
Clean	Ea.	1.98	25		26.98	
Minimum Charge	Job		260		260	
Stain						Includes labor and material to stain exterior type door and trim on both sides.
Install	Ea.	20	48		68	
Minimum Charge	Job		216		216	
Casing Trim						Cost includes material and labor to install 11/16" x 2-1/2" pine ranch style door casing for one side of a double door opening.
Demolish	Ea.		10.90		10.90	
Install	Opng.	37	52		89	
Demolish and Install	Opng.	37	62.90		99.90	
Reinstall	Opng.		41.66		41.66	
Clean	Opng.	.53	8		8.53	
Paint	Ea.	10.20	9.80		20	
Minimum Charge	Job		260		260	
Sliding Screen						Includes material and labor to install sliding screen door, 6' 8" x 3'.
Demolish	Ea.		10.35		10.35	
Install	Ea.	282	26		308	
Demolish and Install	Ea.	282	36.35		318.35	
Reinstall	Ea.		20.83		20.83	
Clean	Ea.	.66	8.35		9.01	
Minimum Charge	Job		260		260	

Screen Door

Screen Door	Unit	Material	Labor	Equip.	Total	Specification
Aluminum						Includes material and labor to install aluminum frame screen door with fiberglass screen cloth, closer, hinges, latch.
Demolish	Ea.		48		48	
Install	Ea.	580	74.50		654.50	
Demolish and Install	Ea.	580	122.50		702.50	
Reinstall	Ea.		59.52		59.52	
Clean	Ea.	3.17	16.65		19.82	
Paint	Ea.	9.80	17.25		27.05	
Minimum Charge	Job		260		260	
Wood						Includes material and labor to install wood screen door with aluminum cloth screen. Frame and hardware not included.
Demolish	Ea.		48		48	
Install	Ea.	735	87		822	
Demolish and Install	Ea.	735	135		870	
Reinstall	Ea.		69.44		69.44	
Clean	Ea.	3.17	16.65		19.82	
Paint	Ea.	9.80	17.25		27.05	
Minimum Charge	Job		260		260	
Shave & Refit						Includes labor to shave and rework door to fit opening at the job site.
Install	Ea.		43.50		43.50	
Minimum Charge	Job		260		260	

Exterior Doors and Windows

Storm Door		Unit	Material	Labor	Equip.	Total	Specification
Good Grade							
	Demolish	Ea.		48		48	Includes material and labor to install
	Install	Ea.	615	74.50		689.50	aluminum frame storm door with
	Demolish and Install	Ea.	615	122.50		737.50	fiberglass screen cloth, upper glass
	Reinstall	Ea.		59.52		59.52	panel, closer, hinges, latch.
	Clean	Ea.	3.17	16.65		19.82	
	Paint	Ea.	9.80	17.25		27.05	
	Minimum Charge	Job		260		260	
Better Grade							
	Demolish	Ea.		48		48	Includes material and labor to install
	Install	Ea.	615	74.50		689.50	aluminum frame storm door with
	Demolish and Install	Ea.	615	122.50		737.50	fiberglass screen cloth, upper and
	Reinstall	Ea.		59.52		59.52	lower glass panel, closer, latch.
	Clean	Ea.	3.17	16.65		19.82	
	Paint	Ea.	9.80	17.25		27.05	
	Minimum Charge	Job		260		260	
Premium Grade							
	Demolish	Ea.		48		48	Includes material and labor to install
	Install	Ea.	615	74.50		689.50	aluminum frame storm door with
	Demolish and Install	Ea.	615	122.50		737.50	fiberglass screen cloth, upper and
	Reinstall	Ea.		59.52		59.52	lower glass panel, closer, latch.
	Clean	Ea.	3.17	16.65		19.82	
	Paint	Ea.	9.80	17.25		27.05	
	Minimum Charge	Job		260		260	
Insulated							
	Demolish	Ea.		48		48	Includes material and labor to install
	Install	Ea.	550	37		587	aluminum frame storm door with
	Demolish and Install	Ea.	550	85		635	fiberglass screen cloth, upper and
	Reinstall	Ea.		29.76		29.76	lower glass panel, closer, latch.
	Clean	Ea.	3.17	16.65		19.82	
	Paint	Ea.	9.80	17.25		27.05	
	Minimum Charge	Job		260		260	

Garage Door		Unit	Material	Labor	Equip.	Total	Specification
9' One Piece Metal							
	Demolish	Ea.		122		122	Includes material and labor to install
	Install	Ea.	2375	130		2505	one piece steel garage door 9' x 7'
	Demolish and Install	Ea.	2375	252		2627	primed including hardware.
	Reinstall	Ea.		104.16		104.16	
	Clean	Ea.	3.96	50		53.96	
	Paint	Ea.	10.85	108		118.85	
	Minimum Charge	Job		260		260	
16' One Piece Metal							
	Demolish	Ea.		145		145	Includes material and labor to install
	Install	Ea.	4050	174		4224	one piece steel garage door 16' x 7'
	Demolish and Install	Ea.	4050	319		4369	primed including hardware.
	Reinstall	Ea.		138.88		138.88	
	Clean	Ea.	7.90	66.50		74.40	
	Paint	Ea.	21.50	123		144.50	
	Minimum Charge	Job		260		260	

Exterior Doors and Windows

Garage Door		Unit	Material	Labor	Equip.	Total	Specification
9' Metal Sectional							
	Demolish	Ea.		122		122	Includes material and labor to install
	Install	Ea.	1325	130		1455	sectional metal garage door 9' x 7'
	Demolish and Install	Ea.	1325	252		1577	primed including hardware.
	Reinstall	Ea.		104.16		104.16	
	Clean	Ea.	4.05	57		61.05	
	Paint	Ea.	21.50	123		144.50	
	Minimum Charge	Job		260		260	
16' Metal Sectional							
	Demolish	Ea.		145		145	Includes material and labor to install
	Install	Ea.	2175	174		2349	sectional metal garage door 16' x 7'
	Demolish and Install	Ea.	2175	319		2494	primed including hardware.
	Reinstall	Ea.		138.88		138.88	
	Clean	Ea.	7.90	66.50		74.40	
	Paint	Ea.	21.50	123		144.50	
	Minimum Charge	Job		260		260	
Metal / Steel Roll Up							
8' x 8' Steel							
	Demolish	Ea.		165		165	Cost includes material and labor to
	Install	Ea.	1125	735		1860	install 8' x 8' steel overhead roll-up,
	Demolish and Install	Ea.	1125	900		2025	chain hoist operated service door.
	Reinstall	Ea.		588.80		588.80	
	Clean	Ea.	4.05	57		61.05	
	Paint	Ea.	23	144		167	
	Minimum Charge	Job		1175		1175	
10' x 10' Steel							
	Demolish	Ea.		193		193	Cost includes material and labor to
	Install	Ea.	1525	840		2365	install 10' x 10' steel overhead roll-up,
	Demolish and Install	Ea.	1525	1033		2558	chain hoist operated service door.
	Reinstall	Ea.		672.91		672.91	
	Clean	Ea.	6.35	80		86.35	
	Paint	Ea.	36.50	144		180.50	
	Minimum Charge	Job		1175		1175	
12' x 12' Steel							
	Demolish	Ea.		231		231	Cost includes material and labor to
	Install	Ea.	1875	980		2855	install 12' x 12' steel overhead roll-up,
	Demolish and Install	Ea.	1875	1211		3086	chain hoist operated service door.
	Reinstall	Ea.		785.07		785.07	
	Clean	Ea.	9.15	89		98.15	
	Paint	Ea.	51.50	173		224.50	
	Minimum Charge	Job		1175		1175	
14' x 14' Steel							
	Demolish	Ea.		289		289	Cost includes material and labor to
	Install	Ea.	2825	1475		4300	install 14' x 14' steel overhead roll-up,
	Demolish and Install	Ea.	2825	1764		4589	chain hoist operated service door.
	Reinstall	Ea.		1177.60		1177.60	
	Clean	Ea.	12.40	114		126.40	
	Paint	Ea.	60.50	216		276.50	
	Minimum Charge	Job		1175		1175	
18' x 18' Steel							
	Demolish	Ea.		320		320	Cost includes material and labor to
	Install	Opng.	4275	1975		6250	install 18' x 18' steel overhead roll-up,
	Demolish and Install	Opng.	4275	2295		6570	chain hoist operated service door.
	Reinstall	Opng.		1570.13		1570.13	
	Clean	Ea.	12.40	66.50		78.90	
	Paint	Ea.	117	430		547	
	Minimum Charge	Job		1175		1175	
Automatic Motor							
	Install	Ea.	1375	52		1427	Includes labor and material to install
	Minimum Charge	Job		260		260	electric opener motor, 1/3 HP.

For customer support on your Contractor's Pricing Guide: Residential Repair & Remodeling, call 800.448.8182.

Exterior Doors and Windows

Garage Door

Garage Door		Unit	Material	Labor	Equip.	Total	Specification
9' One Piece Wood							
	Demolish	Ea.		122		122	Includes material and labor to install
	Install	Ea.	3775	130		3905	one piece wood garage door 9' x 7'
	Demolish and Install	Ea.	3775	252		4027	primed including hardware.
	Reinstall	Ea.		104.16		104.16	
	Clean	Ea.	3.96	50		53.96	
	Paint	Ea.	10.85	108		118.85	
	Minimum Charge	Job		260		260	
9' Wood Sectional							
	Demolish	Ea.		122		122	Includes material and labor to install
	Install	Ea.	2975	130		3105	sectional wood garage door 9' x 7'
	Demolish and Install	Ea.	2975	252		3227	unfinished including hardware.
	Reinstall	Ea.		104.16		104.16	
	Clean	Ea.	3.96	50		53.96	
	Paint	Ea.	10.85	108		118.85	
	Minimum Charge	Job		260		260	
16' Wood Sectional							
	Demolish	Ea.		145		145	Includes material and labor to install
	Install	Ea.	2900	174		3074	sectional fiberglass garage door 16' x
	Demolish and Install	Ea.	2900	319		3219	7' unfinished including hardware.
	Reinstall	Ea.		138.88		138.88	
	Clean	Ea.	7.90	66.50		74.40	
	Paint	Ea.	21.50	123		144.50	
	Minimum Charge	Job		260		260	
Electric Opener							
	Demolish	Ea.		89		89	Includes material and labor to install
	Install	Ea.	1150	97.50		1247.50	electric opener, 1/3 HP economy.
	Demolish and Install	Ea.	1150	186.50		1336.50	
	Reinstall	Ea.		78.17		78.17	
	Clean	Ea.	.15	22		22.15	
	Minimum Charge	Job		260		260	
Springs							
	Install	Pair	106	130		236	Includes labor and material to replace
	Minimum Charge	Job		260		260	springs for garage door.

Commercial Metal Door

Commercial Metal Door		Unit	Material	Labor	Equip.	Total	Specification
Pre-hung Steel Door							
2' 8" x 6' 8"							
	Demolish	Ea.		72.50		72.50	Cost includes material and labor to
	Install	Ea.	865	32.50		897.50	install 2' 8" x 6' 8" pre-hung 1-3/4"
	Demolish and Install	Ea.	865	105		970	18 gauge steel door with steel frame,
	Reinstall	Ea.		32.55		32.55	hinges, aluminum sill,
	Clean	Ea.	3.17	16.65		19.82	weather-stripped.
	Paint	Ea.	12.55	36		48.55	
	Minimum Charge	Job		260		260	
3' x 6' 8"							
	Demolish	Ea.		72.50		72.50	Cost includes material and labor to
	Install	Ea.	1100	34.50		1134.50	install 3' x 6' 8" pre-hung 1-3/4" 18
	Demolish and Install	Ea.	1100	107		1207	gauge steel door with steel frame,
	Reinstall	Ea.		34.72		34.72	hinges, aluminum sill,
	Clean	Ea.	3.17	16.65		19.82	weather-stripped.
	Paint	Ea.	12.55	36		48.55	
	Minimum Charge	Job		260		260	

For customer support on your Contractor's Pricing Guide: Residential Repair & Remodeling, call 800.448.8182.

Exterior Doors and Windows

Commercial Metal Door

	Unit	Material	Labor	Equip.	Total	Specification
3' 6" x 6' 8"						Cost includes material and labor to
Demolish	Ea.		72.50		72.50	install 3' 6" x 6' 8" pre-hung 1-3/4"
Install	Ea.	1175	40		1215	18 gauge steel door with steel frame,
Demolish and Install	Ea.	1175	112.50		1287.50	hinges, aluminum sill,
Reinstall	Ea.		40.06		40.06	weather-stripped.
Clean	Ea.	3.17	16.65		19.82	
Paint	Ea.	12.55	36		48.55	
Minimum Charge	Job		260		260	
4' x 6' 8"						Cost includes material and labor to
Demolish	Ea.		72.50		72.50	install 4' x 6' 8" pre-hung 1-3/4" 18
Install	Ea.	1300	52		1352	gauge steel door with steel frame,
Demolish and Install	Ea.	1300	124.50		1424.50	hinges, aluminum sill,
Reinstall	Ea.		52.08		52.08	weather-stripped.
Clean	Ea.	3.17	16.65		19.82	
Paint	Ea.	12.55	36		48.55	
Minimum Charge	Job		260		260	

Metal Jamb

	Unit	Material	Labor	Equip.	Total	Specification
6' 8" High						Cost includes material and labor to
Demolish	Ea.		29		29	install 18 gauge hollow metal door
Install	Ea.	450	32.50		482.50	frame to fit a 4-1/2" jamb 6' 8" high
Demolish and Install	Ea.	450	61.50		511.50	and up to 3' 6" opening.
Reinstall	Ea.		32.55		32.55	
Clean	Ea.	.66	8.35		9.01	
Paint	Ea.	3.45	8.65		12.10	
Minimum Charge	Job		260		260	
7' High						Cost includes material and labor to
Demolish	Ea.		29		29	install 18 gauge hollow metal door
Install	Ea.	455	34.50		489.50	frame to fit a 4-1/2" jamb 7' high and
Demolish and Install	Ea.	455	63.50		518.50	up to 3' 6" opening.
Reinstall	Ea.		34.72		34.72	
Clean	Ea.	.66	8.35		9.01	
Paint	Ea.	3.45	8.65		12.10	
Minimum Charge	Job		260		260	
8' High						Cost includes material and labor to
Demolish	Ea.		29		29	install 18 gauge hollow metal door
Install	Ea.	273	43.50		316.50	frame to fit a 4-1/2" jamb 8' high and
Demolish and Install	Ea.	273	72.50		345.50	up to 3' 6" opening.
Reinstall	Ea.		43.40		43.40	
Clean	Ea.	.94	10		10.94	
Paint	Ea.	3.86	9.60		13.46	
Minimum Charge	Job		260		260	
9' High						Cost includes material and labor to
Demolish	Ea.		29		29	install 18 gauge hollow metal door
Install	Ea.	340	52		392	frame to fit a 4-1/2" jamb 9' high and
Demolish and Install	Ea.	340	81		421	up to 3' 6" opening.
Reinstall	Ea.		52.08		52.08	
Clean	Ea.	.94	10		10.94	
Paint	Ea.	3.86	9.60		13.46	
Minimum Charge	Job		260		260	

Hardware

	Unit	Material	Labor	Equip.	Total	Specification
Doorknob w / Lock						Includes material and labor to install
Demolish	Ea.		20.50		20.50	keyed entry lock set, bored type
Install	Ea.	283	43.50		326.50	including knob trim, strike and strike
Demolish and Install	Ea.	283	64		347	box for standard commercial
Reinstall	Ea.		34.72		34.72	application.
Clean	Ea.	.32	12.50		12.82	
Minimum Charge	Job		260		260	

For customer support on your Contractor's Pricing Guide: Residential Repair & Remodeling, call 800.448.8182.

Exterior Doors and Windows

Commercial Metal Door		Unit	Material	Labor	Equip.	Total	Specification
Lever Handle							
	Demolish	Ea.		20.50		20.50	Includes material and labor to install
	Install	Ea.	156	52		208	passage lockset, keyless bored type
	Demolish and Install	Ea.	156	72.50		228.50	with lever handle, non-locking
	Reinstall	Ea.		41.66		41.66	standard commercial latchset.
	Clean	Ea.	.32	12.50		12.82	
	Minimum Charge	Job		260		260	
Deadbolt							
	Demolish	Ea.		18.10		18.10	Includes material and labor to install a
	Install	Ea.	136	37		173	deadbolt.
	Demolish and Install	Ea.	136	55.10		191.10	
	Reinstall	Ea.		29.76		29.76	
	Clean	Ea.	.32	12.50		12.82	
	Minimum Charge	Job		260		260	
Panic Device							
	Demolish	Ea.		44.50		44.50	Includes material and labor to install
	Install	Ea.	550	87		637	touch bar, low profile, exit only.
	Demolish and Install	Ea.	550	131.50		681.50	
	Reinstall	Ea.		69.44		69.44	
	Clean	Ea.	.15	22		22.15	
	Minimum Charge	Job		260		260	
Closer							
	Demolish	Ea.		10.85		10.85	Includes material and labor to install
	Install	Ea.	365	87		452	pneumatic heavy duty model for
	Demolish and Install	Ea.	365	97.85		462.85	exterior type doors.
	Reinstall	Ea.		69.44		69.44	
	Clean	Ea.	.15	16.65		16.80	
	Minimum Charge	Job		260		260	
Peep-Hole							
	Demolish	Ea.		9.85		9.85	Includes material and labor to install
	Install	Ea.	15.50	16.30		31.80	peep-hole for door.
	Demolish and Install	Ea.	15.50	26.15		41.65	
	Reinstall	Ea.		13.02		13.02	
	Clean	Ea.	.02	4.98		5	
	Minimum Charge	Job		260		260	
Exit Sign							
	Demolish	Ea.		31.50		31.50	Includes material and labor to install a
	Install	Ea.	121	77.50		198.50	wall mounted interior electric exit sign.
	Demolish and Install	Ea.	121	109		230	
	Reinstall	Ea.		62.08		62.08	
	Clean	Ea.	.02	10.10		10.12	
	Minimum Charge	Job		260		260	
Kickplate							
	Demolish	Ea.		16.05		16.05	Includes material and labor to install
	Install	Ea.	136	34.50		170.50	stainless steel kickplate, 10" x 36".
	Demolish and Install	Ea.	136	50.55		186.55	
	Reinstall	Ea.		27.78		27.78	
	Clean	Ea.	.32	12.50		12.82	
	Minimum Charge	Job		260		260	

Aluminum Window		Unit	Material	Labor	Equip.	Total	Specification
Single Hung Two Light							
2' x 2'							
	Demolish	Ea.		28		28	Cost includes material and labor to
	Install	Ea.	390	107		497	install 2' x 2' single hung, two light,
	Demolish and Install	Ea.	390	135		525	double glazed aluminum window
	Reinstall	Ea.		85.64		85.64	including screen.
	Clean	Ea.	.03	11.55		11.58	
	Minimum Charge	Job		294		294	

For customer support on your Contractor's Pricing Guide: Residential Repair & Remodeling, call 800.448.8182.

Exterior Doors and Windows

Aluminum Window		Unit	Material	Labor	Equip.	Total	Specification
2' x 2' 6"							
	Demolish	Ea.		28		28	Cost includes material and labor to
	Install	Ea.	455	107		562	install 2' x 2' 6" single hung, two
	Demolish and Install	Ea.	455	135		590	light, double glazed aluminum
	Reinstall	Ea.		85.64		85.64	window including screen.
	Clean	Ea.	.03	11.55		11.58	
	Minimum Charge	Job		294		294	
3' x 1' 6"							
	Demolish	Ea.		28		28	Cost includes material and labor to
	Install	Ea.	340	118		458	install 3' x 1' 6" single hung, two
	Demolish and Install	Ea.	340	146		486	light, double glazed aluminum
	Reinstall	Ea.		94.21		94.21	window including screen.
	Clean	Ea.	.02	10.10		10.12	
	Minimum Charge	Job		294		294	
3' x 2'							
	Demolish	Ea.		28		28	Cost includes material and labor to
	Install	Ea.	640	118		758	install 3' x 2' single hung, two light,
	Demolish and Install	Ea.	640	146		786	double glazed aluminum window
	Reinstall	Ea.		94.21		94.21	including screen.
	Clean	Ea.	.02	10.10		10.12	
	Minimum Charge	Job		294		294	
3' x 2' 6"							
	Demolish	Ea.		28		28	Cost includes material and labor to
	Install	Ea.	450	118		568	install 3' x 2' 6" single hung, two
	Demolish and Install	Ea.	450	146		596	light, double glazed aluminum
	Reinstall	Ea.		94.21		94.21	window including screen.
	Clean	Ea.	.03	11.55		11.58	
	Minimum Charge	Job		294		294	
3' x 3'							
	Demolish	Ea.		28		28	Cost includes material and labor to
	Install	Ea.	500	118		618	install 3' x 3' single hung, two light,
	Demolish and Install	Ea.	500	146		646	double glazed aluminum window
	Reinstall	Ea.		94.21		94.21	including screen.
	Clean	Ea.	.03	11.55		11.58	
	Minimum Charge	Job		294		294	
3' x 3' 6"							
	Demolish	Ea.		28		28	Cost includes material and labor to
	Install	Ea.	540	118		658	install 3' x 3' 6" single hung, two
	Demolish and Install	Ea.	540	146		686	light, double glazed aluminum
	Reinstall	Ea.		94.21		94.21	window including screen.
	Clean	Ea.	.04	13.50		13.54	
	Minimum Charge	Job		294		294	
3' x 4'							
	Demolish	Ea.		28		28	Cost includes material and labor to
	Install	Ea.	570	118		688	install 3' x 4' single hung, two light,
	Demolish and Install	Ea.	570	146		716	double glazed aluminum window
	Reinstall	Ea.		94.21		94.21	including screen.
	Clean	Ea.	.04	13.50		13.54	
	Minimum Charge	Job		294		294	
3' x 5'							
	Demolish	Ea.		40.50		40.50	Cost includes material and labor to
	Install	Ea.	645	118		763	install 3' x 5' single hung, two light,
	Demolish and Install	Ea.	645	158.50		803.50	double glazed aluminum window
	Reinstall	Ea.		94.21		94.21	including screen.
	Clean	Ea.	.06	16.20		16.26	
	Minimum Charge	Job		294		294	

For customer support on your Contractor's Pricing Guide: Residential Repair & Remodeling, call 800.448.8182.

Exterior Doors and Windows

Aluminum Window		Unit	Material	Labor	Equip.	Total	Specification
3' x 6'							
	Demolish	Ea.		40.50		40.50	Cost includes material and labor to
	Install	Ea.	695	131		826	install 3' x 6' single hung, two light,
	Demolish and Install	Ea.	695	171.50		866.50	double glazed aluminum window
	Reinstall	Ea.		104.68		104.68	including screen.
	Clean	Ea.	.07	19.05		19.12	
	Minimum Charge	Job		294		294	

Double Hung Two Light

		Unit	Material	Labor	Equip.	Total	Specification
2' x 2'							
	Demolish	Ea.		28		28	Cost includes material and labor to
	Install	Ea.	144	107		251	install 2' x 2' double hung, two light,
	Demolish and Install	Ea.	144	135		279	double glazed aluminum window
	Reinstall	Ea.		85.64		85.64	including screen.
	Clean	Ea.	.02	10.10		10.12	
	Minimum Charge	Job		294		294	
2' x 2' 6"							
	Demolish	Ea.		28		28	Cost includes material and labor to
	Install	Ea.	505	107		612	install 2' x 2' 6" double hung, two
	Demolish and Install	Ea.	505	135		640	light, double glazed aluminum
	Reinstall	Ea.		85.64		85.64	window including screen.
	Clean	Ea.	.02	10.10		10.12	
	Minimum Charge	Job		294		294	
3' x 1' 6"							
	Demolish	Ea.		28		28	Cost includes material and labor to
	Install	Ea.	330	118		448	install 3' x 1' 6" double hung, two
	Demolish and Install	Ea.	330	146		476	light, double glazed aluminum
	Reinstall	Ea.		94.21		94.21	window including screen.
	Clean	Ea.	.02	10.10		10.12	
	Minimum Charge	Job		294		294	
3' x 2'							
	Demolish	Ea.		28		28	Cost includes material and labor to
	Install	Ea.	720	118		838	install 3' x 2' double hung, two light,
	Demolish and Install	Ea.	720	146		866	double glazed aluminum window
	Reinstall	Ea.		94.21		94.21	including screen.
	Clean	Ea.	.02	10.10		10.12	
	Minimum Charge	Job		294		294	
3' x 2' 6"							
	Demolish	Ea.		28		28	Cost includes material and labor to
	Install	Ea.	705	118		823	install 3' x 2' 6" double hung, two
	Demolish and Install	Ea.	705	146		851	light, double glazed aluminum
	Reinstall	Ea.		94.21		94.21	window including screen.
	Clean	Ea.	.03	11.55		11.58	
	Minimum Charge	Job		294		294	
3' x 3'							
	Demolish	Ea.		28		28	Cost includes material and labor to
	Install	Ea.	820	118		938	install 3' x 3' double hung, two light,
	Demolish and Install	Ea.	820	146		966	double glazed aluminum window
	Reinstall	Ea.		94.21		94.21	including screen.
	Clean	Ea.	.03	11.55		11.58	
	Minimum Charge	Job		294		294	
3' x 3' 6"							
	Demolish	Ea.		28		28	Cost includes material and labor to
	Install	Ea.	865	118		983	install 3' x 3' 6" double hung, two
	Demolish and Install	Ea.	865	146		1011	light, double glazed aluminum
	Reinstall	Ea.		94.21		94.21	window including screen.
	Clean	Ea.	.04	13.50		13.54	
	Minimum Charge	Job		294		294	

For customer support on your Contractor's Pricing Guide: Residential Repair & Remodeling, call 800.448.8182.

Exterior Doors and Windows

Aluminum Window		Unit	Material	Labor	Equip.	Total	Specification
3' x 4'							
	Demolish	Ea.		28		28	Cost includes material and labor to
	Install	Ea.	945	118		1063	install 3' x 4' double hung, two light,
	Demolish and Install	Ea.	945	146		1091	double glazed aluminum window
	Reinstall	Ea.		94.21		94.21	including screen.
	Clean	Ea.	.04	13.50		13.54	
	Minimum Charge	Job		294		294	
3' x 5'							
	Demolish	Ea.		40.50		40.50	Cost includes material and labor to
	Install	Ea.	1050	118		1168	install 3' x 5' double hung, two light,
	Demolish and Install	Ea.	1050	158.50		1208.50	double glazed aluminum window
	Reinstall	Ea.		94.21		94.21	including screen.
	Clean	Ea.	.06	16.20		16.26	
	Minimum Charge	Job		294		294	
3' x 6'							
	Demolish	Ea.		40.50		40.50	Cost includes material and labor to
	Install	Ea.	1000	131		1131	install 3' x 6' double hung, two light,
	Demolish and Install	Ea.	1000	171.50		1171.50	double glazed aluminum window
	Reinstall	Ea.		104.68		104.68	including screen.
	Clean	Ea.	.07	19.05		19.12	
	Minimum Charge	Job		294		294	
Sliding Sash							
1' 6" x 3'							
	Demolish	Ea.		28		28	Cost includes material and labor to
	Install	Ea.	370	107		477	install 1' 6" x 3' sliding sash, two
	Demolish and Install	Ea.	370	135		505	light, double glazed aluminum
	Reinstall	Ea.		85.64		85.64	window including screen.
	Clean	Ea.	.02	10.10		10.12	
	Minimum Charge	Job		294		294	
2' x 2'							
	Demolish	Ea.		28		28	Cost includes material and labor to
	Install	Ea.	330	118		448	install 2' x 2' sliding sash, two light,
	Demolish and Install	Ea.	330	146		476	double glazed aluminum window
	Reinstall	Ea.		94.21		94.21	including screen.
	Clean	Ea.	.02	10.10		10.12	
	Minimum Charge	Job		294		294	
3' x 2'							
	Demolish	Ea.		28		28	Cost includes material and labor to
	Install	Ea.	390	118		508	install 3' x 2' sliding sash, two light,
	Demolish and Install	Ea.	390	146		536	double glazed aluminum window
	Reinstall	Ea.		94.21		94.21	including screen.
	Clean	Ea.	.02	10.10		10.12	
	Minimum Charge	Job		294		294	
3' x 2' 6"							
	Demolish	Ea.		28		28	Cost includes material and labor to
	Install	Ea.	440	118		558	install 3' x 2' 6" sliding sash, two
	Demolish and Install	Ea.	440	146		586	light, double glazed aluminum
	Reinstall	Ea.		94.21		94.21	window including screen.
	Clean	Ea.	.03	11.55		11.58	
	Minimum Charge	Job		294		294	
3' x 3'							
	Demolish	Ea.		28		28	Cost includes material and labor to
	Install	Ea.	495	118		613	install 3' x 3' sliding sash, two light,
	Demolish and Install	Ea.	495	146		641	double glazed aluminum window
	Reinstall	Ea.		94.21		94.21	including screen.
	Clean	Ea.	.03	11.55		11.58	
	Minimum Charge	Job		294		294	

For customer support on your Contractor's Pricing Guide: Residential Repair & Remodeling, call 800.448.8182.

Exterior Doors and Windows

Aluminum Window		Unit	Material	Labor	Equip.	Total	Specification
3' x 3' 6"							
	Demolish	Ea.		28		28	Cost includes material and labor to
	Install	Ea.	550	118		668	install 3' x 3' 6" sliding sash, two
	Demolish and Install	Ea.	550	146		696	light, double glazed aluminum
	Reinstall	Ea.		94.21		94.21	window including screen.
	Clean	Ea.	.04	13.50		13.54	
	Minimum Charge	Job		294		294	
3' x 4'							
	Demolish	Ea.		28		28	Cost includes material and labor to
	Install	Ea.	600	131		731	install 3' x 4' sliding sash, two light,
	Demolish and Install	Ea.	600	159		759	double glazed aluminum window
	Reinstall	Ea.		104.68		104.68	including screen.
	Clean	Ea.	.04	13.50		13.54	
	Minimum Charge	Job		294		294	
3' x 5'							
	Demolish	Ea.		40.50		40.50	Cost includes material and labor to
	Install	Ea.	690	131		821	install 3' x 5' sliding sash, two light,
	Demolish and Install	Ea.	690	171.50		861.50	double glazed aluminum window
	Reinstall	Ea.		104.68		104.68	including screen.
	Clean	Ea.	.06	16.20		16.26	
	Minimum Charge	Job		294		294	
4' x 5'							
	Demolish	Ea.		40.50		40.50	Cost includes material and labor to
	Install	Ea.	800	168		968	install 4' x 5' sliding sash, two light,
	Demolish and Install	Ea.	800	208.50		1008.50	double glazed aluminum window
	Reinstall	Ea.		134.58		134.58	including screen.
	Clean	Ea.	.10	23		23.10	
	Minimum Charge	Job		294		294	
5' x 5'							
	Demolish	Ea.		40.50		40.50	Cost includes material and labor to
	Install	Ea.	895	168		1063	install 5' x 5' sliding sash, two light,
	Demolish and Install	Ea.	895	208.50		1103.50	double glazed aluminum window
	Reinstall	Ea.		134.58		134.58	including screen.
	Clean	Ea.	.10	23		23.10	
	Minimum Charge	Job		294		294	
Awning Sash							
26" x 26", 2 Light							
	Demolish	Ea.		28		28	Cost includes material and labor to
	Install	Ea.	395	131		526	install 26" x 26" two light awning,
	Demolish and Install	Ea.	395	159		554	double glazed aluminum window
	Reinstall	Ea.		104.68		104.68	including screen.
	Clean	Ea.	.03	11.55		11.58	
	Minimum Charge	Job		294		294	
26" x 37", 2 Light							
	Demolish	Ea.		28		28	Cost includes material and labor to
	Install	Ea.	655	131		786	install 26" x 37" two light awning,
	Demolish and Install	Ea.	655	159		814	double glazed aluminum window
	Reinstall	Ea.		104.68		104.68	including screen.
	Clean	Ea.	.03	11.55		11.58	
	Minimum Charge	Job		294		294	
38" x 26", 3 Light							
	Demolish	Ea.		28		28	Cost includes material and labor to
	Install	Ea.	715	131		846	install 38" x 26" three light awning,
	Demolish and Install	Ea.	715	159		874	double glazed aluminum window
	Reinstall	Ea.		104.68		104.68	including screen.
	Clean	Ea.	.04	13.50		13.54	
	Minimum Charge	Job		294		294	

110

For customer support on your Contractor's Pricing Guide: Residential Repair & Remodeling, call 800.448.8182.

Exterior Doors and Windows

Aluminum Window		Unit	Material	Labor	Equip.	Total	Specification
38″ x 37″, 3 Light							
	Demolish	Ea.		28		28	Cost includes material and labor to
	Install	Ea.	865	147		1012	install 38″ x 37″ three light awning,
	Demolish and Install	Ea.	865	175		1040	double glazed aluminum window
	Reinstall	Ea.		117.76		117.76	including screen.
	Clean	Ea.	.04	13.50		13.54	
	Minimum Charge	Job		294		294	
38″ x 53″, 3 Light							
	Demolish	Ea.		40.50		40.50	Cost includes material and labor to
	Install	Ea.	885	147		1032	install 38″ x 53″ three light awning,
	Demolish and Install	Ea.	885	187.50		1072.50	double glazed aluminum window
	Reinstall	Ea.		117.76		117.76	including screen.
	Clean	Ea.	.06	16.20		16.26	
	Minimum Charge	Job		294		294	
50″ x 37″, 4 Light							
	Demolish	Ea.		40.50		40.50	Cost includes material and labor to
	Install	Ea.	800	147		947	install 50″ x 37″ four light awning,
	Demolish and Install	Ea.	800	187.50		987.50	double glazed aluminum window
	Reinstall	Ea.		117.76		117.76	including screen.
	Clean	Ea.	.06	16.20		16.26	
	Minimum Charge	Job		294		294	
50″ x 53″, 4 Light							
	Demolish	Ea.		40.50		40.50	Cost includes material and labor to
	Install	Ea.	1450	168		1618	install 50″ x 53″ four light awning,
	Demolish and Install	Ea.	1450	208.50		1658.50	double glazed aluminum window
	Reinstall	Ea.		134.58		134.58	including screen.
	Clean	Ea.	.07	19.05		19.12	
	Minimum Charge	Job		294		294	
63″ x 37″, 5 Light							
	Demolish	Ea.		40.50		40.50	Cost includes material and labor to
	Install	Ea.	1025	168		1193	install 63″ x 37″ five light awning,
	Demolish and Install	Ea.	1025	208.50		1233.50	double glazed aluminum window
	Reinstall	Ea.		134.58		134.58	including screen.
	Clean	Ea.	.07	19.05		19.12	
	Minimum Charge	Job		294		294	
63″ x 53″, 5 Light							
	Demolish	Ea.		40.50		40.50	Cost includes material and labor to
	Install	Ea.	1600	168		1768	install 63″ x 53″ five light awning,
	Demolish and Install	Ea.	1600	208.50		1808.50	double glazed aluminum window
	Reinstall	Ea.		134.58		134.58	including screen.
	Clean	Ea.	.10	23		23.10	
	Minimum Charge	Job		294		294	
Mullion Bars							
	Install	Ea.	16.05	47		63.05	Includes labor and material to install
	Minimum Charge	Job		294		294	mullions for sliding aluminum window.
Bay							
4′ 8″ x 6′ 4″							
	Demolish	Ea.		29.50		29.50	Cost includes material and labor to
	Install	Ea.	2300	130		2430	install 4′ 8″ x 6′ 4″ angle bay double
	Demolish and Install	Ea.	2300	159.50		2459.50	hung unit, pine frame with aluminum
	Reinstall	Ea.		104.16		104.16	cladding, picture center sash,
	Clean	Ea.	.14	27		27.14	hardware, screens.
	Minimum Charge	Job		294		294	

111

For customer support on your Contractor's Pricing Guide: Residential Repair & Remodeling, call 800.448.8182.

Exterior Doors and Windows

Aluminum Window		Unit	Material	Labor	Equip.	Total	Specification
4′ 8″ x 7′ 2″							
	Demolish	Ea.		29.50		29.50	Cost includes material and labor to
	Install	Ea.	2375	130		2505	install 4′ 8″ x 7′ 2″ angle bay double
	Demolish and Install	Ea.	2375	159.50		2534.50	hung unit, pine frame with aluminum
	Reinstall	Ea.		104.16		104.16	cladding, picture center sash,
	Clean	Ea.	.14	27		27.14	hardware, screens.
	Minimum Charge	Job		294		294	
4′ 8″ x 7′ 10″							
	Demolish	Ea.		29.50		29.50	Cost includes material and labor to
	Install	Ea.	2475	130		2605	install 4′ 8″ x 7′ 10″ angle bay
	Demolish and Install	Ea.	2475	159.50		2634.50	double hung unit, pine frame with
	Reinstall	Ea.		104.16		104.16	aluminum cladding, picture center
	Clean	Ea.	.14	27		27.14	sash, hardware, screens.
	Minimum Charge	Job		294		294	
4′ 8″ x 8′ 6″							
	Demolish	Ea.		34		34	Cost includes material and labor to
	Install	Ea.	3300	149		3449	install 4′ 8″ x 8′ 6″ angle bay double
	Demolish and Install	Ea.	3300	183		3483	hung unit, pine frame with aluminum
	Reinstall	Ea.		119.04		119.04	cladding, picture center sash,
	Clean	Ea.	.02	36		36.02	hardware, screens.
	Minimum Charge	Job		294		294	
4′ 8″ x 8′ 10″							
	Demolish	Ea.		34		34	Cost includes material and labor to
	Install	Ea.	3350	149		3499	install 4′ 8″ x 8′ 10″ angle bay
	Demolish and Install	Ea.	3350	183		3533	double hung unit, pine frame with
	Reinstall	Ea.		119.04		119.04	aluminum cladding, picture center
	Clean	Ea.	.02	36		36.02	sash, hardware, screens.
	Minimum Charge	Job		294		294	
4′ 8″ x 9′ 2″							
	Demolish	Ea.		34		34	Cost includes material and labor to
	Install	Ea.	3925	260		4185	install 4′ 8″ x 9′ 2″ angle bay double
	Demolish and Install	Ea.	3925	294		4219	hung unit, pine frame with aluminum
	Reinstall	Ea.		208.32		208.32	cladding, picture center sash,
	Clean	Ea.	.02	36		36.02	hardware, screens.
	Minimum Charge	Job		294		294	
4′ 8″ x 9′ 10″							
	Demolish	Ea.		34		34	Cost includes material and labor to
	Install	Ea.	4000	260		4260	install 4′ 8″ x 9′ 10″ angle bay
	Demolish and Install	Ea.	4000	294		4294	double hung unit, pine frame with
	Reinstall	Ea.		208.32		208.32	aluminum cladding, picture center
	Clean	Ea.	.02	36		36.02	sash, hardware, screens.
	Minimum Charge	Job		294		294	
Casing							
	Demolish	Ea.		12.05		12.05	Cost includes material and labor to
	Install	Opng.	60.50	40		100.50	install 11/16″ x 2-1/2″ pine ranch
	Demolish and Install	Opng.	60.50	52.05		112.55	style window casing.
	Reinstall	Opng.		32.05		32.05	
	Clean	Opng.	.45	4		4.45	
	Paint	Ea.	10.20	8.65		18.85	
	Minimum Charge	Job		294		294	
Storm Window							
2′ x 3′ 6″							
	Demolish	Ea.		16.45		16.45	Includes material and labor to install
	Install	Ea.	525	34.50		559.50	residential aluminum storm window
	Demolish and Install	Ea.	525	50.95		575.95	with mill finish, 2′ x 3′ 6″ high.
	Reinstall	Ea.		27.78		27.78	
	Clean	Ea.	.02	10.10		10.12	
	Minimum Charge	Job		294		294	

112

For customer support on your Contractor's Pricing Guide: Residential Repair & Remodeling, call 800.448.8182.

Exterior Doors and Windows

Aluminum Window		Unit	Material	Labor	Equip.	Total	Specification
2' 6" x 5'							
	Demolish	Ea.		16.45		16.45	Includes material and labor to install
	Install	Ea.	720	37		757	residential aluminum storm window
	Demolish and Install	Ea.	720	53.45		773.45	with mill finish, 2' 6" x 5' high.
	Reinstall	Ea.		29.76		29.76	
	Clean	Ea.	.04	13.50		13.54	
	Minimum Charge	Job		294		294	
4' x 6'							
	Demolish	Ea.		31.50		31.50	Includes material and labor to install
	Install	Ea.	870	41.50		911.50	residential aluminum storm window, 4'
	Demolish and Install	Ea.	870	73		943	x 6' high.
	Reinstall	Ea.		33.33		33.33	
	Clean	Ea.	.10	20		20.10	
	Minimum Charge	Job		294		294	
Hardware							
Weatherstripping							
	Install	L.F.	.62	4.34		4.96	Includes labor and material to install
	Minimum Charge	Job		174		174	vinyl V strip weatherstripping for
							double-hung window.
Trim Set							
	Demolish	Ea.		12.05		12.05	Cost includes material and labor to
	Install	Opng.	60.50	40		100.50	install 11/16" x 2-1/2" pine ranch
	Demolish and Install	Opng.	60.50	52.05		112.55	style window casing.
	Reinstall	Opng.		32.05		32.05	
	Clean	Opng.	.45	4		4.45	
	Paint	Ea.	10.20	8.65		18.85	
	Minimum Charge	Job		260		260	

Vinyl Window		Unit	Material	Labor	Equip.	Total	Specification
Double Hung							
2' x 2' 6"							
	Demolish	Ea.		20		20	Cost includes material and labor to
	Install	Ea.	450	69.50		519.50	install 2' x 2' 6" double hung, two
	Demolish and Install	Ea.	450	89.50		539.50	light, vinyl window with screen.
	Reinstall	Ea.		55.55		55.55	
	Clean	Ea.	.02	10.10		10.12	
	Minimum Charge	Job		174		174	
2' x 3' 6"							
	Demolish	Ea.		20		20	Cost includes material and labor to
	Install	Ea.	395	74.50		469.50	install 2' x 3' 6" double hung, two
	Demolish and Install	Ea.	395	94.50		489.50	light, vinyl window with screen.
	Reinstall	Ea.		59.52		59.52	
	Clean	Ea.	.03	11.55		11.58	
	Minimum Charge	Job		174		174	
2' 6" x 4' 6"							
	Demolish	Ea.		20		20	Cost includes material and labor to
	Install	Ea.	480	80		560	install 2' 6" x 4' 6" double hung, two
	Demolish and Install	Ea.	480	100		580	light, vinyl window with screen.
	Reinstall	Ea.		64.10		64.10	
	Clean	Ea.	.04	13.50		13.54	
	Minimum Charge	Job		174		174	
3' x 4'							
	Demolish	Ea.		20		20	Includes material and labor to install
	Install	Ea.	495	104		599	medium size double hung, 3' x 4',
	Demolish and Install	Ea.	495	124		619	two light, vinyl window with screen.
	Reinstall	Ea.		83.33		83.33	
	Clean	Ea.	.04	13.50		13.54	
	Minimum Charge	Job		174		174	

For customer support on your Contractor's Pricing Guide: Residential Repair & Remodeling, call 800.448.8182.

Exterior Doors and Windows

Vinyl Window

Vinyl Window	Unit	Material	Labor	Equip.	Total	Specification
3' x 4' 6"						
Demolish	Ea.		24.50		24.50	Cost includes material and labor to
Install	Ea.	545	116		661	install 3' x 4' 6" double hung, two
Demolish and Install	Ea.	545	140.50		685.50	light, vinyl window with screen.
Reinstall	Ea.		92.59		92.59	
Clean	Ea.	.06	16.20		16.26	
Minimum Charge	Job		174		174	
4' x 4' 6"						
Demolish	Ea.		24.50		24.50	Cost includes material and labor to
Install	Ea.	555	130		685	install 4' x 4' 6" double hung, two
Demolish and Install	Ea.	555	154.50		709.50	light, vinyl window with screen.
Reinstall	Ea.		104.16		104.16	
Clean	Ea.	.07	19.05		19.12	
Minimum Charge	Job		174		174	
4' x 6'						
Demolish	Ea.		24.50		24.50	Includes material and labor to install
Install	Ea.	625	149		774	large double hung, 4' x 6', two light,
Demolish and Install	Ea.	625	173.50		798.50	vinyl window with screen.
Reinstall	Ea.		119.04		119.04	
Clean	Ea.	.10	23		23.10	
Minimum Charge	Job		174		174	
Trim Set						
Demolish	Ea.		12.05		12.05	Cost includes material and labor to
Install	Opng.	60.50	40		100.50	install 11/16" x 2-1/2" pine ranch
Demolish and Install	Opng.	60.50	52.05		112.55	style window casing.
Reinstall	Opng.		32.05		32.05	
Clean	Opng.	.45	4		4.45	
Paint	Ea.	10.20	8.65		18.85	
Minimum Charge	Job		294		294	

Wood Window

Wood Window	Unit	Material	Labor	Equip.	Total	Specification
Double Hung						
2' 2" x 3' 4"						
Demolish	Ea.		20		20	Cost includes material and labor to
Install	Ea.	485	69.50		554.50	install 2' 2" x 3' 4" double hung, two
Demolish and Install	Ea.	485	89.50		574.50	light, double glazed wood window
Reinstall	Ea.		55.55		55.55	with screen.
Clean	Ea.	.03	11.55		11.58	
Paint	Ea.	6.40	18		24.40	
Minimum Charge	Job		260		260	
2' 2" x 4' 4"						
Demolish	Ea.		20		20	Cost includes material and labor to
Install	Ea.	590	74.50		664.50	install 2' 2" x 4' 4" double hung, two
Demolish and Install	Ea.	590	94.50		684.50	light, double glazed wood window
Reinstall	Ea.		59.52		59.52	with screen.
Clean	Ea.	.03	11.55		11.58	
Paint	Ea.	6.40	18		24.40	
Minimum Charge	Job		260		260	
2' 6" x 3' 4"						
Demolish	Ea.		20		20	Cost includes material and labor to
Install	Ea.	530	80		610	install 2' 6" x 3' 4" double hung, two
Demolish and Install	Ea.	530	100		630	light, double glazed wood window
Reinstall	Ea.		64.10		64.10	with screen.
Clean	Ea.	.04	13.50		13.54	
Paint	Ea.	8.15	21.50		29.65	
Minimum Charge	Job		260		260	

Exterior Doors and Windows

Wood Window		Unit	Material	Labor	Equip.	Total	Specification
2' 6" x 4'							
	Demolish	Ea.		20		20	Cost includes material and labor to
	Install	Ea.	590	87		677	install 2' 6" x 4' double hung, two
	Demolish and Install	Ea.	590	107		697	light, double glazed wood window
	Reinstall	Ea.		69.44		69.44	with screen.
	Clean	Ea.	.04	13.50		13.54	
	Paint	Ea.	8.15	21.50		29.65	
	Minimum Charge	Job		260		260	
2' 6" x 4' 8"							
	Demolish	Ea.		20		20	Cost includes material and labor to
	Install	Ea.	650	87		737	install 2' 6" x 4' 8" double hung, two
	Demolish and Install	Ea.	650	107		757	light, double glazed wood window
	Reinstall	Ea.		69.44		69.44	with screen.
	Clean	Ea.	.04	13.50		13.54	
	Paint	Ea.	8.15	21.50		29.65	
	Minimum Charge	Job		260		260	
2' 10" x 3' 4"							
	Demolish	Ea.		20		20	Cost includes material and labor to
	Install	Ea.	830	104		934	install 2' 10" x 3' 4" double hung, two
	Demolish and Install	Ea.	830	124		954	light, double glazed wood window
	Reinstall	Ea.		83.33		83.33	with screen.
	Clean	Ea.	.04	13.50		13.54	
	Paint	Ea.	8.15	21.50		29.65	
	Minimum Charge	Job		260		260	
2' 10" x 4'							
	Demolish	Ea.		20		20	Cost includes material and labor to
	Install	Ea.	640	104		744	install 2' 10" x 4' double hung, two
	Demolish and Install	Ea.	640	124		764	light, double glazed wood window
	Reinstall	Ea.		83.33		83.33	with screen.
	Clean	Ea.	.04	13.50		13.54	
	Paint	Ea.	8.15	21.50		29.65	
	Minimum Charge	Job		260		260	
3' 7" x 3' 4"							
	Demolish	Ea.		20		20	Cost includes material and labor to
	Install	Ea.	1150	116		1266	install 3' 7" x 3' 4" double hung, two
	Demolish and Install	Ea.	1150	136		1286	light, double glazed wood window
	Reinstall	Ea.		92.59		92.59	with screen.
	Clean	Ea.	.04	13.50		13.54	
	Paint	Ea.	8.15	21.50		29.65	
	Minimum Charge	Job		260		260	
3' 7" x 5' 4"							
	Demolish	Ea.		24.50		24.50	Cost includes material and labor to
	Install	Ea.	1425	116		1541	install 3' 7" x 5' 4" double hung, two
	Demolish and Install	Ea.	1425	140.50		1565.50	light, double glazed wood window
	Reinstall	Ea.		92.59		92.59	with screen.
	Clean	Ea.	.10	23		23.10	
	Paint	Ea.	13.60	36		49.60	
	Minimum Charge	Job		260		260	
4' 3" x 5' 4"							
	Demolish	Ea.		24.50		24.50	Cost includes material and labor to
	Install	Ea.	920	130		1050	install 4' 3" x 5' 4" double hung, two
	Demolish and Install	Ea.	920	154.50		1074.50	light, double glazed wood window
	Reinstall	Ea.		104.16		104.16	with screen.
	Clean	Ea.	.10	23		23.10	
	Paint	Ea.	64.50	43		107.50	
	Minimum Charge	Job		260		260	

For customer support on your Contractor's Pricing Guide: Residential Repair & Remodeling, call 800.448.8182.

Exterior Doors and Windows

Wood Window	Unit	Material	Labor	Equip.	Total	Specification
Casement						
2' x 3' 4"						
Demolish	Ea.		20		20	Cost includes material and labor to
Install	Ea.	1075	52		1127	install 2' x 3' 4" casement window
Demolish and Install	Ea.	1075	72		1147	with 3/4" insulated glass, screens,
Reinstall	Ea.		41.66		41.66	weatherstripping, hardware.
Clean	Ea.	.03	11.55		11.58	
Paint	Ea.	6.40	18		24.40	
Minimum Charge	Job		520		520	
2' x 4'						
Demolish	Ea.		20		20	Cost includes material and labor to
Install	Ea.	1325	58		1383	install 2' x 4' casement window with
Demolish and Install	Ea.	1325	78		1403	3/4" insulated glass, screens,
Reinstall	Ea.		46.29		46.29	weatherstripping, hardware.
Clean	Ea.	.03	11.55		11.58	
Paint	Ea.	6.40	18		24.40	
Minimum Charge	Job		520		520	
2' x 5'						
Demolish	Ea.		20		20	Cost includes material and labor to
Install	Ea.	1350	61.50		1411.50	install 2' x 5' casement window with
Demolish and Install	Ea.	1350	81.50		1431.50	3/4" insulated glass, screens,
Reinstall	Ea.		49.02		49.02	weatherstripping, hardware.
Clean	Ea.	.04	13.50		13.54	
Paint	Ea.	8.15	21.50		29.65	
Minimum Charge	Job		520		520	
2' x 6'						
Demolish	Ea.		20		20	Cost includes material and labor to
Install	Ea.	1525	65		1590	install 2' x 6' casement window with
Demolish and Install	Ea.	1525	85		1610	3/4" insulated glass, screens,
Reinstall	Ea.		52.08		52.08	weatherstripping, hardware.
Clean	Ea.	.04	13.50		13.54	
Paint	Ea.	8.15	21.50		29.65	
Minimum Charge	Job		520		520	
4' x 3' 4"						
Demolish	Ea.		24.50		24.50	Cost includes material and labor to
Install	Ea.	1900	69.50		1969.50	install 4' x 3' 4" casement windows
Demolish and Install	Ea.	1900	94		1994	with 3/4" insulated glass, screens,
Reinstall	Ea.		55.55		55.55	weatherstripping, hardware.
Clean	Ea.	.04	13.50		13.54	
Paint	Ea.	32	21.50		53.50	
Minimum Charge	Job		520		520	
4' x 4'						
Demolish	Ea.		24.50		24.50	Cost includes material and labor to
Install	Ea.	1800	69.50		1869.50	install 4' x 4' casement windows with
Demolish and Install	Ea.	1800	94		1894	3/4" insulated glass, screens,
Reinstall	Ea.		55.55		55.55	weatherstripping, hardware.
Clean	Ea.	.07	19.05		19.12	
Paint	Ea.	38	27		65	
Minimum Charge	Job		520		520	
4' x 5'						
Demolish	Ea.		24.50		24.50	Cost includes material and labor to
Install	Ea.	2275	74.50		2349.50	install 4' x 5' casement windows with
Demolish and Install	Ea.	2275	99		2374	3/4" insulated glass, screens,
Reinstall	Ea.		59.52		59.52	weatherstripping, hardware.
Clean	Ea.	.10	23		23.10	
Paint	Ea.	49.50	36		85.50	
Minimum Charge	Job		520		520	

Exterior Doors and Windows

Wood Window		Unit	Material	Labor	Equip.	Total	Specification
4' x 6'							
	Demolish	Ea.		24.50		24.50	Cost includes material and labor to
	Install	Ea.	2450	87		2537	install 4' x 6' casement windows with
	Demolish and Install	Ea.	2450	111.50		2561.50	3/4" insulated glass, screens,
	Reinstall	Ea.		69.44		69.44	weatherstripping, hardware.
	Clean	Ea.	.10	23		23.10	
	Paint	Ea.	58.50	43		101.50	
	Minimum Charge	Job		520		520	
Casement Bow							
4' 8" x 8' 1"							
	Demolish	Ea.		34		34	Includes material and labor to install
	Install	Ea.	5450	195		5645	casement bow window unit with 4
	Demolish and Install	Ea.	5450	229		5679	lights, pine frame, 1/2" insulated
	Reinstall	Ea.		156.24		156.24	glass, weatherstripping, hardware,
	Clean	Ea.	.14	27		27.14	screens.
	Paint	Ea.	108	86.50		194.50	
	Minimum Charge	Job		520		520	
4' 8" x 10'							
	Demolish	Ea.		34		34	Includes material and labor to install
	Install	Ea.	5225	260		5485	casement bow window unit with 5
	Demolish and Install	Ea.	5225	294		5519	lights, pine frame, 1/2" insulated
	Reinstall	Ea.		208.32		208.32	glass, weatherstripping, hardware,
	Clean	Ea.	.02	36		36.02	screens.
	Paint	Ea.	134	108		242	
	Minimum Charge	Job		520		520	
5' 4" x 8' 1"							
	Demolish	Ea.		34		34	Includes material and labor to install
	Install	Ea.	5000	195		5195	casement bow window unit with 4
	Demolish and Install	Ea.	5000	229		5229	lights, pine frame, 1/2" insulated
	Reinstall	Ea.		156.24		156.24	glass, weatherstripping, hardware,
	Clean	Ea.	.02	36		36.02	screens.
	Paint	Ea.	123	86.50		209.50	
	Minimum Charge	Job		520		520	
5' 4" x 10'							
	Demolish	Ea.		34		34	Includes material and labor to install
	Install	Ea.	5150	260		5410	casement bow window unit with 5
	Demolish and Install	Ea.	5150	294		5444	lights, pine frame, 1/2" insulated
	Reinstall	Ea.		208.32		208.32	glass, weatherstripping, hardware,
	Clean	Ea.	.02	36		36.02	screens.
	Paint	Ea.	155	108		263	
	Minimum Charge	Job		520		520	
6' x 8' 1"							
	Demolish	Ea.		34		34	Includes material and labor to install
	Install	Ea.	5225	260		5485	casement bow window unit with 5
	Demolish and Install	Ea.	5225	294		5519	lights, pine frame, 1/2" insulated
	Reinstall	Ea.		208.32		208.32	glass, weatherstripping, hardware,
	Clean	Ea.	.02	36		36.02	screens.
	Paint	Ea.	141	108		249	
	Minimum Charge	Job		520		520	
6' x 10'							
	Demolish	Ea.		34		34	Includes material and labor to install
	Install	Ea.	5600	260		5860	casement bow window unit with 5
	Demolish and Install	Ea.	5600	294		5894	lights, pine frame, 1/2" insulated
	Reinstall	Ea.		208.32		208.32	glass, weatherstripping, hardware,
	Clean	Ea.	.02	36		36.02	screens.
	Paint	Ea.	175	144		319	
	Minimum Charge	Job		520		520	

For customer support on your Contractor's Pricing Guide: Residential Repair & Remodeling, call 800.448.8182.

Exterior Doors and Windows

Wood Window

	Unit	Material	Labor	Equip.	Total	Specification
Casing						
Demolish	Ea.		12.05		12.05	Cost includes material and labor to
Install	Opng.	60.50	40		100.50	install 11/16" x 2-1/2" pine ranch
Demolish and Install	Opng.	60.50	52.05		112.55	style window casing.
Reinstall	Opng.		32.05		32.05	
Clean	Opng.	.45	4		4.45	
Paint	Ea.	10.20	8.65		18.85	
Minimum Charge	Job		294		294	
Trim Set						
Demolish	Ea.		9.20		9.20	Cost includes material and labor to
Install	Opng.	60.50	40		100.50	install 11/16" x 2-1/2" pine ranch
Demolish and Install	Opng.	60.50	49.20		109.70	style window casing.
Reinstall	Opng.		32.05		32.05	
Clean	Opng.	.45	4		4.45	
Paint	Ea.	10.20	8.65		18.85	
Minimum Charge	Job		260		260	
Wood Picture Window						
3' 6" x 4'						
Demolish	Ea.		24.50		24.50	Includes material and labor to install a
Install	Ea.	305	87		392	wood framed picture window
Demolish and Install	Ea.	305	111.50		416.50	including exterior trim.
Clean	Ea.	.10	23		23.10	
Paint	Ea.	6.20	33		39.20	
Minimum Charge	Job		520		520	
4' x 4' 6"						
Demolish	Ea.		24.50		24.50	Includes material and labor to install a
Install	Ea.	590	94.50		684.50	wood framed picture window
Demolish and Install	Ea.	590	119		709	including exterior trim.
Clean	Ea.	.10	23		23.10	
Paint	Ea.	6.20	33		39.20	
Minimum Charge	Job		520		520	
5' x 4'						
Demolish	Ea.		24.50		24.50	Includes material and labor to install a
Install	Ea.	560	94.50		654.50	wood framed picture window
Demolish and Install	Ea.	560	119		679	including exterior trim.
Clean	Ea.	.10	23		23.10	
Paint	Ea.	13.20	54		67.20	
Minimum Charge	Job		520		520	
6' x 4' 6"						
Demolish	Ea.		24.50		24.50	Includes material and labor to install a
Install	Ea.	840	104		944	wood framed picture window
Demolish and Install	Ea.	840	128.50		968.50	including exterior trim.
Clean	Ea.	.10	23		23.10	
Paint	Ea.	13.20	54		67.20	
Minimum Charge	Job		520		520	

Window Screens

	Unit	Material	Labor	Equip.	Total	Specification
Small						
Demolish	Ea.		4.62		4.62	Includes material and labor to install
Install	Ea.	43.50	15.80		59.30	small window screen, 4 S.F.
Demolish and Install	Ea.	43.50	20.42		63.92	
Reinstall	Ea.		12.63		12.63	
Clean	Ea.	.32	8.35		8.67	
Minimum Charge	Job		130		130	

For customer support on your Contractor's Pricing Guide: Residential Repair & Remodeling, call 800.448.8182.

Exterior Doors and Windows

Window Screens

		Unit	Material	Labor	Equip.	Total	Specification
Standard Size							
	Demolish	Ea.		4.62		4.62	Includes material and labor to install
	Install	Ea.	73	15.80		88.80	medium window screen, 12 S.F.
	Demolish and Install	Ea.	73	20.42		93.42	
	Reinstall	Ea.		12.63		12.63	
	Clean	Ea.	.32	8.35		8.67	
	Minimum Charge	Job		130		130	
Large							
	Demolish	Ea.		4.62		4.62	Includes material and labor to install
	Install	Ea.	98	15.80		113.80	large window screen, 16 S.F.
	Demolish and Install	Ea.	98	20.42		118.42	
	Reinstall	Ea.		12.63		12.63	
	Clean	Ea.	.32	8.35		8.67	
	Minimum Charge	Job		130		130	

Window Glass

		Unit	Material	Labor	Equip.	Total	Specification
Plate, Clear, 1/4″							
	Demolish	S.F.		2.22		2.22	Includes material and labor to install
	Install	S.F.	16.15	8.40		24.55	clear glass.
	Demolish and Install	S.F.	16.15	10.62		26.77	
	Reinstall	S.F.		6.73		6.73	
	Clean	S.F.	.02	.23		.25	
	Minimum Charge	Job		252		252	
Insulated							
	Demolish	S.F.		2.22		2.22	Includes material and labor to install
	Install	S.F.	44.50	14.40		58.90	insulated glass.
	Demolish and Install	S.F.	44.50	16.62		61.12	
	Reinstall	S.F.		11.54		11.54	
	Clean	S.F.	.02	.23		.25	
	Minimum Charge	Job		252		252	
Tinted							
	Demolish	S.F.		2.22		2.22	Includes material and labor to install
	Install	S.F.	9.70	7.75		17.45	tinted glass.
	Demolish and Install	S.F.	9.70	9.97		19.67	
	Reinstall	S.F.		6.21		6.21	
	Clean	S.F.	.02	.23		.25	
	Minimum Charge	Job		252		252	
Obscure / Florentine							
	Demolish	S.F.		2.22		2.22	Includes material and labor to install
	Install	S.F.	40.50	7.20		47.70	obscure glass.
	Demolish and Install	S.F.	40.50	9.42		49.92	
	Reinstall	S.F.		5.77		5.77	
	Clean	S.F.	.02	.23		.25	
	Minimum Charge	Job		252		252	
Laminated Safety							
	Demolish	S.F.		2.22		2.22	Includes material and labor to install
	Install	S.F.	37.50	11.20		48.70	clear laminated glass.
	Demolish and Install	S.F.	37.50	13.42		50.92	
	Reinstall	S.F.		8.97		8.97	
	Clean	S.F.	.02	.23		.25	
	Minimum Charge	Job		252		252	

For customer support on your Contractor's Pricing Guide: Residential Repair & Remodeling, call 800.448.8182.

Exterior Doors and Windows

Window Glass

Window Glass	Unit	Material	Labor	Equip.	Total	Specification
Wired						
Demolish	S.F.		2.22		2.22	Includes material and labor to install
Install	S.F.	32	7.50		39.50	rough obscure wire glass.
Demolish and Install	S.F.	32	9.72		41.72	
Reinstall	S.F.		5.98		5.98	
Clean	S.F.	.02	.23		.25	
Minimum Charge	Job		252		252	
Tempered						
Demolish	S.F.		2.22		2.22	Includes material and labor to install
Install	S.F.	16.15	8.40		24.55	clear glass.
Demolish and Install	S.F.	16.15	10.62		26.77	
Reinstall	S.F.		6.73		6.73	
Clean	S.F.	.02	.23		.25	
Minimum Charge	Job		252		252	
Stained						
Demolish	S.F.		2.22		2.22	Includes material and labor to install
Install	S.F.	83	8.40		91.40	stained glass.
Demolish and Install	S.F.	83	10.62		93.62	
Reinstall	S.F.		6.73		6.73	
Clean	S.F.	.02	.23		.25	
Minimum Charge	Job		252		252	
Etched						
Demolish	S.F.		2.22		2.22	Includes material and labor to install
Install	S.F.	69	8.40		77.40	etched glass.
Demolish and Install	S.F.	69	10.62		79.62	
Reinstall	S.F.		6.73		6.73	
Clean	S.F.	.02	.23		.25	
Minimum Charge	Job		252		252	
Wall Mirror						
Demolish	S.F.		2.22		2.22	Includes material and labor to install
Install	S.F.	28	8.10		36.10	distortion-free float glass mirror, 1/4"
Demolish and Install	S.F.	28	10.32		38.32	thick, cut and polished edges.
Reinstall	S.F.		6.46		6.46	
Clean	S.F.	.02	.23		.25	
Minimum Charge	Job		252		252	
Bullet Resistant						
Demolish	S.F.		2.22		2.22	Includes material and labor to install
Install	S.F.	415	101		516	bullet resistant clear laminated glass.
Demolish and Install	S.F.	415	103.22		518.22	
Reinstall	S.F.		80.77		80.77	
Clean	S.F.	.02	.23		.25	
Minimum Charge	Job		252		252	

Storefront System

Storefront System	Unit	Material	Labor	Equip.	Total	Specification
Clear Glass						
Stub Wall to 8'						
Demolish	S.F.		3.47		3.47	Includes material and labor to install
Install	S.F.	32.50	14.10		46.60	commercial grade 4-1/2" section,
Demolish and Install	S.F.	32.50	17.57		50.07	anodized aluminum with clear glass in
Reinstall	S.F.		11.27		11.27	upper section and safety glass in
Clean	S.F.		.54		.54	lower section.
Minimum Charge	Job		294		294	

120

For customer support on your Contractor's Pricing Guide: Residential Repair & Remodeling, call 800.448.8182.

Exterior Doors and Windows

Storefront System	Unit	Material	Labor	Equip.	Total	Specification
Floor to 10'						
Demolish	S.F.		3.47		3.47	Includes material and labor to install
Install	S.F.	33.50	14.10		47.60	commercial grade 4-1/2" section,
Demolish and Install	S.F.	33.50	17.57		51.07	anodized aluminum with clear glass in
Reinstall	S.F.		11.27		11.27	upper section and safety glass in
Clean	S.F.		.54		.54	lower section.
Minimum Charge	Job		294		294	
Polished Plate Glass						
Stub Wall to 9'						
Demolish	S.F.		3.47		3.47	Cost includes material and labor to
Install	S.F.	20.50	14.10		34.60	install 4' x 9' polished plate glass
Demolish and Install	S.F.	20.50	17.57		38.07	window built on 8" stub wall.
Reinstall	S.F.		11.27		11.27	
Clean	S.F.		.54		.54	
Minimum Charge	Job		294		294	
Stub Wall to 13'						
Demolish	S.F.		3.47		3.47	Cost includes material and labor to
Install	S.F.	23	14.10		37.10	install 4' x 13' polished plate glass
Demolish and Install	S.F.	23	17.57		40.57	window built on 8" stub wall.
Reinstall	S.F.		11.27		11.27	
Clean	S.F.		.54		.54	
Minimum Charge	Job		294		294	
Floor to 9'						
Demolish	S.F.		3.47		3.47	Cost includes material and labor to
Install	S.F.	24	14.10		38.10	install 4' x 9' polished plate glass
Demolish and Install	S.F.	24	17.57		41.57	window built on floor.
Reinstall	S.F.		11.27		11.27	
Clean	S.F.		.54		.54	
Minimum Charge	Job		294		294	
Floor to 13'						
Demolish	S.F.		3.47		3.47	Cost includes material and labor to
Install	S.F.	26	14.10		40.10	install 4' x 13' polished plate glass
Demolish and Install	S.F.	26	17.57		43.57	window built on floor.
Reinstall	S.F.		11.27		11.27	
Clean	S.F.		.54		.54	
Minimum Charge	Job		294		294	
Clear Plate Glass						
Floor to 9'						
Demolish	S.F.		3.47		3.47	Cost includes material and labor to
Install	S.F.	34	14.10		48.10	install 4' x 9' clear plate glass window
Demolish and Install	S.F.	34	17.57		51.57	built on floor.
Reinstall	S.F.		11.27		11.27	
Clean	S.F.		.54		.54	
Minimum Charge	Job		294		294	
Floor to 11'						
Demolish	S.F.		3.47		3.47	Cost includes material and labor to
Install	S.F.	33.50	14.10		47.60	install 4' x 11' clear plate glass
Demolish and Install	S.F.	33.50	17.57		51.07	window built on floor.
Reinstall	S.F.		11.27		11.27	
Clean	S.F.		.54		.54	
Minimum Charge	Job		294		294	
Floor to 13'						
Demolish	S.F.		3.47		3.47	Cost includes material and labor to
Install	S.F.	33.50	14.10		47.60	install 4' x 13' clear plate glass
Demolish and Install	S.F.	33.50	17.57		51.07	window built on floor.
Reinstall	S.F.		11.27		11.27	
Clean	S.F.		.54		.54	
Minimum Charge	Job		294		294	

For customer support on your Contractor's Pricing Guide: Residential Repair & Remodeling, call 800.448.8182.

Exterior Doors and Windows

Storefront System		Unit	Material	Labor	Equip.	Total	Specification
Entrance							
Narrow Stile 3' x 7"							
	Demolish	Ea.		29		29	Cost includes material and labor to
	Install	Ea.	5275	345		5620	install 3' x 7' narrow stile door, center
	Demolish and Install	Ea.	5275	374		5649	pivot concealed closer.
	Reinstall	Ea.		277.76		277.76	
	Clean	Ea.	3.96	50		53.96	
	Minimum Charge	Job		294		294	
Tempered Glass							
	Demolish	Ea.		29		29	Cost includes material and labor to
	Install	Ea.	6375	635		7010	install 3' x 7' tempered glass door,
	Demolish and Install	Ea.	6375	664		7039	center pivot concealed closer.
	Reinstall	Ea.		509.23		509.23	
	Clean	Ea.	3.96	50		53.96	
	Minimum Charge	Job		294		294	
Heavy Section							
	Demolish	Ea.		29		29	Includes material and labor to install
	Install	Ea.	3175	590		3765	wide stile door with 3' 6" x 7'
	Demolish and Install	Ea.	3175	619		3794	opening.
	Reinstall	Ea.		471.04		471.04	
	Clean	Ea.	.79	16.65		17.44	
	Minimum Charge	Job		1000		1000	

Curtain Wall System		Unit	Material	Labor	Equip.	Total	Specification
Regular Weight							
Bronze Anodized							
	Demolish	S.F.		3.47		3.47	Includes material and labor to install
	Install	S.F.	53.50	14.10		67.60	structural curtain wall system with
	Demolish and Install	S.F.	53.50	17.57		71.07	bronze anodized frame of regular
	Clean	S.F.		.54		.54	weight, float glass.
	Minimum Charge	Job		260		260	
Black Anodized							
	Demolish	S.F.		3.47		3.47	Includes material and labor to install
	Install	S.F.	58	14.10		72.10	structural curtain wall system with
	Demolish and Install	S.F.	58	17.57		75.57	black anodized frame of regular
	Clean	S.F.		.54		.54	weight, float glass.
	Minimum Charge	Job		260		260	
Heavy Weight							
Bronze Anodized							
	Demolish	S.F.		3.47		3.47	Includes material and labor to install
	Install	S.F.	53.50	14.10		67.60	structural curtain wall system with
	Demolish and Install	S.F.	53.50	17.57		71.07	bronze anodized frame of regular
	Clean	S.F.		.54		.54	weight, float glass.
	Minimum Charge	Job		260		260	
Black Anodized							
	Demolish	S.F.		3.47		3.47	Includes material and labor to install
	Install	S.F.	55	14.10		69.10	structural curtain wall system with
	Demolish and Install	S.F.	55	17.57		72.57	black anodized frame of heavy
	Clean	S.F.		.54		.54	weight, float glass.
	Minimum Charge	Job		260		260	
Glass							
1" Insulated							
	Install	S.F.	27	13.45		40.45	Cost includes labor and material to
	Minimum Charge	Job		252		252	install 1" insulated glass.

Exterior Doors and Windows

Skylight		Unit	Material	Labor	Equip.	Total	Specification
Dome							
22″ x 22″							
	Demolish	Ea.		102		102	Cost includes material and labor to
	Install	Ea.	305	161		466	install 22″ x 22″ fixed dome skylight.
	Demolish and Install	Ea.	305	263		568	
	Reinstall	Ea.		129.02		129.02	
	Clean	Ea.	.35	14.25		14.60	
	Minimum Charge	Job		970		970	
22″ x 46″							
	Demolish	Ea.		102		102	Cost includes material and labor to
	Install	Ea.	370	194		564	install 22″ x 46″ fixed dome skylight.
	Demolish and Install	Ea.	370	296		666	
	Reinstall	Ea.		154.83		154.83	
	Clean	Ea.		22		22	
	Minimum Charge	Job		970		970	
30″ x 30″							
	Demolish	Ea.		102		102	Cost includes material and labor to
	Install	Ea.	495	161		656	install 30″ x 30″ fixed dome skylight.
	Demolish and Install	Ea.	495	263		758	
	Reinstall	Ea.		129.02		129.02	
	Clean	Ea.		22		22	
	Minimum Charge	Job		970		970	
30″ x 46″							
	Demolish	Ea.		102		102	Cost includes material and labor to
	Install	Ea.	505	194		699	install 30″ x 46″ fixed dome skylight.
	Demolish and Install	Ea.	505	296		801	
	Reinstall	Ea.		154.83		154.83	
	Clean	Ea.		22		22	
	Minimum Charge	Job		970		970	
Fixed							
22″ x 27″							
	Demolish	Ea.		102		102	Includes material and labor to install
	Install	Ea.	400	161		561	double glazed skylight, 22″ x 27″,
	Demolish and Install	Ea.	400	263		663	fixed.
	Reinstall	Ea.		129.02		129.02	
	Clean	Ea.	.35	14.25		14.60	
	Minimum Charge	Job		970		970	
22″ x 46″							
	Demolish	Ea.		102		102	Includes material and labor to install
	Install	Ea.	390	194		584	double glazed skylight, 22″ x 46″,
	Demolish and Install	Ea.	390	296		686	fixed.
	Reinstall	Ea.		154.83		154.83	
	Clean	Ea.		22		22	
	Minimum Charge	Job		970		970	
44″ x 46″							
	Demolish	Ea.		102		102	Includes material and labor to install
	Install	Ea.	645	194		839	double glazed skylight, 44″ x 46″,
	Demolish and Install	Ea.	645	296		941	fixed.
	Reinstall	Ea.		154.83		154.83	
	Clean	Ea.		25		25	
	Minimum Charge	Job		970		970	
Operable							
22″ x 27″							
	Demolish	Ea.		102		102	Includes material and labor to install
	Install	Ea.	730	161		891	double glazed skylight, 22″ x 27″,
	Demolish and Install	Ea.	730	263		993	operable.
	Reinstall	Ea.		129.02		129.02	
	Clean	Ea.	.35	14.25		14.60	
	Minimum Charge	Job		970		970	

For customer support on your Contractor's Pricing Guide: Residential Repair & Remodeling, call 800.448.8182.

Exterior Doors and Windows

Skylight

	Unit	Material	Labor	Equip.	Total	Specification
22" x 46"						
Demolish	Ea.		102		102	Includes material and labor to install
Install	Ea.	880	194		1074	double glazed skylight, 22" x 46",
Demolish and Install	Ea.	880	296		1176	operable.
Reinstall	Ea.		154.83		154.83	
Clean	Ea.		22		22	
Minimum Charge	Job		970		970	
44" x 46"						
Demolish	Ea.		102		102	Includes material and labor to install
Install	Ea.	1275	194		1469	double glazed skylight, 44" x 46",
Demolish and Install	Ea.	1275	296		1571	operable.
Reinstall	Ea.		154.83		154.83	
Clean	Ea.		25		25	
Minimum Charge	Job		970		970	

Industrial Steel Window

	Unit	Material	Labor	Equip.	Total	Specification
100% Fixed						
Demolish	S.F.		2.85		2.85	Includes material and labor to install
Install	S.F.	36	5.90		41.90	commercial grade steel windows,
Demolish and Install	S.F.	36	8.75		44.75	including glazing.
Reinstall	S.F.		4.71		4.71	
Clean	Ea.	.14	1.69		1.83	
Paint	S.F.	.35	.96		1.31	
Minimum Charge	Job		294		294	
50% Vented						
Demolish	S.F.		2.85		2.85	Includes material and labor to install
Install	S.F.	69.50	5.90		75.40	commercial grade steel windows,
Demolish and Install	S.F.	69.50	8.75		78.25	including glazing.
Reinstall	S.F.		4.71		4.71	
Clean	Ea.	.14	1.69		1.83	
Paint	S.F.	.35	.96		1.31	
Minimum Charge	Job		294		294	

For customer support on your Contractor's Pricing Guide: Residential Repair & Remodeling, call 800.448.8182.

Masonry

Brick Veneer

Concrete Masonry Units

Masonry Wall		Unit	Material	Labor	Equip.	Total	Specification
Unreinforced CMU							
4" x 8" x 16"							
	Demolish	S.F.		.16	.44	.60	Includes material and labor to install
	Install	S.F.	2.62	5.25		7.87	normal weight 4" x 8" x 16" block.
	Demolish and Install	S.F.	2.62	5.41	.44	8.47	
	Clean	S.F.	.12	.61		.73	
	Paint	S.F.	.48	.98		1.46	
	Minimum Charge	Job		259		259	
6" x 8" x 16"							
	Demolish	S.F.		.21	.59	.80	Includes material and labor to install
	Install	S.F.	2.84	5.50		8.34	normal weight 6" x 8" x 16" block.
	Demolish and Install	S.F.	2.84	5.71	.59	9.14	
	Clean	S.F.	.12	.61		.73	
	Paint	S.F.	.48	.98		1.46	
	Minimum Charge	Job		259		259	
8" x 8" x 16"							
	Demolish	S.F.		.42	1.18	1.60	Includes material and labor to install
	Install	S.F.	3.14	6.05		9.19	normal weight 8" x 8" x 16" block.
	Demolish and Install	S.F.	3.14	6.47	1.18	10.79	
	Clean	S.F.	.12	.61		.73	
	Paint	S.F.	.48	.98		1.46	
	Minimum Charge	Job		259		259	
12" x 8" x 16"							
	Demolish	S.F.		.42	1.18	1.60	Includes material and labor to install
	Install	S.F.	4.74	9.20		13.94	normal weight 12" x 8" x 16" block.
	Demolish and Install	S.F.	4.74	9.62	1.18	15.54	
	Clean	S.F.	.12	.61		.73	
	Paint	S.F.	.48	.98		1.46	
	Minimum Charge	Job		259		259	
Reinforced CMU							
4" x 8" x 16"							
	Demolish	S.F.		2.47	1.37	3.84	Includes material and labor to install
	Install	S.F.	2.88	5.40		8.28	normal weight 4" x 8" x 16" block,
	Demolish and Install	S.F.	2.88	7.87	1.37	12.12	mortar and 1/2" re-bar installed
	Clean	S.F.	.12	.61		.73	vertically every 24" O.C.
	Paint	S.F.	.48	.98		1.46	
	Minimum Charge	Job		475		475	
6" x 8" x 16"							
	Demolish	S.F.		2.60	.92	3.52	Includes material and labor to install
	Install	S.F.	3.06	5.65		8.71	normal weight 6" x 8" x 16" block,
	Demolish and Install	S.F.	3.06	8.25	.92	12.23	mortar and 1/2" re-bar installed
	Clean	S.F.	.12	.61		.73	vertically every 24" O.C.
	Paint	S.F.	.48	.98		1.46	
	Minimum Charge	Job		475		475	

For customer support on your Contractor's Pricing Guide: Residential Repair & Remodeling, call 800.448.8182.

Masonry

Masonry Wall

Masonry Wall		Unit	Material	Labor	Equip.	Total	Specification
8" x 8" x 16"							Includes material and labor to install
	Demolish	S.F.		2.74	.97	3.71	normal weight 8" x 8" x 16" block,
	Install	S.F.	3.37	6.15		9.52	mortar and 1/2" re-bar installed
	Demolish and Install	S.F.	3.37	8.89	.97	13.23	vertically every 24" O.C.
	Clean	S.F.	.12	.61		.73	
	Paint	S.F.	.48	.98		1.46	
	Minimum Charge	Job		475		475	
12" x 8" x 16"							Includes material and labor to install
	Demolish	S.F.		2.83	1	3.83	normal weight 12" x 8" x 16" block,
	Install	S.F.	4.99	9.50		14.49	mortar and 1/2" re-bar installed
	Demolish and Install	S.F.	4.99	12.33	1	18.32	vertically every 24" O.C.
	Clean	S.F.	.12	.61		.73	
	Paint	S.F.	.48	.98		1.46	
	Minimum Charge	Job		475		475	

Structural Brick

4" Single Row

		Unit	Material	Labor	Equip.	Total	Specification
	Demolish	S.F.		3.55		3.55	Cost includes material and labor to
	Install	S.F.	7	11.25		18.25	install 4" red brick wall, running bond
	Demolish and Install	S.F.	7	14.80		21.80	with 3/8" concave joints, mortar and
	Clean	S.F.	.12	.53		.65	ties.
	Paint	S.F.	.48	.98		1.46	
	Minimum Charge	Job		475		475	

8" Double Row

		Unit	Material	Labor	Equip.	Total	Specification
	Demolish	S.F.		3.55		3.55	Cost includes material and labor to
	Install	S.F.	16.95	17.95		34.90	install 8" red brick wall, running bond,
	Demolish and Install	S.F.	16.95	21.50		38.45	with 3/8" concave joints, mortar and
	Clean	S.F.	.12	.53		.65	wall ties.
	Paint	S.F.	.48	.98		1.46	
	Minimum Charge	Job		475		475	

12" Triple Row

		Unit	Material	Labor	Equip.	Total	Specification
	Demolish	S.F.		4.93		4.93	Cost includes material and labor to
	Install	S.F.	25.50	25.50		51	install 12" red brick wall, running
	Demolish and Install	S.F.	25.50	30.43		55.93	bond, with 3/8" concave joints,
	Clean	S.F.	.12	.53		.65	mortar and wall ties.
	Paint	S.F.	.48	.98		1.46	
	Minimum Charge	Job		475		475	

Brick Veneer

4" Red Common

		Unit	Material	Labor	Equip.	Total	Specification
	Demolish	S.F.		3.55		3.55	Cost includes material and labor to
	Install	S.F.	8.10	10.55		18.65	install 4" common red face brick,
	Demolish and Install	S.F.	8.10	14.10		22.20	running bond with 3/8" concave
	Clean	S.F.	.12	.53		.65	joints, mortar and wall ties.
	Paint	S.F.	.48	.98		1.46	
	Minimum Charge	Job		259		259	

4" Used

		Unit	Material	Labor	Equip.	Total	Specification
	Demolish	S.F.		3.55		3.55	Cost includes material and labor to
	Install	S.F.	7	11.25		18.25	install 4" red brick wall, running bond
	Demolish and Install	S.F.	7	14.80		21.80	with 3/8" concave joints, mortar and
	Clean	S.F.	.12	.53		.65	ties.
	Paint	S.F.	.48	.98		1.46	
	Minimum Charge	Job		259		259	

For customer support on your Contractor's Pricing Guide: Residential Repair & Remodeling, call 800.448.8182.

Masonry

Masonry Wall

	Unit	Material	Labor	Equip.	Total	Specification
6" Jumbo						
Demolish	S.F.		3.55		3.55	Cost includes material and labor to
Install	S.F.	6.25	5.55		11.80	install 6" red jumbo brick veneer,
Demolish and Install	S.F.	6.25	9.10		15.35	running bond with 3/8" concave
Clean	S.F.	.12	.53		.65	joints, mortar and ties.
Paint	S.F.	.48	.98		1.46	
Minimum Charge	Job		259		259	
Norman (11-1/2")						
Demolish	S.F.		3.55		3.55	Cost includes material and labor to
Install	S.F.	7.20	7.55		14.75	install 4" norman red brick wall,
Demolish and Install	S.F.	7.20	11.10		18.30	running bond with 3/8" concave
Clean	S.F.	.12	.53		.65	joints, mortar and ties.
Paint	S.F.	.48	.98		1.46	
Minimum Charge	Job		259		259	
Glazed						
Demolish	S.F.		3.55		3.55	Includes material, labor and
Install	S.F.	25	11.55		36.55	equipment to install glazed brick.
Demolish and Install	S.F.	25	15.10		40.10	
Clean	S.F.	.12	.53		.65	
Paint	S.F.	.48	.98		1.46	
Minimum Charge	Job		259		259	
Slumpstone						
6" x 4"						
Demolish	S.F.		.16	.44	.60	Cost includes material and labor to
Install	S.F.	7.55	5.95		13.50	install 6" x 4" x 16" concrete slump
Demolish and Install	S.F.	7.55	6.11	.44	14.10	block.
Clean	S.F.	.12	.61		.73	
Paint	S.F.	.48	.98		1.46	
Minimum Charge	Job		259		259	
6" x 6"						
Demolish	S.F.		.21	.59	.80	Cost includes material and labor to
Install	S.F.	5.70	6.15		11.85	install 6" x 6" x 16" concrete slump
Demolish and Install	S.F.	5.70	6.36	.59	12.65	block.
Clean	S.F.	.12	.61		.73	
Paint	S.F.	.48	.98		1.46	
Minimum Charge	Job		259		259	
8" x 6"						
Demolish	S.F.		.21	.59	.80	Cost includes material and labor to
Install	S.F.	8.70	6.35		15.05	install 8" x 6" x 16" concrete slump
Demolish and Install	S.F.	8.70	6.56	.59	15.85	block.
Clean	S.F.	.12	.61		.73	
Paint	S.F.	.48	.98		1.46	
Minimum Charge	Job		259		259	
12" x 4"						
Demolish	S.F.		.16	.44	.60	Cost includes material and labor to
Install	S.F.	15.20	7.30		22.50	install 12" x 4" x 16" concrete slump
Demolish and Install	S.F.	15.20	7.46	.44	23.10	block.
Clean	S.F.	.12	.61		.73	
Paint	S.F.	.48	.98		1.46	
Minimum Charge	Job		259		259	
12" x 6"						
Demolish	S.F.		.21	.59	.80	Cost includes material and labor to
Install	S.F.	16.50	7.95		24.45	install 12" x 6" x 16" concrete slump
Demolish and Install	S.F.	16.50	8.16	.59	25.25	block.
Clean	S.F.	.12	.61		.73	
Paint	S.F.	.48	.98		1.46	
Minimum Charge	Job		259		259	

For customer support on your Contractor's Pricing Guide: Residential Repair & Remodeling, call 800.448.8182.

Masonry

Masonry Wall		Unit	Material	Labor	Equip.	Total	Specification
Stucco							
Three Coat							
	Demolish	S.Y.		3.06		3.06	Includes material and labor to install
	Install	S.Y.	6.40	12	.67	19.07	stucco on exterior masonry walls.
	Demolish and Install	S.Y.	6.40	15.06	.67	22.13	
	Clean	S.Y.		5.55		5.55	
	Paint	S.F.	.34	.74		1.08	
	Minimum Charge	Job		254		254	
Three Coat w / Wire Mesh							
	Demolish	S.Y.		5.55		5.55	Includes material and labor to install
	Install	S.Y.	12.05	46.50	2.12	60.67	stucco on wire mesh over wood
	Demolish and Install	S.Y.	12.05	52.05	2.12	66.22	framing.
	Clean	S.Y.		5.55		5.55	
	Paint	S.F.	.34	.74		1.08	
	Minimum Charge	Job		254		254	
Fill Cracks							
	Install	S.F.	.52	4.06		4.58	Includes labor and material to fill in
	Minimum Charge	Job		254		254	hairline cracks up to 1/8" with filler, including sand and fill per S.F.
Furring							
1" x 2"							
	Demolish	L.F.		.18		.18	Cost includes material and labor to
	Install	L.F.	.20	1.05		1.25	install 1" x 2" furring on masonry wall.
	Demolish and Install	L.F.	.20	1.23		1.43	
	Minimum Charge	Job		260		260	
2" x 2"							
	Demolish	L.F.		.19		.19	Cost includes material and labor to
	Install	L.F.	1.97	1.16		3.13	install 2" x 2" furring on masonry wall.
	Demolish and Install	L.F.	1.97	1.35		3.32	
	Minimum Charge	Job		260		260	
2" x 4"							
	Demolish	L.F.		.21		.21	Cost includes material and labor to
	Install	L.F.	1.09	1.32		2.41	install 2" x 4" furring on masonry wall.
	Demolish and Install	L.F.	1.09	1.53		2.62	
	Reinstall	L.F.		1.05		1.05	
	Minimum Charge	Job		260		260	
Maintenance / Cleaning							
Repoint							
	Install	S.F.	1.36	6.45		7.81	Includes labor and material to cut and
	Minimum Charge	Hr.		64.50		64.50	repoint brick, hard mortar, running bond.
Acid Etch							
	Install	S.F.	.08	.92		1	Includes labor and material to acid
	Minimum Charge	Job		200	70.50	270.50	wash smooth brick.
Pressure Wash							
	Clean	S.F.		1.07	.36	1.43	Includes labor and material to high
	Minimum Charge	Job		200	70.50	270.50	pressure water wash masonry, average.
Waterproof							
	Install	S.F.	.65	.98		1.63	Includes labor and material to install
	Minimum Charge	Job		244		244	masonry waterproofing.

128

For customer support on your Contractor's Pricing Guide: Residential Repair & Remodeling, call 800.448.8182.

Masonry

Chimney		Unit	Material	Labor	Equip.	Total	Specification
Demo & Haul Exterior							
	Demolish	C.F.		7.25		7.25	Includes minimum labor and
	Minimum Charge	Job		222		222	equipment to remove masonry and
							haul debris to truck or dumpster.
Demo & Haul Interior							
	Demolish	C.F.		10		10	Includes minimum labor and
	Minimum Charge	Job		222		222	equipment to remove masonry and
							haul debris to truck or dumpster.
Install New (w / flue)							
Brick							
	Install	V.L.F.	83	59.50		142.50	Includes material and labor to install
	Minimum Charge	Job		259		259	brick chimney with flue. Foundation
							not included.
Natural Stone							
	Install	V.L.F.	91.50	95		186.50	Includes labor and material to install
	Minimum Charge	Job		259		259	natural stone chimney.
Install From Roof Up							
Brick							
	Install	V.L.F.	83	59.50		142.50	Includes material and labor to install
	Minimum Charge	Job		259		259	brick chimney with flue. Foundation
							not included.
Natural Stone							
	Install	V.L.F.	91.50	79.50		171	Includes labor and material to install
	Minimum Charge	Job		259		259	natural stone chimney.
Install Base Pad							
	Install	Ea.	530	290	1.49	821.49	Includes labor and material to replace
	Minimum Charge	Job		1275	36	1311	spread footing, 5' square x 16" deep.
Metal Cap							
	Install	Ea.	94.50	32.50		127	Includes labor and materials to install
	Minimum Charge	Job		260		260	bird and squirrel screens in a chimney
							flue.

Fireplace		Unit	Material	Labor	Equip.	Total	Specification
Common Brick							
30"							
	Demolish	Ea.		1575		1575	Includes material and labor to install a
	Install	Ea.	4200	4750		8950	complete unit which includes common
	Demolish and Install	Ea.	4200	6325		10525	brick, foundation, damper, flue lining
	Minimum Charge	Job		259		259	stack to 15' high and a 30" fire box.
36"							
	Demolish	Ea.		1575		1575	Includes material and labor to install a
	Install	Ea.	4475	5300		9775	complete unit which includes common
	Demolish and Install	Ea.	4475	6875		11350	brick, foundation, damper, flue lining
	Minimum Charge	Job		259		259	stack to 15' high and a 36" fire box.
42"							
	Demolish	Ea.		1575		1575	Includes material and labor to install a
	Install	Ea.	4850	5950		10800	complete unit which includes common
	Demolish and Install	Ea.	4850	7525		12375	brick, foundation, damper, flue lining
	Minimum Charge	Job		259		259	stack to 15' high and a 42" fire box.

For customer support on your Contractor's Pricing Guide: Residential Repair & Remodeling, call 800.448.8182.

Masonry

Fireplace

	Unit	Material	Labor	Equip.	Total	Specification
48"						
Demolish	Ea.		1575		1575	Includes material and labor to install a
Install	Ea.	5125	6775		11900	complete unit which includes common
Demolish and Install	Ea.	5125	8350		13475	brick, foundation, damper, flue lining
Minimum Charge	Job		259		259	stack to 15' high and a 48" fire box.
Raised Hearth						
Brick Hearth						
Demolish	S.F.		6.95		6.95	Includes material and labor to install a
Install	S.F.	2.60	39.50		42.10	brick fireplace hearth based on 30" x
Demolish and Install	S.F.	2.60	46.45		49.05	29" opening.
Clean	S.F.	.12	.53		.65	
Minimum Charge	Job		259		259	
Marble Hearth / Facing						
Demolish	Ea.		44.50		44.50	Includes material and labor to install
Install	Ea.	330	159		489	marble fireplace hearth.
Demolish and Install	Ea.	330	203.50		533.50	
Reinstall	Ea.		127.58		127.58	
Clean	Ea.	40	8		48	
Minimum Charge	Job		259		259	
Breast (face) Brick						
Demolish	S.F.		6.95		6.95	Includes material and labor to install
Install	S.F.	7.10	11		18.10	standard red brick, running bond.
Demolish and Install	S.F.	7.10	17.95		25.05	
Clean	S.F.		1.07	.36	1.43	
Minimum Charge	Hr.		64.50		64.50	
Fire Box						
Demolish	Ea.		355		355	Includes material and labor to install
Install	Ea.	790	475		1265	fireplace box only (110 brick).
Demolish and Install	Ea.	790	830		1620	
Clean	Ea.	.12	20		20.12	
Minimum Charge	Hr.		64.50		64.50	
Natural Stone						
Demolish	S.F.		6.95		6.95	Includes material and labor to install
Install	S.F.	19	20		39	high priced stone, split or rock face.
Demolish and Install	S.F.	19	26.95		45.95	
Clean	S.F.	.12	1.19		1.31	
Minimum Charge	Job		475		475	
Slumpstone Veneer						
Demolish	S.F.		.21	.59	.80	Cost includes material and labor to
Install	S.F.	8.70	6.35		15.05	install 8" x 6" x 16" concrete slump
Demolish and Install	S.F.	8.70	6.56	.59	15.85	block.
Clean	S.F.	.12	.61		.73	
Paint	S.F.	.48	.98		1.46	
Minimum Charge	Job		259		259	
Prefabricated						
Firebox						
Demolish	Ea.		63.50		63.50	Includes material and labor to install
Install	Ea.	2100	475		2575	up to 43" radiant heat, zero
Demolish and Install	Ea.	2100	538.50		2638.50	clearance, prefabricated fireplace.
Reinstall	Ea.		378.76		378.76	Flues, doors or blowers not included.
Clean	Ea.	1.58	25		26.58	
Minimum Charge	Job		260		260	

Masonry

Fireplace	Unit	Material	Labor	Equip.	Total	Specification
Flue						
Demolish	V.L.F.		7.40		7.40	Includes material and labor to install
Install	V.L.F.	50.50	19.65		70.15	prefabricated chimney vent, 8"
Demolish and Install	V.L.F.	50.50	27.05		77.55	exposed metal flue.
Reinstall	V.L.F.		15.74		15.74	
Clean	L.F.	.64	5		5.64	
Minimum Charge	Job		370		370	
Blower						
Demolish	Ea.		5.90		5.90	Includes material and labor to install
Install	Ea.	211	35.50		246.50	blower for metal fireplace flue.
Demolish and Install	Ea.	211	41.40		252.40	
Reinstall	Ea.		28.40		28.40	
Clean	Ea.	3.17	50		53.17	
Minimum Charge	Job		370		370	
Accessories						
Wood Mantle						
Demolish	L.F.		7.40		7.40	Includes material and labor to install
Install	L.F.	19.35	14.90		34.25	wood fireplace mantel.
Demolish and Install	L.F.	19.35	22.30		41.65	
Reinstall	L.F.		14.88		14.88	
Clean	S.F.		.50		.50	
Paint	L.F.	.20	11.75		11.95	
Minimum Charge	Job		260		260	
Colonial Style						
Demolish	Ea.		64.50		64.50	Includes material and labor to install
Install	Opng.	3350	260		3610	wood fireplace mantel.
Demolish and Install	Opng.	3350	324.50		3674.50	
Reinstall	Opng.		260.40		260.40	
Clean	Ea.		.90		.90	
Paint	Ea.	38.50	162		200.50	
Minimum Charge	Job		260		260	
Modern Design						
Demolish	Ea.		64.50		64.50	Includes material and labor to install
Install	Opng.	1000	104		1104	wood fireplace mantel.
Demolish and Install	Opng.	1000	168.50		1168.50	
Reinstall	Opng.		104.16		104.16	
Clean	Ea.		.90		.90	
Paint	Ea.	38.50	162		200.50	
Minimum Charge	Job		260		260	
Glass Door Unit						
Demolish	Ea.		9.25		9.25	Includes material and labor to install a
Install	Ea.	820	129		949	fireplace glass door unit.
Demolish and Install	Ea.	820	138.25		958.25	
Reinstall	Ea.		129.30		129.30	
Clean	Ea.	.39	15.35		15.74	
Minimum Charge	Job		260		260	
Fire Screen						
Demolish	Ea.		4.62		4.62	Includes material and labor to install a
Install	Ea.	385	5.40		390.40	fireplace fire screen.
Demolish and Install	Ea.	385	10.02		395.02	
Reinstall	Ea.		5.39		5.39	
Clean	Ea.	.19	13.30		13.49	
Minimum Charge	Job		260		260	

For customer support on your Contractor's Pricing Guide: Residential Repair & Remodeling, call 800.448.8182.

Siding

Shingles

Board & Batten

Texture 1-11

Exterior Siding

	Unit	Material	Labor	Equip.	Total	Specification
Trim						
1" x 2"						
Demolish	L.F.		.37		.37	Cost includes material and labor to
Install	L.F.	.20	1.58		1.78	install 1" x 2" pine trim.
Demolish and Install	L.F.	.20	1.95		2.15	
Clean	L.F.	.01	.27		.28	
Paint	L.F.	.07	.67		.74	
Minimum Charge	Job		260		260	
1" x 3"						
Demolish	L.F.		.37		.37	Cost includes material and labor to
Install	L.F.	.36	1.80		2.16	install 1" x 3" pine trim.
Demolish and Install	L.F.	.36	2.17		2.53	
Clean	L.F.	.01	.27		.28	
Paint	L.F.	.07	.67		.74	
Minimum Charge	Job		260		260	
1" x 4"						
Demolish	L.F.		.37		.37	Cost includes material and labor to
Install	L.F.	.83	2.08		2.91	install 1" x 4" pine trim.
Demolish and Install	L.F.	.83	2.45		3.28	
Clean	L.F.	.01	.27		.28	
Paint	L.F.	.07	.67		.74	
Minimum Charge	Job		260		260	
1" x 6"						
Demolish	L.F.		.37		.37	Cost includes material and labor to
Install	L.F.	1.14	2.08		3.22	install 1" x 6" pine trim.
Demolish and Install	L.F.	1.14	2.45		3.59	
Clean	L.F.	.01	.27		.28	
Paint	L.F.	.24	.67		.91	
Minimum Charge	Job		260		260	
1" x 8"						
Demolish	L.F.		.40		.40	Cost includes material and labor to
Install	L.F.	.90	2.60		3.50	install 1" x 8" pine trim.
Demolish and Install	L.F.	.90	3		3.90	
Clean	L.F.	.02	.29		.31	
Paint	L.F.	.24	.67		.91	
Minimum Charge	Job		260		260	
Aluminum						
8" Non-Insulated						
Demolish	S.F.		1		1	Includes material and labor to install
Install	S.F.	3.59	2.12		5.71	horizontal, colored clapboard, 8" to
Demolish and Install	S.F.	3.59	3.12		6.71	10" wide, plain.
Reinstall	S.F.		2.12		2.12	
Clean	S.F.		.40		.40	
Minimum Charge	Job		260		260	

Siding

Exterior Siding	Unit	Material	Labor	Equip.	Total	Specification
8" Insulated						
Demolish	S.F.		1		1	Includes material and labor to install
Install	S.F.	3.95	2.12		6.07	horizontal, colored insulated
Demolish and Install	S.F.	3.95	3.12		7.07	clapboard, 8" wide, plain.
Reinstall	S.F.		2.12		2.12	
Clean	S.F.		.40		.40	
Minimum Charge	Job		260		260	
12" Non-Insulated						
Demolish	S.F.		1		1	Includes material and labor to install
Install	S.F.	4.18	1.82		6	horizontal, colored clapboard, plain.
Demolish and Install	S.F.	4.18	2.82		7	
Reinstall	S.F.		1.82		1.82	
Clean	S.F.		.40		.40	
Minimum Charge	Job		260		260	
12" Insulated						
Demolish	S.F.		1		1	Includes material and labor to install
Install	S.F.	3.80	1.82		5.62	horizontal, colored insulated
Demolish and Install	S.F.	3.80	2.82		6.62	clapboard, plain.
Reinstall	S.F.		1.82		1.82	
Clean	S.F.		.40		.40	
Minimum Charge	Job		260		260	
Clapboard 8" to 10"						
Demolish	S.F.		1		1	Includes material and labor to install
Install	S.F.	3.59	2.12		5.71	horizontal, colored clapboard, 8" to
Demolish and Install	S.F.	3.59	3.12		6.71	10" wide, plain.
Reinstall	S.F.		2.12		2.12	
Clean	S.F.		.40		.40	
Minimum Charge	Job		260		260	
Corner Strips						
Demolish	L.F.		.52		.52	Includes material and labor to install
Install	V.L.F.	1.35	2.12		3.47	corners for horizontal lapped
Demolish and Install	V.L.F.	1.35	2.64		3.99	aluminum siding.
Reinstall	V.L.F.		2.12		2.12	
Clean	L.F.		.27		.27	
Minimum Charge	Job		260		260	
Vinyl						
Non-Insulated						
Demolish	S.F.		.78		.78	Cost includes material and labor to
Install	S.F.	3.43	1.89		5.32	install 10" non-insulated solid vinyl
Demolish and Install	S.F.	3.43	2.67		6.10	siding.
Reinstall	S.F.		1.52		1.52	
Clean	S.F.		.40		.40	
Minimum Charge	Job		260		260	
Insulated						
Demolish	S.F.		.78		.78	Cost includes material and labor to
Install	S.F.	5.45	2.60		8.05	install insulated solid vinyl siding
Demolish and Install	S.F.	5.45	3.38		8.83	panels.
Reinstall	S.F.		2.08		2.08	
Clean	S.F.		.40		.40	
Minimum Charge	Job		260		260	
Corner Strips						
Demolish	L.F.		.49		.49	Cost includes material and labor to
Install	L.F.	3.11	1.49		4.60	install corner posts for vinyl siding
Demolish and Install	L.F.	3.11	1.98		5.09	panels.
Reinstall	L.F.		1.19		1.19	
Clean	L.F.		.27		.27	
Minimum Charge	Job		260		260	

For customer support on your Contractor's Pricing Guide: Residential Repair & Remodeling, call 800.448.8182.

Siding

Exterior Siding	Unit	Material	Labor	Equip.	Total	Specification
Fiber Cement Siding						
4' x 8' Panel						
Demolish	S.F.		.78		.78	Cost includes material and labor to
Install	S.F.	1.01	1.14		2.15	install 4' x 8' fiber cement 7/16"
Demolish and Install	S.F.	1.01	1.92		2.93	thick.
Clean	S.F.		.40		.40	
Paint	S.F.	.70	1.52		2.22	
Minimum Charge	Job		260		260	
Lap						
Demolish	S.F.		.77		.77	Cost includes material and labor to
Install	S.F.	1.55	2.06		3.61	install 8" or 12" fiber cement lap
Demolish and Install	S.F.	1.55	2.83		4.38	siding 7/16" thick with wood grain
Clean	S.F.		.40		.40	finish.
Paint	S.F.	.70	1.52		2.22	
Minimum Charge	Job		260		260	
Plywood Panel						
Douglas Fir (T-1-11)						
Demolish	S.F.		.61		.61	Includes material and labor to install
Install	S.F.	1.56	1.62		3.18	Douglas fir 4' x 8' plywood exterior
Demolish and Install	S.F.	1.56	2.23		3.79	siding plain or patterns.
Reinstall	S.F.		1.62		1.62	
Clean	S.F.		.40		.40	
Paint	S.F.	.70	1.52		2.22	
Minimum Charge	Job		260		260	
Redwood						
Demolish	S.F.		.61		.61	Includes material and labor to install
Install	S.F.	2.83	1.62		4.45	redwood 4' x 8' plywood exterior
Demolish and Install	S.F.	2.83	2.23		5.06	siding plain or patterns.
Reinstall	S.F.		1.62		1.62	
Clean	S.F.		.40		.40	
Paint	S.F.	.70	1.52		2.22	
Minimum Charge	Job		260		260	
Cedar						
Demolish	S.F.		.61		.61	Cost includes material and labor to
Install	S.F.	3.55	1.62		5.17	install 4' x 8' cedar plywood exterior
Demolish and Install	S.F.	3.55	2.23		5.78	siding plain or patterns.
Reinstall	S.F.		1.62		1.62	
Clean	S.F.		.40		.40	
Paint	S.F.	.70	1.52		2.22	
Minimum Charge	Job		260		260	
Tongue and Groove						
Fir, Tongue and Groove						
Demolish	S.F.		1.11		1.11	Includes material and labor to install
Install	S.F.	7.80	1.58		9.38	fir tongue and groove exterior siding
Demolish and Install	S.F.	7.80	2.69		10.49	1" x 8".
Reinstall	S.F.		1.26		1.26	
Clean	S.F.		.40		.40	
Paint	S.F.	.70	1.52		2.22	
Minimum Charge	Job		260		260	
Pine, Tongue and Groove						
Demolish	S.F.		1.11		1.11	Includes labor and material to install
Install	S.F.	1.89	1.58		3.47	Pine tongue and groove exterior
Demolish and Install	S.F.	1.89	2.69		4.58	siding 1" x 8".
Reinstall	S.F.		1.58		1.58	
Clean	S.F.		.40		.40	
Paint	S.F.	.70	1.52		2.22	
Minimum Charge	Job		260		260	

For customer support on your Contractor's Pricing Guide: Residential Repair & Remodeling, call 800.448.8182.

Siding

Exterior Siding	Unit	Material	Labor	Equip.	Total	Specification
Cedar, Tongue and Groove						
Demolish	S.F.		1.11		1.11	Includes labor and material to install
Install	S.F.	12.05	2.37		14.42	cedar tongue and groove exterior
Demolish and Install	S.F.	12.05	3.48		15.53	siding 1″ x 8″.
Reinstall	S.F.		1.89		1.89	
Clean	S.F.		.40		.40	
Paint	S.F.	.70	1.52		2.22	
Minimum Charge	Job		260		260	
Redwood, Tongue and Groove						
Demolish	S.F.		1.11		1.11	Includes labor and material to install
Install	S.F.	9.65	2.37		12.02	redwood tongue and groove exterior
Demolish and Install	S.F.	9.65	3.48		13.13	siding 1″ x 4″.
Reinstall	S.F.		2.37		2.37	
Clean	S.F.		.40		.40	
Paint	S.F.	.70	1.52		2.22	
Minimum Charge	Job		260		260	

Board and Batten

Fir

	Unit	Material	Labor	Equip.	Total	Specification
Demolish	S.F.		1.11		1.11	Cost includes material and labor to
Install	S.F.	.32	1.58		1.90	install 1″ x 8″ fir boards with 1″ x 3″
Demolish and Install	S.F.	.32	2.69		3.01	battens.
Reinstall	S.F.		1.26		1.26	
Clean	S.F.		.40		.40	
Paint	S.F.	.70	1.52		2.22	
Minimum Charge	Job		260		260	

Pine

	Unit	Material	Labor	Equip.	Total	Specification
Demolish	S.F.		1.11		1.11	Cost includes material and labor to
Install	S.F.	4.48	1.39		5.87	install 1″ x 10″ pine boards with 1″ x
Demolish and Install	S.F.	4.48	2.50		6.98	3″ battens.
Reinstall	S.F.		1.11		1.11	
Clean	S.F.		.40		.40	
Paint	S.F.	.70	1.52		2.22	
Minimum Charge	Job		260		260	

Cedar

	Unit	Material	Labor	Equip.	Total	Specification
Demolish	S.F.		1.11		1.11	Cost includes material and labor to
Install	S.F.	14.25	1.39		15.64	install 1″ x 10″ cedar boards with 1″ x
Demolish and Install	S.F.	14.25	2.50		16.75	3″ battens.
Reinstall	S.F.		1.11		1.11	
Clean	S.F.		.40		.40	
Paint	S.F.	.70	1.52		2.22	
Minimum Charge	Job		260		260	

Redwood

	Unit	Material	Labor	Equip.	Total	Specification
Demolish	S.F.		1.11		1.11	Cost includes material and labor to
Install	S.F.	8.05	1.39		9.44	install 1″ x 10″ redwood boards with
Demolish and Install	S.F.	8.05	2.50		10.55	1″ x 3″ battens.
Reinstall	S.F.		1.11		1.11	
Clean	S.F.		.40		.40	
Paint	S.F.	.70	1.52		2.22	
Minimum Charge	Job		260		260	

Wood Shingle

	Unit	Material	Labor	Equip.	Total	Specification
Demolish	Sq.		102		102	Includes material and labor to install
Install	Sq.	450	231		681	wood shingles/shakes.
Demolish and Install	Sq.	450	333		783	
Clean	Sq.		42		42	
Paint	Sq.	43	136		179	
Minimum Charge	Job		260		260	

For customer support on your Contractor's Pricing Guide: Residential Repair & Remodeling, call 800.448.8182.

Siding

Exterior Siding	Unit	Material	Labor	Equip.	Total	Specification
Synthetic Stucco System						
Demolish	S.F.		3.70		3.70	Cost includes material and labor to
Install	S.F.	3.17	8.95	.50	12.62	install 2 coats of adhesive mixture,
Demolish and Install	S.F.	3.17	12.65	.50	16.32	glass fiber mesh, 1" insulation board,
Clean	S.F.		.36		.36	and textured finish coat.
Paint	S.F.	.34	.74		1.08	
Minimum Charge	Job		1200	66.50	1266.50	

For customer support on your Contractor's Pricing Guide: Residential Repair & Remodeling, call 800.448.8182.

Site Work

Front-end Loader　　　　　　　　Bulldozer

Grading		Unit	Material	Labor	Equip.	Total	Specification
Hand (fine)							Includes labor and material to install fine grade for slab on grade, hand grade.
	Install	S.F.		.80		.80	
	Minimum Charge	Job		200		200	
Machine (fine)							Includes labor and material to install fine grade for roadway, base or leveling course.
	Install	S.Y.		.70	.93	1.63	
	Minimum Charge	Job		475	635	1110	

Asphalt Paving		Unit	Material	Labor	Equip.	Total	Specification
Seal Coat							Includes labor and material to install sealcoat for a small area.
	Install	S.F.	.17	.14		.31	
	Minimum Charge	Job		200		200	
Wearing Course							Cost includes labor and material to install 1" thick wearing course.
	Install	S.F.	98	.51	.38	98.89	
	Minimum Charge	Job		200		200	

Masonry Pavers		Unit	Material	Labor	Equip.	Total	Specification
Natural Concrete							
Sand Base							Includes material and labor to install concrete pavers on sand base.
	Demolish	S.F.		4.44		4.44	
	Install	S.F.	6.90	8.60		15.50	
	Demolish and Install	S.F.	6.90	13.04		19.94	
	Reinstall	S.F.		6.90		6.90	
	Clean	S.F.		1.60		1.60	
	Minimum Charge	Job		259		259	
Mortar Base							Includes material and labor to install concrete pavers on mortar base.
	Demolish	S.F.		7.40	1.80	9.20	
	Install	S.F.	6.05	7.20		13.25	
	Demolish and Install	S.F.	6.05	14.60	1.80	22.45	
	Reinstall	S.F.		5.75		5.75	
	Clean	S.F.		1.45		1.45	
	Minimum Charge	Job		259		259	

Site Work

Masonry Pavers

Masonry Pavers	Unit	Material	Labor	Equip.	Total	Specification
Adobe						
Sand Base						
Demolish	S.F.		4.44		4.44	Includes material and labor to install
Install	S.F.	3.23	5.75		8.98	adobe pavers on sand base.
Demolish and Install	S.F.	3.23	10.19		13.42	
Reinstall	S.F.		4.60		4.60	
Clean	S.F.		1.60		1.60	
Minimum Charge	Job		259		259	
Mortar Base						
Demolish	S.F.		7.40	1.80	9.20	Includes material and labor to install
Install	S.F.	3.63	7.20		10.83	adobe pavers on mortar base.
Demolish and Install	S.F.	3.63	14.60	1.80	20.03	
Reinstall	S.F.		5.75		5.75	
Clean	S.F.		1.45		1.45	
Minimum Charge	Job		259		259	
Standard Brick						
Sand Base						
Demolish	S.F.		4.44		4.44	Includes material and labor to install
Install	S.F.	5.10	9.50		14.60	standard brick pavers on sand base.
Demolish and Install	S.F.	5.10	13.94		19.04	
Reinstall	S.F.		7.62		7.62	
Clean	S.F.		1.60		1.60	
Minimum Charge	Job		259		259	
Mortar Base						
Demolish	S.F.		7.40	1.80	9.20	Includes material and labor to install
Install	S.F.	5.45	3.45		8.90	standard brick pavers on mortar base.
Demolish and Install	S.F.	5.45	10.85	1.80	18.10	
Reinstall	S.F.		2.76		2.76	
Clean	S.F.		1.45		1.45	
Minimum Charge	Job		259		259	
Deluxe Brick						
Sand Base						
Demolish	S.F.		4.44		4.44	Includes material and labor to install
Install	S.F.	6.40	7.20		13.60	deluxe brick pavers on sand base.
Demolish and Install	S.F.	6.40	11.64		18.04	
Reinstall	S.F.		5.75		5.75	
Clean	S.F.		1.60		1.60	
Minimum Charge	Job		259		259	
Mortar Base						
Demolish	S.F.		7.40	1.80	9.20	Includes material and labor to install
Install	S.F.	6.15	12.95		19.10	deluxe brick pavers on mortar base.
Demolish and Install	S.F.	6.15	20.35	1.80	28.30	
Reinstall	S.F.		10.34		10.34	
Clean	S.F.		1.45		1.45	
Minimum Charge	Job		259		259	

Fencing

Fencing	Unit	Material	Labor	Equip.	Total	Specification
Block w / Footing						
4″						
Demolish	S.F.		.16	.44	.60	Cost includes material and labor to
Install	S.F.	2.51	5.65		8.16	install 4″ non-reinforced concrete
Demolish and Install	S.F.	2.51	5.81	.44	8.76	block.
Clean	S.F.	.24	1.33	.45	2.02	
Paint	S.F.	.54	.82		1.36	
Minimum Charge	Job		259		259	

140

For customer support on your Contractor's Pricing Guide: Residential Repair & Remodeling, call 800.448.8182.

Site Work

Fencing

	Unit	Material	Labor	Equip.	Total	Specification
4" Reinforced						
Demolish	S.F.		2.47	1.37	3.84	Cost includes material and labor to
Install	S.F.	2.77	5.70		8.47	install 4" reinforced concrete block.
Demolish and Install	S.F.	2.77	8.17	1.37	12.31	
Clean	S.F.	.24	1.33	.45	2.02	
Paint	S.F.	.54	.82		1.36	
Minimum Charge	Job		259		259	
6"						
Demolish	S.F.		.21	.59	.80	Cost includes material and labor to
Install	S.F.	2.74	6.05		8.79	install 6" non-reinforced concrete
Demolish and Install	S.F.	2.74	6.26	.59	9.59	block.
Clean	S.F.	.24	1.33	.45	2.02	
Paint	S.F.	.54	.82		1.36	
Minimum Charge	Job		259		259	
6" Reinforced						
Demolish	S.F.		2.60	.92	3.52	Cost includes material and labor to
Install	S.F.	2.95	6.15		9.10	install 6" reinforced concrete block.
Demolish and Install	S.F.	2.95	8.75	.92	12.62	
Clean	S.F.	.24	1.33	.45	2.02	
Paint	S.F.	.54	.82		1.36	
Minimum Charge	Job		259		259	
Cap for Block Wall						
Demolish	L.F.		2.78		2.78	Includes labor and equipment to install
Install	L.F.	8.05	12.70		20.75	precast concrete coping.
Demolish and Install	L.F.	8.05	15.48		23.53	
Clean	L.F.	.03	4.16		4.19	
Paint	S.F.	.54	.82		1.36	
Minimum Charge	Job		259		259	
Block Pilaster						
Demolish	Ea.		191		191	Includes material and labor to install
Install	Ea.	168	190		358	block pilaster for fence.
Demolish and Install	Ea.	168	381		549	
Reinstall	Ea.		190.40		190.40	
Paint	Ea.	13.25	19.20		32.45	
Minimum Charge	Job		259		259	
Block w / o Footing						
4"						
Demolish	S.F.		.16	.44	.60	Cost includes material and labor to
Install	S.F.	1.73	4.33		6.06	install 4" non-reinforced hollow
Demolish and Install	S.F.	1.73	4.49	.44	6.66	lightweight block.
Clean	S.F.	.24	1.33	.45	2.02	
Paint	S.F.	.54	.82		1.36	
Minimum Charge	Job		259		259	
4" Reinforced						
Demolish	S.F.		2.47	1.37	3.84	Cost includes material and labor to
Install	S.F.	1.91	5.55		7.46	install 4" reinforced hollow lightweight
Demolish and Install	S.F.	1.91	8.02	1.37	11.30	block.
Clean	S.F.	.24	1.33	.45	2.02	
Paint	S.F.	.54	.82		1.36	
Minimum Charge	Job		259		259	
6"						
Demolish	S.F.		.21	.59	.80	Cost includes material and labor to
Install	S.F.	3.04	4.76		7.80	install 6" non-reinforced hollow
Demolish and Install	S.F.	3.04	4.97	.59	8.60	lightweight block.
Clean	S.F.	.24	1.33	.45	2.02	
Paint	S.F.	.54	.82		1.36	
Minimum Charge	Job		259		259	

For customer support on your Contractor's Pricing Guide: Residential Repair & Remodeling, call 800.448.8182.

Site Work

Fencing

	Unit	Material	Labor	Equip.	Total	Specification
6″ Reinforced						
Demolish	S.F.		2.60	.92	3.52	Cost includes material and labor to
Install	S.F.	3.22	6		9.22	install 6″ reinforced hollow lightweight
Demolish and Install	S.F.	3.22	8.60	.92	12.74	block.
Clean	S.F.	.24	1.33	.45	2.02	
Paint	S.F.	.54	.82		1.36	
Minimum Charge	Job		259		259	
Cap for Block Wall						
Demolish	L.F.		2.78		2.78	Includes labor and equipment to install
Install	L.F.	8.05	12.70		20.75	precast concrete coping.
Demolish and Install	L.F.	8.05	15.48		23.53	
Clean	L.F.	.03	4.16		4.19	
Paint	S.F.	.54	.82		1.36	
Minimum Charge	Job		259		259	
Block Pilaster						
Demolish	Ea.		191		191	Includes material and labor to install
Install	Ea.	168	190		358	block pilaster for fence.
Demolish and Install	Ea.	168	381		549	
Reinstall	Ea.		190.40		190.40	
Paint	Ea.	13.25	19.20		32.45	
Minimum Charge	Job		259		259	

Chain Link

3′ High w / o Top Rail

	Unit	Material	Labor	Equip.	Total	Specification
Demolish	L.F.		3.47	.85	4.32	Includes material and labor to install
Install	L.F.	6.65	2.45	1	10.10	residential galvanized fence 3′ high
Demolish and Install	L.F.	6.65	5.92	1.85	14.42	with 1-3/8″ steel post 10′ O.C. set
Reinstall	L.F.		1.96	.80	2.76	without concrete.
Clean	L.F.	.10	1.45		1.55	
Minimum Charge	Job		610		610	

4′ High w / o Top Rail

	Unit	Material	Labor	Equip.	Total	Specification
Demolish	L.F.		3.47	.85	4.32	Includes material and labor to install
Install	L.F.	7.55	3.03	1.24	11.82	residential galvanized fence 4′ high
Demolish and Install	L.F.	7.55	6.50	2.09	16.14	with 1-3/8″ steel post 10′ O.C. set
Reinstall	L.F.		2.43	.99	3.42	without concrete.
Clean	L.F.	.13	1.60		1.73	
Minimum Charge	Job		610		610	

5′ High w / o Top Rail

	Unit	Material	Labor	Equip.	Total	Specification
Demolish	L.F.		3.47	.85	4.32	Includes material and labor to install
Install	L.F.	8.85	3.97	1.62	14.44	residential galvanized fence 5′ high
Demolish and Install	L.F.	8.85	7.44	2.47	18.76	with 1-3/8″ steel post 10′ O.C. set
Reinstall	L.F.		3.17	1.29	4.46	without concrete.
Clean	L.F.	.15	1.78		1.93	
Minimum Charge	Job		610		610	

3′ High w / Top Rail

	Unit	Material	Labor	Equip.	Total	Specification
Demolish	L.F.		3.69	.90	4.59	Includes material and labor to install
Install	L.F.	9.20	2.58	1.05	12.83	residential galvanized fence 3′ high
Demolish and Install	L.F.	9.20	6.27	1.95	17.42	with 1-3/8″ steel post 10′ O.C. set
Reinstall	L.F.		2.06	.84	2.90	without concrete.
Clean	L.F.	.10	1.48		1.58	
Minimum Charge	Job		610		610	

4′ High w / Top Rail

	Unit	Material	Labor	Equip.	Total	Specification
Demolish	L.F.		3.69	.90	4.59	Includes material and labor to install
Install	L.F.	10.05	3.22	1.31	14.58	residential galvanized fence 4′ high
Demolish and Install	L.F.	10.05	6.91	2.21	19.17	with 1-5/8″ steel post 10′ O.C. set
Reinstall	L.F.		2.58	1.05	3.63	without concrete and 1-5/8″ top rail
Clean	L.F.	.13	1.63		1.76	with sleeves.
Minimum Charge	Job		610		610	

For customer support on your Contractor's Pricing Guide: Residential Repair & Remodeling, call 800.448.8182.

Site Work

Fencing

	Unit	Material	Labor	Equip.	Total	Specification
5' High w / Top Rail						Includes material and labor to install
Demolish	L.F.		3.69	.90	4.59	residential galvanized fence 5' high
Install	L.F.	11.40	4.30	1.75	17.45	with 1-5/8" steel post 10' O.C. set
Demolish and Install	L.F.	11.40	7.99	2.65	22.04	without concrete and 1-5/8" top rail
Reinstall	L.F.		3.44	1.40	4.84	with sleeves.
Clean	L.F.	.15	1.82		1.97	
Minimum Charge	Job		610		610	
6' High w / Top Rail						Includes material and labor to install
Demolish	L.F.		3.69	.90	4.59	residential galvanized fence 6' high
Install	L.F.	12.75	6.45	2.63	21.83	with 2" steel post 10' O.C. set without
Demolish and Install	L.F.	12.75	10.14	3.53	26.42	concrete and 1-3/8" top rail with
Reinstall	L.F.		5.16	2.10	7.26	sleeves.
Clean	L.F.	.19	2.05		2.24	
Minimum Charge	Job		610		610	
Barbed Wire						Cost includes material and labor to
Demolish	L.F.		2.06		2.06	install 3 strand galvanized barbed
Install	L.F.	.66	.80	.39	1.85	wire fence.
Demolish and Install	L.F.	.66	2.86	.39	3.91	
Reinstall	L.F.		.64	.32	.96	
Minimum Charge	Job		610	390	1000	
6' Redwood						Cost includes material and labor to
Demolish	L.F.		13.60		13.60	install 6' high redwood fencing.
Install	L.F.	43.50	10.30	4.20	58	
Demolish and Install	L.F.	43.50	23.90	4.20	71.60	
Reinstall	L.F.		8.25	3.36	11.61	
Clean	L.F.	.19	1.86		2.05	
Paint	L.F.	1.75	4.54		6.29	
Minimum Charge	Job		610		610	
6' Red Cedar						Cost includes material and labor to
Demolish	L.F.		13.60		13.60	install 6' cedar fence.
Install	L.F.	28.50	10.30	4.20	43	
Demolish and Install	L.F.	28.50	23.90	4.20	56.60	
Reinstall	L.F.		8.25	3.36	11.61	
Clean	L.F.	.19	1.86		2.05	
Paint	L.F.	1.75	4.54		6.29	
Minimum Charge	Job		610		610	
6' Shadowbox						Cost includes material and labor to
Demolish	L.F.		13.60		13.60	install 6' pressure treated pine fence.
Install	L.F.	47.50	8.60	3.50	59.60	
Demolish and Install	L.F.	47.50	22.20	3.50	73.20	
Reinstall	L.F.		6.87	2.80	9.67	
Clean	L.F.	.19	1.86		2.05	
Paint	L.F.	1.75	4.54		6.29	
Minimum Charge	Job		610		610	
3' Picket						Includes material, labor and
Demolish	L.F.		13.60		13.60	equipment to install pressure treated
Install	L.F.	17.10	9.20	3.75	30.05	pine picket fence.
Demolish and Install	L.F.	17.10	22.80	3.75	43.65	
Reinstall	L.F.		7.36	3	10.36	
Clean	L.F.	.10	1.60		1.70	
Paint	L.F.	.87	2.47		3.34	
Minimum Charge	Job		610		610	

Site Work

Fencing

	Unit	Material	Labor	Equip.	Total	Specification
Wrought Iron						
Demolish	L.F.		29.50		29.50	Cost includes material and labor to
Install	L.F.	51.50	16.35		67.85	install 5' high wrought iron fence with
Demolish and Install	L.F.	51.50	45.85		97.35	pickets at 4-1/2" O.C.
Reinstall	L.F.		13.06		13.06	
Clean	L.F.		1.60		1.60	
Paint	L.F.	7.15	5.10		12.25	
Minimum Charge	Job		610		610	
Post						
Dirt Based Wood						
Demolish	Ea.		34		34	Includes material and labor to install
Install	Ea.	52.50	122		174.50	wood post set in earth.
Demolish and Install	Ea.	52.50	156		208.50	
Reinstall	Ea.		97.96		97.96	
Clean	Ea.		5		5	
Paint	Ea.	1.93	1.73		3.66	
Minimum Charge	Job		610		610	
Concrete Based Wood						
Demolish	Ea.		49	12	61	Includes material and labor to install
Install	Ea.	53.50	153		206.50	wood post set in concrete.
Demolish and Install	Ea.	53.50	202	12	267.50	
Clean	Ea.		5		5	
Paint	Ea.	1.93	1.73		3.66	
Minimum Charge	Job		610		610	
Block Pilaster						
Demolish	Ea.		191		191	Includes material and labor to install
Install	Ea.	168	190		358	block pilaster for fence.
Demolish and Install	Ea.	168	381		549	
Reinstall	Ea.		190.40		190.40	
Paint	Ea.	13.25	19.20		32.45	
Minimum Charge	Job		259		259	
Gate						
Chain Link						
Demolish	Ea.		45.50		45.50	Cost includes material and labor to
Install	Ea.	163	68		231	install 3' wide chain link pass gate
Demolish and Install	Ea.	163	113.50		276.50	and hardware.
Reinstall	Ea.		54.42		54.42	
Clean	Ea.		5.55		5.55	
Minimum Charge	Job		610		610	
Redwood						
Demolish	Ea.		39		39	Includes material, labor and
Install	Ea.	670	68		738	equipment to install redwood pass
Demolish and Install	Ea.	670	107		777	gate and hardware.
Reinstall	Ea.		54.42		54.42	
Clean	Ea.		5.55		5.55	
Paint	Ea.	2.88	8.65		11.53	
Minimum Charge	Job		610		610	
Red Cedar						
Demolish	Ea.		39		39	Cost includes material and labor to
Install	Ea.	340	68		408	install 3' wide red cedar pass gate
Demolish and Install	Ea.	340	107		447	and hardware.
Reinstall	Ea.		54.42		54.42	
Clean	Ea.		5.55		5.55	
Paint	Ea.	2.88	8.65		11.53	
Minimum Charge	Job		610		610	

For customer support on your Contractor's Pricing Guide: Residential Repair & Remodeling, call 800.448.8182.

Site Work

Fencing	Unit	Material	Labor	Equip.	Total	Specification
Wrought Iron						
Demolish	Ea.		29.50		29.50	Includes material, labor and
Install	Ea.	425	153		578	equipment to install wrought iron gate
Demolish and Install	Ea.	425	182.50		607.50	with hardware.
Reinstall	Ea.		122.45		122.45	
Clean	Ea.		5.55		5.55	
Paint	Ea.	14.25	8.65		22.90	
Minimum Charge	Job		610		610	

For customer support on your Contractor's Pricing Guide: Residential Repair & Remodeling, call 800.448.8182.

Rough Mechanical

Fire-extinguishing System

Space Heater

Hot-air Furnace

Plumbing

	Unit	Material	Labor	Equip.	Total	Specification
Components						
3/4" Supply						
Demolish	L.F.		3.26		3.26	Includes material and labor to install
Install	L.F.	5.50	7.75		13.25	type L copper tubing including
Demolish and Install	L.F.	5.50	11.01		16.51	couplings and hangers at 10' O.C.
Reinstall	L.F.		7.73		7.73	
Minimum Charge	Job		294		294	
1/2" Fixture Supply						
Install	Ea.	70	196		266	Includes labor and materials to install supply pipe only to a plumbing fixture.
2" Single Vent Stack						
Demolish	Ea.		54.50		54.50	Includes material and labor to install
Install	Ea.	54	294		348	PVC vent stack pipe with couplings at
Demolish and Install	Ea.	54	348.50		402.50	10' O.C.
Reinstall	Ea.		293.60		293.60	
Minimum Charge	Job		294		294	
4" Drain						
Demolish	L.F.		4.35		4.35	Includes material and labor to install
Install	L.F.	14.30	9.60		23.90	DWV type PVC pipe with couplings at
Demolish and Install	L.F.	14.30	13.95		28.25	10' O.C. and 3 strap hangers per 10'
Reinstall	L.F.		9.61		9.61	run.
Minimum Charge	Job		294		294	
1-1/2" Drain						
Demolish	L.F.		3.26		3.26	Includes material and labor to install
Install	L.F.	4.82	6.25		11.07	DWV type PVC pipe with couplings at
Demolish and Install	L.F.	4.82	9.51		14.33	10' O.C. and 3 strap hangers per 10'
Reinstall	L.F.		6.25		6.25	run.
Minimum Charge	Job		294		294	
Maintenance						
Flush Out / Unclog						
Clean	Job		294		294	Includes 4 hour minimum labor charge
Minimum Charge	Job		294		294	for a plumber.
Vent Cap						
Large						
Demolish	Ea.		14.85		14.85	Includes labor and material to install a
Install	Ea.	153	26.50		179.50	PVC plumbing pipe vent cap.
Demolish and Install	Ea.	153	41.35		194.35	
Clean	Ea.	.44	9.10		9.54	
Paint	Ea.	.55	9.80		10.35	
Minimum Charge	Job		147		147	

Rough Mechanical

Plumbing	Unit	Material	Labor	Equip.	Total	Specification
Medium						
Demolish	Ea.		14.85		14.85	Includes labor and material to install a
Install	Ea.	142	25.50		167.50	PVC plumbing pipe vent cap.
Demolish and Install	Ea.	142	40.35		182.35	
Clean	Ea.	.44	9.10		9.54	
Paint	Ea.	.55	9.80		10.35	
Minimum Charge	Job		147		147	
Small						
Demolish	Ea.		14.85		14.85	Includes material and labor to install a
Install	Ea.	11.20	24.50		35.70	PVC plumbing pipe vent cap.
Demolish and Install	Ea.	11.20	39.35		50.55	
Clean	Ea.	.44	9.10		9.54	
Paint	Ea.	.55	9.80		10.35	
Minimum Charge	Job		147		147	

Plumbing	Unit	Material	Labor	Equip.	Total	Specification
Sump Pump						
1/4 HP						
Demolish	Ea.		58.50		58.50	Includes material and labor to install a
Install	Ea.	209	92		301	submersible sump pump.
Demolish and Install	Ea.	209	150.50		359.50	
Reinstall	Ea.		73.40		73.40	
Clean	Ea.	1.11	25		26.11	
Minimum Charge	Job		294		294	
1/3 HP						
Demolish	Ea.		58.50		58.50	Includes material and labor to install a
Install	Ea.	212	98		310	submersible sump pump.
Demolish and Install	Ea.	212	156.50		368.50	
Reinstall	Ea.		78.29		78.29	
Clean	Ea.	1.11	25		26.11	
Minimum Charge	Job		294		294	
1/2 HP						
Demolish	Ea.		58.50		58.50	Includes material and labor to install a
Install	Ea.	315	109		424	submersible sump pump.
Demolish and Install	Ea.	315	167.50		482.50	
Reinstall	Ea.		86.99		86.99	
Clean	Ea.	1.11	25		26.11	
Minimum Charge	Job		294		294	
Clean and Service						
Clean	Ea.		36.50		36.50	Includes labor to clean and service a
Minimum Charge	Job		294		294	sump pump.

Water Heater	Unit	Material	Labor	Equip.	Total	Specification
Electric						
30 Gallon						
Demolish	Ea.		184		184	Includes material and labor to install a
Install	Ea.	925	267		1192	30 gallon electric water heater.
Demolish and Install	Ea.	925	451		1376	
Reinstall	Ea.		213.53		213.53	
Clean	Ea.	1.11	25		26.11	
Minimum Charge	Job		294		294	

For customer support on your Contractor's Pricing Guide: Residential Repair & Remodeling, call 800.448.8182.

Rough Mechanical

Water Heater

	Unit	Material	Labor	Equip.	Total	Specification
40 Gallon						
Demolish	Ea.		184		184	Includes material and labor to install a
Install	Ea.	785	294		1079	40 gallon electric water heater.
Demolish and Install	Ea.	785	478		1263	
Reinstall	Ea.		234.88		234.88	
Clean	Ea.	1.11	25		26.11	
Minimum Charge	Job		294		294	
52 Gallon						
Demolish	Ea.		184		184	Includes material and labor to install a
Install	Ea.	1025	294		1319	52 gallon electric water heater.
Demolish and Install	Ea.	1025	478		1503	
Reinstall	Ea.		234.88		234.88	
Clean	Ea.	1.11	25		26.11	
Minimum Charge	Job		294		294	
82 Gallon						
Demolish	Ea.		184		184	Includes material and labor to install
Install	Ea.	1600	365		1965	an 82 gallon electric water heater.
Demolish and Install	Ea.	1600	549		2149	
Reinstall	Ea.		293.60		293.60	
Clean	Ea.	1.11	25		26.11	
Minimum Charge	Job		294		294	

Gas-fired

	Unit	Material	Labor	Equip.	Total	Specification
30 Gallon Gas-fired						
Demolish	Ea.		226		226	Includes material and labor to install a
Install	Ea.	1150	294		1444	30 gallon gas water heater.
Demolish and Install	Ea.	1150	520		1670	
Reinstall	Ea.		234.88		234.88	
Clean	Ea.	1.11	25		26.11	
Minimum Charge	Job		294		294	
30 Gallon Gas-fired, Flue						
Install	Ea.	59.50	84		143.50	Includes labor and materials to install
Minimum Charge	Job		294		294	a flue for a water heater (up to 50 gallon size).
40 Gallon Gas-fired						
Demolish	Ea.		226		226	Includes material and labor to install a
Install	Ea.	885	310		1195	40 gallon gas water heater.
Demolish and Install	Ea.	885	536		1421	
Reinstall	Ea.		247.24		247.24	
Clean	Ea.	1.11	25		26.11	
Minimum Charge	Job		294		294	
40 Gallon Gas-fired, Flue						
Install	Ea.	59.50	84		143.50	Includes labor and materials to install
Minimum Charge	Job		294		294	a flue for a water heater (up to 50 gallon size).
50 Gallon Gas-fired						
Demolish	Ea.		226		226	Includes material and labor to install a
Install	Ea.	895	325		1220	50 gallon gas water heater.
Demolish and Install	Ea.	895	551		1446	
Reinstall	Ea.		260.98		260.98	
Clean	Ea.	1.11	25		26.11	
Minimum Charge	Job		294		294	
50 Gallon Gas-fired, Flue						
Install	Ea.	59.50	84		143.50	Includes labor and materials to install
Minimum Charge	Job		294		294	a flue for a water heater (up to 50 gallon size).

For customer support on your Contractor's Pricing Guide: Residential Repair & Remodeling, call 800.448.8182.

Rough Mechanical

Water Heater

Water Heater	Unit	Material	Labor	Equip.	Total	Specification
Solar Hot Water System						
Water Heater						
Demolish	Ea.		147		147	Includes material and labor to install a
Install	Ea.	1675	390		2065	solar hot water heat exchanger
Demolish and Install	Ea.	1675	537		2212	storage tank, 100 gallon.
Reinstall	Ea.		313.17		313.17	
Minimum Charge	Job		315		315	
Solar Panel						
Demolish	S.F.		2.15		2.15	Includes material and labor to install
Install	S.F.	38.50	3.86		42.36	solar panel.
Demolish and Install	S.F.	38.50	6.01		44.51	
Reinstall	S.F.		3.09		3.09	
Minimum Charge	Job		315		315	
Repair Plumbing						
Install	Job		315		315	Includes 4 hour minimum labor charge
Minimum Charge	Job		315		315	for a steamfitter or pipefitter.
Insulating Wrap						
Demolish	Ea.		6.80		6.80	Includes material and labor to install
Install	Ea.	21.50	18.35		39.85	insulating wrap around a 100 gallon
Demolish and Install	Ea.	21.50	25.15		46.65	solar hot water heat exchanger
Reinstall	Ea.		14.68		14.68	storage tank.
Minimum Charge	Job		294		294	
Water Softener						
Demolish	Ea.		109		109	Includes material and labor to install a
Install	Ea.	2975	264		3239	14 GPM water softener.
Demolish and Install	Ea.	2975	373		3348	
Reinstall	Ea.		211.36		211.36	
Clean	Ea.	1.11	25		26.11	
Minimum Charge	Ea.		294		294	

Ductwork

Ductwork	Unit	Material	Labor	Equip.	Total	Specification
Galvanized, Square or Rectangular						
6"						
Demolish	L.F.		2.63		2.63	Includes material and labor to install
Install	L.F.	15.55	12.60		28.15	rectangular ductwork with a greatest
Demolish and Install	L.F.	15.55	15.23		30.78	dimension of 6".
Reinstall	L.F.		10.10		10.10	
Clean	L.F.		3.03		3.03	
Minimum Charge	Job		284		284	
8"						
Demolish	L.F.		2.63		2.63	Includes material and labor to install
Install	L.F.	20.50	16.25		36.75	rectangular ductwork with a greatest
Demolish and Install	L.F.	20.50	18.88		39.38	dimension of 8".
Reinstall	L.F.		12.98		12.98	
Clean	L.F.		4.04		4.04	
Minimum Charge	Job		284		284	
10"						
Demolish	L.F.		2.63		2.63	Includes material and labor to install
Install	L.F.	26	21		47	rectangular ductwork with a greatest
Demolish and Install	L.F.	26	23.63		49.63	dimension of 10".
Reinstall	L.F.		16.83		16.83	
Clean	L.F.		5.05		5.05	
Minimum Charge	Job		284		284	

For customer support on your Contractor's Pricing Guide: Residential Repair & Remodeling, call 800.448.8182.

Rough Mechanical

Ductwork	Unit	Material	Labor	Equip.	Total	Specification
Galvanized, Spiral						
4"						
Demolish	L.F.		3.16		3.16	Cost includes material and labor to
Install	L.F.	7.10	2.84		9.94	install 4" diameter spiral pre-formed
Demolish and Install	L.F.	7.10	6		13.10	steel galvanized ductwork.
Reinstall	L.F.		2.27		2.27	
Clean	L.F.		1.59		1.59	
Minimum Charge	Job		284		284	
6"						
Demolish	L.F.		3.16		3.16	Cost includes material and labor to
Install	L.F.	10.05	3.65		13.70	install 6" diameter spiral pre-formed
Demolish and Install	L.F.	10.05	6.81		16.86	steel galvanized ductwork.
Reinstall	L.F.		2.92		2.92	
Clean	L.F.		2.38		2.38	
Minimum Charge	Job		284		284	
8"						
Demolish	L.F.		3.16		3.16	Cost includes material and labor to
Install	L.F.	11.45	5.10		16.55	install 8" diameter spiral pre-formed
Demolish and Install	L.F.	11.45	8.26		19.71	steel galvanized ductwork.
Reinstall	L.F.		4.09		4.09	
Clean	L.F.		3.17		3.17	
Minimum Charge	Job		284		284	
10"						
Demolish	L.F.		3.94		3.94	Cost includes material and labor to
Install	L.F.	14.15	6.40		20.55	install 10" diameter spiral pre-formed
Demolish and Install	L.F.	14.15	10.34		24.49	steel galvanized ductwork.
Reinstall	L.F.		5.11		5.11	
Clean	L.F.		3.97		3.97	
Minimum Charge	Job		284		284	
12"						
Demolish	L.F.		5.25		5.25	Cost includes material and labor to
Install	L.F.	15.70	8.50		24.20	install 12" diameter spiral pre-formed
Demolish and Install	L.F.	15.70	13.75		29.45	steel galvanized ductwork.
Reinstall	L.F.		6.82		6.82	
Clean	L.F.		4.76		4.76	
Minimum Charge	Job		284		284	
16"						
Demolish	L.F.		10.50		10.50	Cost includes material and labor to
Install	L.F.	26	17.05		43.05	install 16" diameter spiral pre-formed
Demolish and Install	L.F.	26	27.55		53.55	steel galvanized ductwork.
Reinstall	L.F.		13.64		13.64	
Clean	L.F.		6.35		6.35	
Minimum Charge	Job		284		284	
Flexible Fiberglass						
6" Non-insulated						
Demolish	L.F.		2.43		2.43	Cost includes material and labor to
Install	L.F.	3.64	3.65		7.29	install 6" diameter non-insulated
Demolish and Install	L.F.	3.64	6.08		9.72	flexible fiberglass fabric on corrosion
Reinstall	L.F.		2.92		2.92	resistant metal ductwork.
Clean	L.F.		2.38		2.38	
Minimum Charge	Job		284		284	
8" Non-insulated						
Demolish	L.F.		3.51		3.51	Cost includes material and labor to
Install	L.F.	4.80	5.10		9.90	install 8" diameter non-insulated
Demolish and Install	L.F.	4.80	8.61		13.41	flexible fiberglass fabric on corrosion
Reinstall	L.F.		4.09		4.09	resistant metal ductwork.
Clean	L.F.		3.17		3.17	
Minimum Charge	Job		284		284	

Rough Mechanical

Ductwork

Ductwork	Unit	Material	Labor	Equip.	Total	Specification
10″ Non-insulated						
Demolish	L.F.		4.51		4.51	Cost includes material and labor to
Install	L.F.	5.95	6.40		12.35	install 10″ diameter non-insulated
Demolish and Install	L.F.	5.95	10.91		16.86	flexible fiberglass fabric on corrosion
Reinstall	L.F.		5.11		5.11	resistant metal ductwork.
Clean	L.F.		3.97		3.97	
Minimum Charge	Job		284		284	
12″ Non-insulated						
Demolish	L.F.		6.30		6.30	Cost includes material and labor to
Install	L.F.	7.45	8.50		15.95	install 12″ diameter non-insulated
Demolish and Install	L.F.	7.45	14.80		22.25	flexible fiberglass fabric on corrosion
Reinstall	L.F.		6.82		6.82	resistant metal ductwork.
Clean	L.F.		4.76		4.76	
Minimum Charge	Job		284		284	
4″ Insulated						
Demolish	L.F.		1.86		1.86	Cost includes material and labor to
Install	L.F.	4.57	3.01		7.58	install 4″ diameter insulated flexible
Demolish and Install	L.F.	4.57	4.87		9.44	fiberglass fabric on corrosion resistant
Reinstall	L.F.		2.41		2.41	metal ductwork.
Clean	L.F.		1.59		1.59	
Minimum Charge	Job		284		284	
6″ Insulated						
Demolish	L.F.		2.43		2.43	Cost includes material and labor to
Install	L.F.	5.40	3.93		9.33	install 6″ diameter insulated flexible
Demolish and Install	L.F.	5.40	6.36		11.76	fiberglass fabric on corrosion resistant
Reinstall	L.F.		3.15		3.15	metal ductwork.
Clean	L.F.		2.38		2.38	
Minimum Charge	Job		284		284	
8″ Insulated						
Demolish	L.F.		3.51		3.51	Cost includes material and labor to
Install	L.F.	6.60	5.70		12.30	install 8″ diameter insulated flexible
Demolish and Install	L.F.	6.60	9.21		15.81	fiberglass fabric on corrosion resistant
Reinstall	L.F.		4.55		4.55	metal ductwork.
Clean	L.F.		3.17		3.17	
Minimum Charge	Job		284		284	
10″ Insulated						
Demolish	L.F.		4.51		4.51	Cost includes material and labor to
Install	L.F.	8.20	7.30		15.50	install 10″ diameter insulated flexible
Demolish and Install	L.F.	8.20	11.81		20.01	fiberglass fabric on corrosion resistant
Reinstall	L.F.		5.85		5.85	metal ductwork.
Clean	L.F.		3.97		3.97	
Minimum Charge	Job		284		284	
12″ Insulated						
Demolish	L.F.		6.30		6.30	Cost includes material and labor to
Install	L.F.	9.75	10.25		20	install 12″ diameter insulated flexible
Demolish and Install	L.F.	9.75	16.55		26.30	fiberglass fabric on corrosion resistant
Reinstall	L.F.		8.18		8.18	metal ductwork.
Clean	L.F.		4.76		4.76	
Minimum Charge	Job		284		284	

Rough Mechanical

Heating	Unit	Material	Labor	Equip.	Total	Specification
Hot Water Baseboard						
Baseboard						
Demolish	L.F.		10.70		10.70	Includes material and labor to install
Install	L.F.	14.20	18.80		33	baseboard radiation.
Demolish and Install	L.F.	14.20	29.50		43.70	
Reinstall	L.F.		15.03		15.03	
Clean	L.F.	1.11	6.25		7.36	
Minimum Charge	Job		294		294	
3'						
Demolish	Ea.		32.50		32.50	Includes material and labor to install
Install	Ea.	28.50	62.50		91	fin tube baseboard radiation in 3'
Demolish and Install	Ea.	28.50	95		123.50	sections.
Reinstall	Ea.		50.08		50.08	
Clean	Ea.	.32	19.05		19.37	
Minimum Charge	Job		294		294	
4'						
Demolish	Ea.		43.50		43.50	Includes material and labor to install
Install	Ea.	38	70.50		108.50	fin tube baseboard radiation in 4'
Demolish and Install	Ea.	38	114		152	sections.
Reinstall	Ea.		56.34		56.34	
Clean	Ea.	.64	25		25.64	
Minimum Charge	Job		294		294	
5'						
Demolish	Ea.		54.50		54.50	Includes material and labor to install
Install	Ea.	47.50	80.50		128	fin tube baseboard radiation in 5'
Demolish and Install	Ea.	47.50	135		182.50	sections.
Reinstall	Ea.		64.39		64.39	
Clean	Ea.	.64	30.50		31.14	
Minimum Charge	Job		294		294	
6'						
Demolish	Ea.		65		65	Includes material and labor to install
Install	Ea.	56.50	94		150.50	fin tube baseboard radiation in 6'
Demolish and Install	Ea.	56.50	159		215.50	sections.
Reinstall	Ea.		75.13		75.13	
Clean	Ea.	.64	36.50		37.14	
Minimum Charge	Job		294		294	
7'						
Demolish	Ea.		72.50		72.50	Includes material and labor to install
Install	Ea.	66	117		183	fin tube baseboard radiation in 7'
Demolish and Install	Ea.	66	189.50		255.50	sections.
Reinstall	Ea.		93.95		93.95	
Clean	Ea.	.79	44.50		45.29	
Minimum Charge	Job		294		294	
9'						
Demolish	Ea.		109		109	Includes material and labor to install
Install	Ea.	85.50	168		253.50	fin tube baseboard radiation in 9'
Demolish and Install	Ea.	85.50	277		362.50	sections.
Reinstall	Ea.		134.22		134.22	
Clean	Ea.	1.27	57		58.27	
Minimum Charge	Job		294		294	

For customer support on your Contractor's Pricing Guide: Residential Repair & Remodeling, call 800.448.8182.

Rough Mechanical

Heating	Unit	Material	Labor	Equip.	Total	Specification
Cabinet Unit Heater						
50 MBH						
Demolish	Ea.		56.50		56.50	Includes material and labor to install a
Install	Ea.	1950	161		2111	cabinet space heater with an output of
Demolish and Install	Ea.	1950	217.50		2167.50	50 MBH.
Reinstall	Ea.		128.79		128.79	
Clean	Ea.	.79	50		50.79	
Minimum Charge	Job		315		315	
75 MBH						
Demolish	Ea.		56.50		56.50	Includes material and labor to install a
Install	Ea.	1100	188		1288	cabinet space heater with an output of
Demolish and Install	Ea.	1100	244.50		1344.50	75 MBH.
Reinstall	Ea.		150.25		150.25	
Clean	Ea.	.79	50		50.79	
Minimum Charge	Job		315		315	
125 MBH						
Demolish	Ea.		67.50		67.50	Includes material and labor to install a
Install	Ea.	1375	225		1600	cabinet space heater with an output of
Demolish and Install	Ea.	1375	292.50		1667.50	125 MBH.
Reinstall	Ea.		180.30		180.30	
Clean	Ea.	.79	50		50.79	
Minimum Charge	Job		315		315	
175 MBH						
Demolish	Ea.		84.50		84.50	Includes material and labor to install a
Install	Ea.	4000	320		4320	cabinet space heater with an output of
Demolish and Install	Ea.	4000	404.50		4404.50	175 MBH.
Reinstall	Ea.		257.57		257.57	
Clean	Ea.	.79	50		50.79	
Minimum Charge	Job		315		315	

Furnace Unit	Unit	Material	Labor	Equip.	Total	Specification
Wall (gas-fired)						
25 MBH w / Thermostat						
Demolish	Ea.		128		128	Includes material and labor to install a
Install	Ea.	1075	205		1280	gas fired hot air furnace rated at 25
Demolish and Install	Ea.	1075	333		1408	MBH.
Reinstall	Ea.		163.66		163.66	
Clean	Ea.	1.58	111		112.58	
Minimum Charge	Job		315		315	
35 MBH w / Thermostat						
Demolish	Ea.		128		128	Includes material and labor to install a
Install	Ea.	1125	227		1352	gas fired hot air furnace rated at 35
Demolish and Install	Ea.	1125	355		1480	MBH.
Reinstall	Ea.		181.85		181.85	
Clean	Ea.	1.58	111		112.58	
Minimum Charge	Job		315		315	
50 MBH w / Thermostat						
Demolish	Ea.		128		128	Includes material and labor to install a
Install	Ea.	3000	256		3256	gas fired hot air furnace rated at 50
Demolish and Install	Ea.	3000	384		3384	MBH.
Reinstall	Ea.		204.58		204.58	
Clean	Ea.	1.58	111		112.58	
Minimum Charge	Job		315		315	

Rough Mechanical

Furnace Unit	Unit	Material	Labor	Equip.	Total	Specification
60 MBH w / Thermostat						
Demolish	Ea.		128		128	Includes material and labor to install a
Install	Ea.	1950	284		2234	gas fired hot air furnace rated at 60
Demolish and Install	Ea.	1950	412		2362	MBH.
Reinstall	Ea.		227.31		227.31	
Clean	Ea.	1.58	111		112.58	
Minimum Charge	Job		315		315	
Floor (gas-fired)						
30 MBH w / Thermostat						
Demolish	Ea.		128		128	Includes material and labor to install a
Install	Ea.	2525	205		2730	gas fired hot air floor furnace rated at
Demolish and Install	Ea.	2525	333		2858	30 MBH.
Reinstall	Ea.		163.66		163.66	
Clean	Ea.	1.58	111		112.58	
Minimum Charge	Job		315		315	
50 MBH w / Thermostat						
Demolish	Ea.		128		128	Includes material and labor to install a
Install	Ea.	2925	256		3181	gas fired hot air floor furnace rated at
Demolish and Install	Ea.	2925	384		3309	50 MBH.
Reinstall	Ea.		204.58		204.58	
Clean	Ea.	1.58	111		112.58	
Minimum Charge	Job		315		315	
Gas Forced Air						
50 MBH						
Demolish	Ea.		128		128	Includes material and labor to install
Install	Ea.	1325	256		1581	an up flow gas fired hot air furnace
Demolish and Install	Ea.	1325	384		1709	rated at 50 MBH.
Reinstall	Ea.		204.58		204.58	
Clean	Ea.	1.58	111		112.58	
Minimum Charge	Job		315		315	
80 MBH						
Demolish	Ea.		128		128	Includes material and labor to install
Install	Ea.	1600	284		1884	an up flow gas fired hot air furnace
Demolish and Install	Ea.	1600	412		2012	rated at 80 MBH.
Reinstall	Ea.		227.31		227.31	
Clean	Ea.	1.58	111		112.58	
Minimum Charge	Job		315		315	
100 MBH						
Demolish	Ea.		128		128	Includes material and labor to install
Install	Ea.	1600	320		1920	an up flow gas fired hot air furnace
Demolish and Install	Ea.	1600	448		2048	rated at 100 MBH.
Reinstall	Ea.		255.72		255.72	
Clean	Ea.	1.58	111		112.58	
Minimum Charge	Job		315		315	
120 MBH						
Demolish	Ea.		128		128	Includes material and labor to install
Install	Ea.	1625	340		1965	an up flow gas fired hot air furnace
Demolish and Install	Ea.	1625	468		2093	rated at 120 MBH.
Reinstall	Ea.		272.77		272.77	
Clean	Ea.	1.58	111		112.58	
Minimum Charge	Job		315		315	

Rough Mechanical

Furnace Unit		Unit	Material	Labor	Equip.	Total	Specification
Electric Forced Air							
31.4 MBH							
	Demolish	Ea.		128		128	Includes material and labor to install
	Install	Ea.	2725	296		3021	an electric hot air furnace rated at
	Demolish and Install	Ea.	2725	424		3149	31.4 MBH.
	Reinstall	Ea.		237.01		237.01	
	Clean	Ea.	1.58	111		112.58	
	Minimum Charge	Job		315		315	
47.1 MBH							
	Demolish	Ea.		128		128	Includes material and labor to install
	Install	Ea.	3100	310		3410	an electric hot air furnace rated at
	Demolish and Install	Ea.	3100	438		3538	47.1 MBH.
	Reinstall	Ea.		248.04		248.04	
	Clean	Ea.	1.58	111		112.58	
	Minimum Charge	Job		315		315	
62.9 MBH							
	Demolish	Ea.		128		128	Includes material and labor to install
	Install	Ea.	3300	335		3635	an electric hot air furnace rated at
	Demolish and Install	Ea.	3300	463		3763	62.9 MBH.
	Reinstall	Ea.		266.64		266.64	
	Clean	Ea.	1.58	111		112.58	
	Minimum Charge	Job		315		315	
78.5 MBH							
	Demolish	Ea.		128		128	Includes material and labor to install
	Install	Ea.	4175	340		4515	an electric hot air furnace rated at
	Demolish and Install	Ea.	4175	468		4643	78.5 MBH.
	Reinstall	Ea.		273.48		273.48	
	Clean	Ea.	1.58	111		112.58	
	Minimum Charge	Job		315		315	
110 MBH							
	Demolish	Ea.		128		128	Includes material and labor to install
	Install	Ea.	5925	360		6285	an electric hot air furnace rated at
	Demolish and Install	Ea.	5925	488		6413	110 MBH.
	Reinstall	Ea.		288.26		288.26	
	Clean	Ea.	1.58	111		112.58	
	Minimum Charge	Job		315		315	
A/C Junction Box							
	Install	Ea.	85.50	155		240.50	Includes labor and materials to install
	Minimum Charge	Job		315		315	a heating system emergency shut-off switch. Cost includes up to 40 feet of non-metallic cable.

Boiler		Unit	Material	Labor	Equip.	Total	Specification
Gas-Fired							
60 MBH							
	Demolish	Ea.		1150		1150	Includes material and labor to install a
	Install	Ea.	4025	2050		6075	gas fired cast iron boiler rated at 60
	Demolish and Install	Ea.	4025	3200		7225	MBH.
	Reinstall	Ea.		1639.10		1639.10	
	Clean	Ea.	3.17	66.50		69.67	
	Minimum Charge	Job		315		315	

156

For customer support on your Contractor's Pricing Guide: Residential Repair & Remodeling, call 800.448.8182.

Rough Mechanical

Boiler

Boiler	Unit	Material	Labor	Equip.	Total	Specification
Gas-Fired						
125 MBH						
Demolish	Ea.		3125		3125	Includes material and labor to install a
Install	Ea.	4400	2500		6900	gas fired cast iron boiler rated at 125
Demolish and Install	Ea.	4400	5625		10025	MBH.
Reinstall	Ea.		2003.34		2003.34	
Clean	Ea.	3.17	66.50		69.67	
Minimum Charge	Job		315		315	
Gas-Fired						
400 MBH						
Demolish	Ea.		3125		3125	Includes material and labor to install a
Install	Ea.	5875	3525		9400	gas fired cast iron boiler rated at 400
Demolish and Install	Ea.	5875	6650		12525	MBH.
Reinstall	Ea.		3521.50		3521.50	
Clean	Ea.	3.17	66.50		69.67	
Minimum Charge	Job		315		315	

Heat Pump

Heat Pump	Unit	Material	Labor	Equip.	Total	Specification
2 Ton						
Demolish	Ea.		315		315	Includes material and labor to install a
Install	Ea.	2000	565		2565	2 ton heat pump.
Demolish and Install	Ea.	2000	880		2880	
Reinstall	Ea.		450.75		450.75	
Clean	Ea.	3.17	50		53.17	
Minimum Charge	Job		315		315	
3 Ton						
Demolish	Ea.		520		520	Includes material and labor to install a
Install	Ea.	2450	940		3390	3 ton heat pump.
Demolish and Install	Ea.	2450	1460		3910	
Reinstall	Ea.		751.25		751.25	
Clean	Ea.	3.17	50		53.17	
Minimum Charge	Job		315		315	
4 Ton						
Demolish	Ea.		785		785	Includes material and labor to install a
Install	Ea.	2825	1400		4225	4 ton heat pump.
Demolish and Install	Ea.	2825	2185		5010	
Reinstall	Ea.		1126.88		1126.88	
Clean	Ea.	3.17	50		53.17	
Minimum Charge	Job		315		315	
5 Ton						
Demolish	Ea.		1250		1250	Includes material and labor to install a
Install	Ea.	3100	2250		5350	5 ton heat pump.
Demolish and Install	Ea.	3100	3500		6600	
Reinstall	Ea.		1803.01		1803.01	
Clean	Ea.	3.17	50		53.17	
Minimum Charge	Job		315		315	
Thermostat						
Demolish	Ea.		17.40		17.40	Includes material and labor to install
Install	Ea.	330	78.50		408.50	an electric thermostat, 2 set back.
Demolish and Install	Ea.	330	95.90		425.90	
Reinstall	Ea.		62.60		62.60	
Clean	Ea.	.07	12.50		12.57	
Minimum Charge	Job		315		315	
A/C Junction Box						
Install	Ea.	680	480		1160	Includes labor and materials to install
Minimum Charge	Job		310		310	a heat pump circuit including up to 40 feet of non-metallic cable.

For customer support on your Contractor's Pricing Guide: Residential Repair & Remodeling, call 800.448.8182.

Rough Mechanical

Heating	Unit	Material	Labor	Equip.	Total	Specification
Electric Baseboard						
Demolish	L.F.		8.60		8.60	Includes material and labor to install
Install	L.F.	15.90	20.50		36.40	electric baseboard heating elements.
Demolish and Install	L.F.	15.90	29.10		45	
Reinstall	L.F.		16.55		16.55	
Clean	L.F.	1.11	6.25		7.36	
Minimum Charge	Job		310		310	
2' 6"						
Demolish	Ea.		28.50		28.50	Includes material and labor to install a
Install	Ea.	47	77.50		124.50	2' 6" long section of electric
Demolish and Install	Ea.	47	106		153	baseboard heating elements.
Reinstall	Ea.		62.08		62.08	
Clean	Ea.	.32	15.35		15.67	
Minimum Charge	Job		310		310	
3'						
Demolish	Ea.		31.50		31.50	Includes material and labor to install a
Install	Ea.	55	77.50		132.50	3' long section of electric baseboard
Demolish and Install	Ea.	55	109		164	heating elements.
Reinstall	Ea.		62.08		62.08	
Clean	Ea.	.32	19.05		19.37	
Minimum Charge	Job		310		310	
4'						
Demolish	Ea.		34.50		34.50	Includes material and labor to install a
Install	Ea.	66	92.50		158.50	4' long section of electric baseboard
Demolish and Install	Ea.	66	127		193	heating elements.
Reinstall	Ea.		74.13		74.13	
Clean	Ea.	.13	4.16		4.29	
Minimum Charge	Job		310		310	
5'						
Demolish	Ea.		38.50		38.50	Includes material and labor to install a
Install	Ea.	80	109		189	5' long section of electric baseboard
Demolish and Install	Ea.	80	147.50		227.50	heating elements.
Reinstall	Ea.		87.13		87.13	
Clean	Ea.	.64	30.50		31.14	
Minimum Charge	Job		310		310	
6'						
Demolish	Ea.		43		43	Includes material and labor to install a
Install	Ea.	89	124		213	6' long section of electric baseboard
Demolish and Install	Ea.	89	167		256	heating elements.
Reinstall	Ea.		99.33		99.33	
Clean	Ea.	.13	5		5.13	
Minimum Charge	Job		310		310	
8'						
Demolish	Ea.		57.50		57.50	Includes material and labor to install
Install	Ea.	105	155		260	an 8' long section of electric
Demolish and Install	Ea.	105	212.50		317.50	baseboard heating elements.
Reinstall	Ea.		124.16		124.16	
Clean	Ea.	.13	6.25		6.38	
Minimum Charge	Job		310		310	
10'						
Demolish	Ea.		69		69	Includes material and labor to install a
Install	Ea.	117	188		305	10' long section of electric baseboard
Demolish and Install	Ea.	117	257		374	heating elements.
Reinstall	Ea.		150.50		150.50	
Clean	Ea.	1.27	62.50		63.77	
Minimum Charge	Job		310		310	

For customer support on your Contractor's Pricing Guide: Residential Repair & Remodeling, call 800.448.8182.

Rough Mechanical

Heating

Heating	Unit	Material	Labor	Equip.	Total	Specification
Electric Radiant Wall Panel						
Demolish	Ea.		46		46	Includes material and labor to install
Install	Ea.	420	124		544	an 18" x 23" electric radiant heat wall
Demolish and Install	Ea.	420	170		590	panel.
Reinstall	Ea.		99.33		99.33	
Clean	Ea.	.15	12.50		12.65	
Minimum Charge	Job		310		310	

Heating Control

Heating Control	Unit	Material	Labor	Equip.	Total	Specification
Thermostat						
Demolish	Ea.		17.40		17.40	Includes material and labor to install
Install	Ea.	330	78.50		408.50	an electric thermostat, 2 set back.
Demolish and Install	Ea.	330	95.90		425.90	
Reinstall	Ea.		62.60		62.60	
Clean	Ea.	.07	12.50		12.57	
Minimum Charge	Job		315		315	

Air Conditioning

Air Conditioning	Unit	Material	Labor	Equip.	Total	Specification
Self-contained						
2 Ton						
Demolish	Ea.		545		545	Includes material and labor to install a
Install	Ea.	5400	970		6370	rooftop air conditioning unit, 2 ton.
Demolish and Install	Ea.	5400	1515		6915	No hoisting is included.
Reinstall	Ea.		972.29		972.29	
Clean	Ea.	3.17	50		53.17	
Minimum Charge	Job		315		315	
3 Ton						
Demolish	Ea.		545		545	Includes material and labor to install a
Install	Ea.	4725	1275		6000	rooftop air conditioning unit, 3 ton.
Demolish and Install	Ea.	4725	1820		6545	No hoisting is included.
Reinstall	Ea.		1282		1282	
Clean	Ea.	3.17	50		53.17	
Minimum Charge	Job		315		315	
4 Ton						
Demolish	Ea.		545		545	Includes material and labor to install a
Install	Ea.	5750	1475		7225	rooftop air conditioning unit, 4 ton.
Demolish and Install	Ea.	5750	2020		7770	No hoisting is included.
Reinstall	Ea.		1486.65		1486.65	
Clean	Ea.	3.17	50		53.17	
Minimum Charge	Job		315		315	
5 Ton						
Demolish	Ea.		545		545	Includes material and labor to install a
Install	Ea.	5800	1600		7400	rooftop air conditioning unit, 5 ton.
Demolish and Install	Ea.	5800	2145		7945	No hoisting is included.
Reinstall	Ea.		1605.24		1605.24	
Clean	Ea.	3.17	50		53.17	
Minimum Charge	Job		315		315	
Compressor						
2 Ton						
Demolish	Ea.		520		520	Includes material and labor to install a
Install	Ea.	415	315		730	refrigeration compressor.
Demolish and Install	Ea.	415	835		1250	
Reinstall	Ea.		250.40		250.40	
Clean	Ea.	3.17	50		53.17	
Minimum Charge	Job		315		315	

For customer support on your Contractor's Pricing Guide: Residential Repair & Remodeling, call 800.448.8182.

Rough Mechanical

Air Conditioning

	Unit	Material	Labor	Equip.	Total	Specification
2.5 Ton						
Demolish	Ea.		520		520	Includes material and labor to install a refrigeration compressor.
Install	Ea.	490	340		830	
Demolish and Install	Ea.	490	860		1350	
Reinstall	Ea.		271.29		271.29	
Clean	Ea.	3.17	50		53.17	
Minimum Charge	Job		315		315	
3 Ton						
Demolish	Ea.		520		520	Includes material and labor to install a refrigeration compressor.
Install	Ea.	535	355		890	
Demolish and Install	Ea.	535	875		1410	
Reinstall	Ea.		283.10		283.10	
Clean	Ea.	3.17	50		53.17	
Minimum Charge	Job		315		315	
4 Ton						
Demolish	Ea.		520		520	Includes material and labor to install a refrigeration compressor.
Install	Ea.	680	285		965	
Demolish and Install	Ea.	680	805		1485	
Reinstall	Ea.		227.64		227.64	
Clean	Ea.	3.17	50		53.17	
Minimum Charge	Job		315		315	
5 Ton						
Demolish	Ea.		520		520	Includes material and labor to install a refrigeration compressor.
Install	Ea.	755	390		1145	
Demolish and Install	Ea.	755	910		1665	
Reinstall	Ea.		310.09		310.09	
Clean	Ea.	3.17	50		53.17	
Minimum Charge	Job		315		315	

Condensing Unit

	Unit	Material	Labor	Equip.	Total	Specification
2 Ton Air-cooled						
Demolish	Ea.		149		149	Includes material and labor to install a 2 ton condensing unit.
Install	Ea.	1625	535		2160	
Demolish and Install	Ea.	1625	684		2309	
Reinstall	Ea.		429.29		429.29	
Clean	Ea.	3.17	50		53.17	
Minimum Charge	Job		315		315	
2.5 Ton Air-cooled						
Demolish	Ea.		174		174	Includes material and labor to install a 2.5 ton condensing unit.
Install	Ea.	1725	665		2390	
Demolish and Install	Ea.	1725	839		2564	
Reinstall	Ea.		530.30		530.30	
Clean	Ea.	3.17	50		53.17	
Minimum Charge	Job		315		315	
3 Ton Air-cooled						
Demolish	Ea.		209		209	Includes material and labor to install a 3 ton condensing unit.
Install	Ea.	1925	865		2790	
Demolish and Install	Ea.	1925	1074		2999	
Reinstall	Ea.		693.46		693.46	
Clean	Ea.	3.17	50		53.17	
Minimum Charge	Job		315		315	
4 Ton Air-cooled						
Demolish	Ea.		315		315	Includes material and labor to install a 4 ton condensing unit.
Install	Ea.	2400	1250		3650	
Demolish and Install	Ea.	2400	1565		3965	
Reinstall	Ea.		1001.67		1001.67	
Clean	Ea.	3.17	50		53.17	
Minimum Charge	Job		315		315	

For customer support on your Contractor's Pricing Guide: Residential Repair & Remodeling, call 800.448.8182.

Rough Mechanical

Air Conditioning	Unit	Material	Labor	Equip.	Total	Specification
5 Ton Air-cooled						
Demolish	Ea.		470		470	Includes material and labor to install a
Install	Ea.	2775	1875		4650	5 ton condensing unit.
Demolish and Install	Ea.	2775	2345		5120	
Reinstall	Ea.		1502.51		1502.51	
Clean	Ea.	3.17	50		53.17	
Minimum Charge	Job		315		315	
Air Handler						
1,200 CFM						
Demolish	Ea.		375		375	Includes material and labor to install a
Install	Ea.	6000	865		6865	1200 CFM capacity air handling unit.
Demolish and Install	Ea.	6000	1240		7240	
Reinstall	Ea.		693.46		693.46	
Clean	Ea.	3.17	50		53.17	
Minimum Charge	Job		315		315	
2,000 CFM						
Demolish	Ea.		375		375	Includes material and labor to install a
Install	Ea.	12700	1025		13725	2000 CFM capacity air handling unit.
Demolish and Install	Ea.	12700	1400		14100	
Reinstall	Ea.		819.55		819.55	
Clean	Ea.	3.17	50		53.17	
Minimum Charge	Job		315		315	
3,000 CFM						
Demolish	Ea.		415		415	Includes material and labor to install a
Install	Ea.	13100	1125		14225	3000 CFM capacity air handling unit.
Demolish and Install	Ea.	13100	1540		14640	
Reinstall	Ea.		901.50		901.50	
Clean	Ea.	3.17	50		53.17	
Minimum Charge	Job		315		315	
Evaporative Cooler						
4,300 CFM, 1/3 HP						
Demolish	Ea.		145		145	Includes material and labor to install
Install	Ea.	1325	320		1645	an evaporative cooler.
Demolish and Install	Ea.	1325	465		1790	
Reinstall	Ea.		255.72		255.72	
Clean	Ea.	1.27	50		51.27	
Minimum Charge	Job		315		315	
4,700 CFM, 1/2 HP						
Demolish	Ea.		145		145	Includes material and labor to install
Install	Ea.	1275	330		1605	an evaporative cooler.
Demolish and Install	Ea.	1275	475		1750	
Reinstall	Ea.		263.97		263.97	
Clean	Ea.	1.27	50		51.27	
Minimum Charge	Job		315		315	
5,600 CFM, 3/4 HP						
Demolish	Ea.		145		145	Includes material and labor to install
Install	Ea.	2825	340		3165	an evaporative cooler.
Demolish and Install	Ea.	2825	485		3310	
Reinstall	Ea.		272.77		272.77	
Clean	Ea.	1.27	50		51.27	
Minimum Charge	Job		315		315	
Master Cool Brand						
Demolish	Ea.		145		145	Includes material and labor to install
Install	Ea.	1300	355		1655	an evaporative cooler.
Demolish and Install	Ea.	1300	500		1800	
Reinstall	Ea.		282.17		282.17	
Clean	Ea.	1.27	50		51.27	
Minimum Charge	Job		315		315	

Rough Mechanical

Air Conditioning	Unit	Material	Labor	Equip.	Total	Specification
Motor						
Demolish	Ea.		81.50		81.50	Includes material and labor to install
Install	Ea.	310	248		558	an evaporative cooler motor.
Demolish and Install	Ea.	310	329.50		639.50	
Reinstall	Ea.		198.78		198.78	
Clean	Ea.	1.27	50		51.27	
Minimum Charge	Job		315		315	
Water Pump						
Demolish	Ea.		81.50		81.50	Includes material and labor to install
Install	Ea.	80.50	142		222.50	an evaporative cooler water pump.
Demolish and Install	Ea.	80.50	223.50		304	
Reinstall	Ea.		113.60		113.60	
Clean	Ea.	1.27	50		51.27	
Minimum Charge	Job		315		315	
Fan Belt						
Demolish	Ea.		16.30		16.30	Includes material and labor to install
Install	Ea.	58	28.50		86.50	an evaporative cooler fan belt.
Demolish and Install	Ea.	58	44.80		102.80	
Reinstall	Ea.		22.72		22.72	
Minimum Charge	Job		315		315	
Pads (4 ea.)						
Demolish	Ea.		16.30		16.30	Includes material and labor to install
Install	Ea.	38.50	71		109.50	pre-fabricated roof pads for an
Demolish and Install	Ea.	38.50	87.30		125.80	evaporative cooler.
Reinstall	Ea.		56.80		56.80	
Clean	Ea.	1.27	50		51.27	
Minimum Charge	Job		315		315	
Service call						
Install	Job		284		284	Includes 4 hour minimum labor charge
						for a sheetmetal worker.
Window Mounted						
Low Output BTU						
Install	Ea.	385	65		450	Includes material and labor to install a
Reinstall	Ea.		52.08		52.08	window air conditioning unit.
Clean	Ea.	.47	12.50		12.97	
Minimum Charge	Job		315		315	
Mid-range Output BTU						
Demolish	Ea.		45.50		45.50	Includes material and labor to install a
Install	Ea.	460	87		547	window air conditioning unit.
Demolish and Install	Ea.	460	132.50		592.50	
Reinstall	Ea.		69.44		69.44	
Clean	Ea.	.47	12.50		12.97	
Minimum Charge	Job		315		315	
High Output BTU						
Demolish	Ea.		58		58	Includes material and labor to install a
Install	Ea.	665	87		752	window air conditioning unit.
Demolish and Install	Ea.	665	145		810	
Reinstall	Ea.		69.44		69.44	
Clean	Ea.	.47	12.50		12.97	
Minimum Charge	Job		315		315	
Hotel-type Hot / Cold						
Demolish	Ea.		89.50		89.50	Includes material and labor to install
Install	Ea.	4375	375		4750	packaged air conditioning unit,
Demolish and Install	Ea.	4375	464.50		4839.50	15000 BTU cooling, 13900 BTU
Reinstall	Ea.		300.50		300.50	heating.
Clean	Ea.	.64	25		25.64	
Minimum Charge	Job		315		315	

For customer support on your Contractor's Pricing Guide: Residential Repair & Remodeling, call 800.448.8182.

Rough Mechanical

Air Conditioning

Air Conditioning	Unit	Material	Labor	Equip.	Total	Specification
Thermostat						
Demolish	Ea.		17.40		17.40	Includes material and labor to install
Install	Ea.	330	78.50		408.50	an electric thermostat, 2 set back.
Demolish and Install	Ea.	330	95.90		425.90	
Reinstall	Ea.		62.60		62.60	
Clean	Ea.	.07	12.50		12.57	
Minimum Charge	Job		315		315	
Concrete Pad						
Demolish	S.F.		2.78		2.78	Includes material and labor to install
Install	S.F.	4.59	6.40		10.99	air conditioner pad.
Demolish and Install	S.F.	4.59	9.18		13.77	
Reinstall	S.F.		5.12		5.12	
Clean	S.F.	.07	.80		.87	
Paint	S.F.	.44	.44		.88	
Minimum Charge	Job		315		315	
A/C Junction Box						
Install	Ea.	240	177		417	Includes labor and materials to install
Minimum Charge	Job		310		310	an air conditioning circuit including up to 40 feet of non-metallic cable.

Sprinkler System

Sprinkler System	Unit	Material	Labor	Equip.	Total	Specification
Exposed						
Wet Type						
Demolish	SF Flr.		.26		.26	Includes material and labor to install
Install	SF Flr.	4.48	2.63		7.11	an exposed wet pipe fire sprinkler
Demolish and Install	SF Flr.	4.48	2.89		7.37	system.
Reinstall	SF Flr.		2.63		2.63	
Minimum Charge	Job		294		294	
Dry Type						
Demolish	SF Flr.		.26		.26	Includes material and labor to install
Install	SF Flr.	5.15	2.63		7.78	an exposed dry pipe fire sprinkler
Demolish and Install	SF Flr.	5.15	2.89		8.04	system.
Reinstall	SF Flr.		2.63		2.63	
Minimum Charge	Job		294		294	
Concealed						
Wet Type						
Demolish	SF Flr.		.52		.52	Includes material and labor to install a
Install	SF Flr.	4.48	3.59		8.07	concealed wet pipe fire sprinkler
Demolish and Install	SF Flr.	4.48	4.11		8.59	system.
Reinstall	SF Flr.		3.59		3.59	
Minimum Charge	Job		294		294	
Dry Type						
Demolish	SF Flr.		.52		.52	Includes material and labor to install a
Install	SF Flr.	5.15	3.59		8.74	concealed dry pipe fire sprinkler
Demolish and Install	SF Flr.	5.15	4.11		9.26	system.
Reinstall	SF Flr.		3.59		3.59	
Minimum Charge	Job		294		294	

Rough Electrical

Junction Boxes   Cable   Switch Box

Electrical Per S.F.

	Unit	Material	Labor	Equip.	Total	Specification
Residential						
Wiring, Lighting and Fixtures						Includes labor and material to replace house wiring, lighting fixtures, and built-in appliances. Apply S.F. cost to floor area including garages but not basements.
Install	S.F.	2.04	2.59		4.63	
Minimum Charge	Job		310		310	
Wiring Only						Includes labor and material to replace house wiring only. Apply S.F. cost to floor area including garages but not basements.
Install	S.F.	.25	1.02		1.27	

Main Service

	Unit	Material	Labor	Equip.	Total	Specification
Complete Panel						
100 Amp.						Cost includes material and labor to install 100 Amp. residential service panel including 12 breaker interior panel, 10 single pole breakers and hook-up. Wiring is not included.
Demolish	Ea.		255		255	
Install	Ea.	640	520		1160	
Demolish and Install	Ea.	640	775		1415	
Reinstall	Ea.		417.34		417.34	
Clean	Ea.	.32	100		100.32	
Minimum Charge	Job		310		310	
150 Amp.						Cost includes material and labor to install 150 Amp. residential service, incl. 20 breaker exterior panel, main switch, 1 GFI and 14 single pole breakers, meter socket including hookup. Wiring is not included.
Demolish	Ea.		255		255	
Install	Ea.	1000	605		1605	
Demolish and Install	Ea.	1000	860		1860	
Reinstall	Ea.		482.17		482.17	
Clean	Ea.	.32	100		100.32	
Minimum Charge	Job		310		310	
200 Amp.						Cost includes material and labor to install 200 Amp. residential service, incl. 20 breaker exterior panel, main switch, 1 GFI and 18 single pole breakers, meter socket including hookup. Wiring is not included.
Demolish	Ea.		255		255	
Install	Ea.	1700	690		2390	
Demolish and Install	Ea.	1700	945		2645	
Reinstall	Ea.		551.82		551.82	
Clean	Ea.	.32	100		100.32	
Minimum Charge	Job		310		310	

Rough Electrical

Main Service

	Unit	Material	Labor	Equip.	Total	Specification
Masthead						
Demolish	Ea.		43		43	Includes material and labor to install
Install	Ea.	114	259		373	galvanized steel conduit including
Demolish and Install	Ea.	114	302		416	masthead.
Reinstall	Ea.		206.93		206.93	
Minimum Charge	Job		310		310	
Allowance to Repair						
Install	Job		310		310	Minimum labor charge for residential
Minimum Charge	Job		310		310	wiring.

For customer support on your Contractor's Pricing Guide: Residential Repair & Remodeling, call 800.448.8182.

Insulation

Wall Insulation

Floor Insulation

	Unit	Material	Labor	Equip.	Total	Specification
Kraft Faced Batt						
3-1/2" (R-13)						
Demolish	S.F.		.18		.18	Cost includes material and labor to
Install	S.F.	1.33	.74		2.07	install 3-1/2" (R-13) kraft paper or foil
Demolish and Install	S.F.	1.33	.92		2.25	faced roll and batt fiberglass
Minimum Charge	Job		260		260	insulation, joists 16" O.C.
6" (R-19)						
Demolish	S.F.		.18		.18	Cost includes material and labor to
Install	S.F.	1.56	.87		2.43	install 6" (R-19) kraft paper or foil
Demolish and Install	S.F.	1.56	1.05		2.61	faced roll and batt fiberglass
Minimum Charge	Job		260		260	insulation, joists 16" O.C.
10" (R-30)						
Demolish	S.F.		.18		.18	Cost includes material and labor to
Install	S.F.	1.67	1.04		2.71	install 10" (R-30) kraft paper-faced roll
Demolish and Install	S.F.	1.67	1.22		2.89	and batt fiberglass insulation, joists
Minimum Charge	Job		260		260	16" O.C.
12" (R-38)						
Demolish	S.F.		.18		.18	Cost includes material and labor to
Install	S.F.	1.96	1.10		3.06	install 12" (R-38) kraft paper-faced roll
Demolish and Install	S.F.	1.96	1.28		3.24	and batt fiberglass insulation, joists
Minimum Charge	Job		260		260	16" O.C.
Foil Faced Batt						
3-1/2" (R-13)						
Demolish	S.F.		.18		.18	Cost includes material and labor to
Install	S.F.	1.33	.74		2.07	install 3-1/2" (R-13) kraft paper or foil
Demolish and Install	S.F.	1.33	.92		2.25	faced roll and batt fiberglass
Minimum Charge	Job		260		260	insulation, joists 16" O.C.
6" (R-19)						
Demolish	S.F.		.18		.18	Cost includes material and labor to
Install	S.F.	1.56	.87		2.43	install 6" (R-19) kraft paper or foil
Demolish and Install	S.F.	1.56	1.05		2.61	faced roll and batt fiberglass
Minimum Charge	Job		260		260	insulation, joists 16" O.C.
Unfaced Batt						
3-1/2" (R-13)						
Demolish	S.F.		.18		.18	Cost includes material and labor to
Install	S.F.	.84	.87		1.71	install 3-1/2" (R-13) unfaced roll and
Demolish and Install	S.F.	.84	1.05		1.89	batt fiberglass insulation, joists 16"
Minimum Charge	Job		260		260	O.C.

Insulation

Floor Insulation

Floor Insulation		Unit	Material	Labor	Equip.	Total	Specification
6″ (R-19)							
	Demolish	S.F.		.18		.18	Cost includes material and labor to
	Install	S.F.	1.31	1.04		2.35	install 6″ (R-19) unfaced roll and batt
	Demolish and Install	S.F.	1.31	1.22		2.53	fiberglass insulation, joists 16″ O.C.
	Minimum Charge	Job		260		260	
10″ (R-30)							
	Demolish	S.F.		.18		.18	Cost includes material and labor to
	Install	S.F.	2.21	1.16		3.37	install 10″ (R-30) unfaced roll and batt
	Demolish and Install	S.F.	2.21	1.34		3.55	fiberglass insulation, joists 16″ O.C.
	Minimum Charge	Job		260		260	
12″ (R-38)							
	Demolish	S.F.		.18		.18	Cost includes material and labor to
	Install	S.F.	2.57	1.23		3.80	install 12″ (R-38) unfaced roll and batt
	Demolish and Install	S.F.	2.57	1.41		3.98	fiberglass insulation, joists 16″ O.C.
	Minimum Charge	Job		260		260	
Rigid Board							
1/2″							
	Demolish	S.F.		.18		.18	Includes material and labor to install
	Install	S.F.	.64	.65		1.29	rigid foam insulation board foil faced
	Demolish and Install	S.F.	.64	.83		1.47	two sides.
	Minimum Charge	Job		260		260	
3/4″							
	Demolish	S.F.		.18		.18	Includes material and labor to install
	Install	S.F.	.79	.65		1.44	rigid foam insulation board foil faced
	Demolish and Install	S.F.	.79	.83		1.62	two sides.
	Minimum Charge	Job		260		260	
1-1/2″							
	Demolish	S.F.		.18		.18	Includes material and labor to install
	Install	S.F.	1.54	.69		2.23	rigid foam insulation board foil faced
	Demolish and Install	S.F.	1.54	.87		2.41	two sides.
	Minimum Charge	Job		260		260	
Support Wire							
	Demolish	S.F.		.07		.07	Includes material and labor to install
	Install	S.F.	.06	.14		.20	friction fit support wire.
	Demolish and Install	S.F.	.06	.21		.27	
	Minimum Charge	Job		260		260	
Vapor Barrier							
	Demolish	S.F.		.10		.10	Cost includes material and labor to
	Install	S.F.	.08	.14		.22	install 4 mil polyethylene vapor barrier
	Demolish and Install	S.F.	.08	.24		.32	10′ wide sheets with 6″ overlaps.
	Minimum Charge	Job		260		260	

Wall Insulation

Wall Insulation		Unit	Material	Labor	Equip.	Total	Specification
Kraft Faced Batt							
3-1/2″ (R-13)							
	Demolish	S.F.		.18		.18	Cost includes material and labor to
	Install	S.F.	1.12	.39		1.51	install 3-1/2″ (R-13) kraft paper-faced
	Demolish and Install	S.F.	1.12	.57		1.69	roll and batt fiberglass insulation, joist
	Minimum Charge	Job		260		260	or studs 16″ O.C.
6″ (R-19)							
	Demolish	S.F.		.18		.18	Cost includes material and labor to
	Install	S.F.	1.35	.39		1.74	install 6″ (R-19) kraft paper-faced roll
	Demolish and Install	S.F.	1.35	.57		1.92	and batt fiberglass insulation, joist or
	Minimum Charge	Job		260		260	studs 16″ O.C.

Insulation

Wall Insulation		Unit	Material	Labor	Equip.	Total	Specification
10″ (R-30)							
	Demolish	S.F.		.18		.18	Cost includes material and labor to
	Install	S.F.	1.67	.45		2.12	install 10″ (R-30) kraft paper-faced roll
	Demolish and Install	S.F.	1.67	.63		2.30	and batt fiberglass insulation, joist or
	Minimum Charge	Job		260		260	studs 16″ O.C.
12″ (R-38)							
	Demolish	S.F.		.18		.18	Cost includes material and labor to
	Install	S.F.	1.96	.45		2.41	install 12″ (R-38) kraft paper-faced roll
	Demolish and Install	S.F.	1.96	.63		2.59	and batt fiberglass insulation, joist or
	Minimum Charge	Job		260		260	studs 16″ O.C.
Foil Faced Batt							
3-1/2″ (R-13)							
	Demolish	S.F.		.18		.18	Cost includes material and labor to
	Install	S.F.	.87	.39		1.26	install 3-1/2″ (R-13) foil faced roll and
	Demolish and Install	S.F.	.87	.57		1.44	batt fiberglass insulation, joist or studs
	Minimum Charge	Job		260		260	16″ O.C.
6″ (R-19)							
	Demolish	S.F.		.18		.18	Cost includes material and labor to
	Install	S.F.	1.27	.39		1.66	install 6″ (R-19) foil faced roll and batt
	Demolish and Install	S.F.	1.27	.57		1.84	fiberglass insulation, joist or studs 16″
	Minimum Charge	Job		260		260	O.C.
Unfaced Batt							
3-1/2″ (R-13)							
	Demolish	S.F.		.18		.18	Cost includes material and labor to
	Install	S.F.	.84	.39		1.23	install 3-1/2″ (R-13) unfaced roll and
	Demolish and Install	S.F.	.84	.57		1.41	batt fiberglass insulation, joist or studs
	Minimum Charge	Job		260		260	16″ O.C.
6″ (R-19)							
	Demolish	S.F.		.18		.18	Cost includes material and labor to
	Install	S.F.	1.31	.45		1.76	install 6″ (R-19) unfaced roll and batt
	Demolish and Install	S.F.	1.31	.63		1.94	fiberglass insulation, joist or studs 16″
	Minimum Charge	Job		260		260	O.C.
10″ (R-30)							
	Demolish	S.F.		.18		.18	Cost includes material and labor to
	Install	S.F.	2.21	.45		2.66	install 10″ (R-30) unfaced roll and batt
	Demolish and Install	S.F.	2.21	.63		2.84	fiberglass insulation, joist or studs 16″
	Minimum Charge	Job		260		260	O.C.
12″ (R-38)							
	Demolish	S.F.		.18		.18	Cost includes material and labor to
	Install	S.F.	2.57	.52		3.09	install 12″ (R-38) unfaced roll and batt
	Demolish and Install	S.F.	2.57	.70		3.27	fiberglass insulation, joist or studs 16″
	Minimum Charge	Job		260		260	O.C.
Blown							
R-13							
	Demolish	S.F.		.18		.18	Includes material and labor to install
	Install	S.F.	.48	1.77	1.02	3.27	fiberglass blown-in insulation in wall
	Demolish and Install	S.F.	.48	1.95	1.02	3.45	with wood siding.
	Minimum Charge	Job		610	350	960	
R-19							
	Demolish	S.F.		.18		.18	Includes material and labor to install
	Install	S.F.	.74	2.04	1.17	3.95	fiberglass blown-in insulation in wall
	Demolish and Install	S.F.	.74	2.22	1.17	4.13	with wood siding.
	Minimum Charge	Job		610	350	960	

Insulation

Wall Insulation

Wall Insulation	Unit	Material	Labor	Equip.	Total	Specification
Rigid Board						
1/2"						
Demolish	S.F.		.22		.22	Includes material and labor to install
Install	S.F.	.64	.65		1.29	rigid foam insulation board, foil faced,
Demolish and Install	S.F.	.64	.87		1.51	two sides.
Minimum Charge	Job		260		260	
3/4"						
Demolish	S.F.		.22		.22	Includes material and labor to install
Install	S.F.	.79	.65		1.44	rigid foam insulation board, foil faced,
Demolish and Install	S.F.	.79	.87		1.66	two sides.
Minimum Charge	Job		260		260	
1-1/2"						
Demolish	S.F.		.22		.22	Includes material and labor to install
Install	S.F.	1.54	.71		2.25	rigid foam insulation board, foil faced,
Demolish and Install	S.F.	1.54	.93		2.47	two sides.
Minimum Charge	Job		260		260	
Vapor Barrier						
Demolish	S.F.		.06		.06	Cost includes material and labor to
Install	S.F.	.08	.14		.22	install 4 mil polyethylene vapor barrier
Demolish and Install	S.F.	.08	.20		.28	10' wide sheets with 6" overlaps.
Minimum Charge	Job		260		260	

Ceiling Insulation

Ceiling Insulation	Unit	Material	Labor	Equip.	Total	Specification
Kraft Faced Batt						
3-1/2" (R-13)						
Demolish	S.F.		.18		.18	Cost includes material and labor to
Install	S.F.	1.12	.39		1.51	install 3-1/2" (R-13) kraft paper-faced
Demolish and Install	S.F.	1.12	.57		1.69	roll and batt fiberglass insulation, joist
Minimum Charge	Job		260		260	or studs 16" O.C.
6" (R-19)						
Demolish	S.F.		.18		.18	Cost includes material and labor to
Install	S.F.	1.35	.39		1.74	install 6" (R-19) kraft paper-faced roll
Demolish and Install	S.F.	1.35	.57		1.92	and batt fiberglass insulation, joist or
Minimum Charge	Job		260		260	studs 16" O.C.
10" (R-30)						
Demolish	S.F.		.18		.18	Cost includes material and labor to
Install	S.F.	1.67	.45		2.12	install 10" (R-30) kraft paper-faced roll
Demolish and Install	S.F.	1.67	.63		2.30	and batt fiberglass insulation, joist or
Minimum Charge	Job		260		260	studs 16" O.C.
12" (R-38)						
Demolish	S.F.		.18		.18	Cost includes material and labor to
Install	S.F.	1.96	.45		2.41	install 12" (R-38) kraft paper-faced roll
Demolish and Install	S.F.	1.96	.63		2.59	and batt fiberglass insulation, joist or
Minimum Charge	Job		260		260	studs 16" O.C.
Foil Faced Batt						
3-1/2" (R-13)						
Demolish	S.F.		.18		.18	Cost includes material and labor to
Install	S.F.	.87	.39		1.26	install 3-1/2" (R-13) foil faced roll and
Demolish and Install	S.F.	.87	.57		1.44	batt fiberglass insulation, joist or studs
Minimum Charge	Job		260		260	16" O.C.

170

For customer support on your Contractor's Pricing Guide: Residential Repair & Remodeling, call 800.448.8182.

Insulation

Ceiling Insulation		Unit	Material	Labor	Equip.	Total	Specification
6" (R-19)							
	Demolish	S.F.		.18		.18	Cost includes material and labor to
	Install	S.F.	1.27	.39		1.66	install 6" (R-19) foil faced roll and batt
	Demolish and Install	S.F.	1.27	.57		1.84	fiberglass insulation, joist or studs 16"
	Minimum Charge	Job		260		260	O.C.
Unfaced Batt							
3-1/2" (R-13)							
	Demolish	S.F.		.18		.18	Cost includes material and labor to
	Install	S.F.	.84	.39		1.23	install 3-1/2" (R-13) unfaced roll and
	Demolish and Install	S.F.	.84	.57		1.41	batt fiberglass insulation, joist or studs
	Minimum Charge	Job		260		260	16" O.C.
6" (R-19)							
	Demolish	S.F.		.18		.18	Cost includes material and labor to
	Install	S.F.	1.31	.45		1.76	install 6" (R-19) unfaced roll and batt
	Demolish and Install	S.F.	1.31	.63		1.94	fiberglass insulation, joist or studs 16"
	Minimum Charge	Job		260		260	O.C.
10" (R-30)							
	Demolish	S.F.		.18		.18	Cost includes material and labor to
	Install	S.F.	2.21	.45		2.66	install 10" (R-30) unfaced roll and batt
	Demolish and Install	S.F.	2.21	.63		2.84	fiberglass insulation, joist or studs 16"
	Minimum Charge	Job		260		260	O.C.
12" (R-38)							
	Demolish	S.F.		.18		.18	Cost includes material and labor to
	Install	S.F.	2.57	.52		3.09	install 12" (R-38) unfaced roll and batt
	Demolish and Install	S.F.	2.57	.70		3.27	fiberglass insulation, joist or studs 16"
	Minimum Charge	Job		260		260	O.C.
Blown							
5"							
	Demolish	S.F.		.55		.55	Includes material and labor to install
	Install	S.F.	.33	.32	.18	.83	fiberglass blown-in insulation.
	Demolish and Install	S.F.	.33	.87	.18	1.38	
	Minimum Charge	Job		610	350	960	
6"							
	Demolish	S.F.		.55		.55	Includes material and labor to install
	Install	S.F.	.46	.41	.23	1.10	fiberglass blown-in insulation.
	Demolish and Install	S.F.	.46	.96	.23	1.65	
	Minimum Charge	Job		610	350	960	
9"							
	Demolish	S.F.		.55		.55	Includes material and labor to install
	Install	S.F.	.58	.56	.32	1.46	fiberglass blown-in insulation.
	Demolish and Install	S.F.	.58	1.11	.32	2.01	
	Minimum Charge	Job		610	350	960	
12"							
	Demolish	S.F.		.55		.55	Includes material and labor to install
	Install	S.F.	.79	.82	.47	2.08	fiberglass blown-in insulation.
	Demolish and Install	S.F.	.79	1.37	.47	2.63	
	Minimum Charge	Job		610	350	960	
Rigid Board							
1/2"							
	Demolish	S.F.		.31		.31	Includes material and labor to install
	Install	S.F.	.64	.65		1.29	rigid foam insulation board, foil faced,
	Demolish and Install	S.F.	.64	.96		1.60	two sides.
	Minimum Charge	Job		260		260	

Insulation

Ceiling Insulation

		Unit	Material	Labor	Equip.	Total	Specification
3/4"							
	Demolish	S.F.		.31		.31	Includes material and labor to install
	Install	S.F.	.79	.65		1.44	rigid foam insulation board, foil faced,
	Demolish and Install	S.F.	.79	.96		1.75	two sides.
	Minimum Charge	Job		260		260	
1-1/2"							
	Demolish	S.F.		.31		.31	Includes material and labor to install
	Install	S.F.	1.54	.71		2.25	rigid foam insulation board, foil faced,
	Demolish and Install	S.F.	1.54	1.02		2.56	two sides.
	Minimum Charge	Job		260		260	

Exterior Insulation

		Unit	Material	Labor	Equip.	Total	Specification
Rigid Board							
1/2"							
	Demolish	S.F.		.17		.17	Includes material and labor to install
	Install	S.F.	.64	.65		1.29	rigid foam insulation board foil faced
	Demolish and Install	S.F.	.64	.82		1.46	two sides.
	Minimum Charge	Job		260		260	
3/4"							
	Demolish	S.F.		.17		.17	Includes material and labor to install
	Install	S.F.	.79	.65		1.44	rigid foam insulation board foil faced
	Demolish and Install	S.F.	.79	.82		1.61	two sides.
	Minimum Charge	Job		260		260	
1-1/2"							
	Demolish	S.F.		.17		.17	Includes material and labor to install
	Install	S.F.	1.54	.69		2.23	rigid foam insulation board foil faced
	Demolish and Install	S.F.	1.54	.86		2.40	two sides.
	Minimum Charge	Job		260		260	
Vapor Barrier							
	Demolish	S.F.		.10		.10	Cost includes material and labor to
	Install	S.F.	.08	.14		.22	install 4 mil polyethylene vapor barrier
	Demolish and Install	S.F.	.08	.24		.32	10' wide sheets with 6" overlaps.
	Minimum Charge	Job		260		260	

Roof Insulation Board

		Unit	Material	Labor	Equip.	Total	Specification
Fiberglass							
3/4" Fiberglass							
	Demolish	Sq.		57.50		57.50	Includes material and labor to install
	Install	Sq.	62.50	29.50		92	fiberglass board insulation 3/4" thick,
	Demolish and Install	Sq.	62.50	87		149.50	R-2.80 and C-0.36 values.
	Minimum Charge	Job		260		260	
1" Fiberglass							
	Demolish	Sq.		57.50		57.50	Includes material and labor to install
	Install	Sq.	128	33.50		161.50	fiberglass board insulation 1" thick,
	Demolish and Install	Sq.	128	91		219	R-4.20 and C-0.24 values.
	Minimum Charge	Job		260		260	
1-3/8" Fiberglass							
	Demolish	Sq.		57.50		57.50	Includes material and labor to install
	Install	Sq.	570	36.50		606.50	fiberglass board insulation 1-3/8"
	Demolish and Install	Sq.	570	94		664	thick, R-5.30 and C-0.19 values.
	Minimum Charge	Job		260		260	

For customer support on your Contractor's Pricing Guide: Residential Repair & Remodeling, call 800.448.8182.

Insulation

Roof Insulation Board		Unit	Material	Labor	Equip.	Total	Specification
1-5/8″ Fiberglass							
	Demolish	Sq.		57.50		57.50	Includes material and labor to install
	Install	Sq.	208	38		246	fiberglass board insulation 1-5/8″
	Demolish and Install	Sq.	208	95.50		303.50	thick, R-6.70 and C-0.15 values.
	Minimum Charge	Job		260		260	
2-1/4″ Fiberglass							
	Demolish	Sq.		57.50		57.50	Includes material and labor to install
	Install	Sq.	405	42		447	fiberglass board insulation 2-1/4″
	Demolish and Install	Sq.	405	99.50		504.50	thick, R-8.30 and C-0.12 values.
	Minimum Charge	Job		260		260	

Perlite

		Unit	Material	Labor	Equip.	Total	Specification
1″ Perlite							
	Demolish	Sq.		57.50		57.50	Includes material and labor to install
	Install	Sq.	118	33.50		151.50	perlite board insulation 1″ thick,
	Demolish and Install	Sq.	118	91		209	R-2.80 and C-0.36 values.
	Minimum Charge	Job		260		260	
4″ Perlite							
	Demolish	Sq.		57.50		57.50	Includes material and labor to install
	Install	Sq.	475	64		539	perlite board insulation 4″ thick,
	Demolish and Install	Sq.	475	121.50		596.50	R-10.0 and C-0.10 values.
	Minimum Charge	Job		260		260	
1-1/2″ Perlite							
	Demolish	Sq.		57.50		57.50	Includes material and labor to install
	Install	Sq.	165	36.50		201.50	perlite board insulation 1-1/2″ thick,
	Demolish and Install	Sq.	165	94		259	R-4.20 and C-0.24 values.
	Minimum Charge	Job		260		260	
2″ Perlite							
	Demolish	Sq.		57.50		57.50	Includes material and labor to install
	Install	Sq.	207	44.50		251.50	perlite board insulation 2″ thick,
	Demolish and Install	Sq.	207	102		309	R-5.30 and C-0.19 values.
	Minimum Charge	Job		260		260	
2-1/2″ Perlite							
	Demolish	Sq.		57.50		57.50	Includes material and labor to install
	Install	Sq.	695	50		745	perlite board insulation 2-1/2″ thick,
	Demolish and Install	Sq.	695	107.50		802.50	R-6.70 and C-0.15 values.
	Minimum Charge	Job		260		260	
3″ Perlite							
	Demolish	Sq.		57.50		57.50	Includes material and labor to install
	Install	Sq.	350	53.50		403.50	perlite board insulation 3″ thick,
	Demolish and Install	Sq.	350	111		461	R-8.30 and C-0.12 values.
	Minimum Charge	Job		260		260	

Polystyrene

		Unit	Material	Labor	Equip.	Total	Specification
2″ Polystyrene							
	Demolish	Sq.		57.50		57.50	Includes material and labor to install
	Install	Sq.	72.50	42		114.50	extruded polystyrene board, R5.0/ 2″
	Demolish and Install	Sq.	72.50	99.50		172	in thickness @25PSI compressive
	Minimum Charge	Job		260		260	strength.
1″ Polystyrene							
	Demolish	Sq.		57.50		57.50	Includes material and labor to install
	Install	Sq.	36.50	28.50		65	extruded polystyrene board, R5.0/ 1″
	Demolish and Install	Sq.	36.50	86		122.50	in thickness @25PSI compressive
	Minimum Charge	Job		260		260	strength.
1-1/2″ Polystyrene							
	Demolish	Sq.		57.50		57.50	Includes material and labor to install
	Install	Sq.	54.50	33.50		88	extruded polystyrene board, R5.0/
	Demolish and Install	Sq.	54.50	91		145.50	1-1/2″ in thickness @25PSI
	Minimum Charge	Job		260		260	compressive strength.

For customer support on your Contractor's Pricing Guide: Residential Repair & Remodeling, call 800.448.8182.

Insulation

Roof Insulation Board	Unit	Material	Labor	Equip.	Total	Specification
Urethane						
3/4" Urethane						
Demolish	Sq.		57.50		57.50	Includes material and labor to install
Install	Sq.	93.50	25		118.50	urethane board insulation 3/4" thick,
Demolish and Install	Sq.	93.50	82.50		176	R-5.30 and C-0.19 values.
Minimum Charge	Job		260		260	
1" Urethane						
Demolish	Sq.		57.50		57.50	Includes material and labor to install
Install	Sq.	158	26.50		184.50	urethane board insulation 1" thick,
Demolish and Install	Sq.	158	84		242	R-6.70 and C-0.15 values.
Minimum Charge	Job		260		260	
1-1/2" Urethane						
Demolish	Sq.		57.50		57.50	Includes material and labor to install
Install	Sq.	140	31		171	urethane board insulation 1-1/2"
Demolish and Install	Sq.	140	88.50		228.50	thick, R-10.1 and C-0.09 values.
Minimum Charge	Job		260		260	
2" Urethane						
Demolish	Sq.		57.50		57.50	Includes material and labor to install
Install	Sq.	169	36.50		205.50	urethane board insulation 2" thick,
Demolish and Install	Sq.	169	94		263	R-13.4 and C-0.06 values.
Minimum Charge	Job		260		260	

For customer support on your Contractor's Pricing Guide: Residential Repair & Remodeling, call 800.448.8182.

Walls / Ceilings

Gypsum Wallboard

Tile Ceiling

Gypsum Wallboard Ceiling

Wall Plaster

		Unit	Material	Labor	Equip.	Total	Specification
Interior							
Cement							
	Demolish	S.Y.		7.40		7.40	Includes material and labor to install
	Install	S.Y.	26	32.50	1.80	60.30	cement plaster on interior wood
	Demolish and Install	S.Y.	26	39.90	1.80	67.70	framing with metal lath.
	Clean	S.Y.		3.63		3.63	
	Paint	S.Y.	2.10	5.75		7.85	
	Minimum Charge	Job		254		254	
Acoustic							
	Demolish	S.Y.		7.40		7.40	Includes material and labor to install
	Install	S.Y.	15.45	39.50	1.80	56.75	acoustical cement plaster on interior
	Demolish and Install	S.Y.	15.45	46.90	1.80	64.15	wood framing with metal lath.
	Clean	S.Y.		3.63		3.63	
	Paint	S.Y.	2.10	5.75		7.85	
	Minimum Charge	Job		254		254	
Gypsum							
	Demolish	S.Y.		7.40		7.40	Includes material and labor to install
	Install	S.Y.	16.25	33.50	1.53	51.28	gypsum wall or ceiling plaster on
	Demolish and Install	S.Y.	16.25	40.90	1.53	58.68	interior wood framing with metal lath.
	Clean	S.Y.		3.63		3.63	
	Paint	S.Y.	2.10	5.75		7.85	
	Minimum Charge	Job		254		254	
Stucco							
	Demolish	S.Y.		17.75		17.75	Includes material and labor to install
	Install	S.Y.	12.05	46.50	2.12	60.67	stucco on wire mesh over wood
	Demolish and Install	S.Y.	12.05	64.25	2.12	78.42	framing.
	Clean	S.Y.		3.63		3.63	
	Paint	S.Y.	2.10	5.75		7.85	
	Minimum Charge	Job		254		254	
Thin Coat							
Veneer							
	Demolish	S.Y.		7.40		7.40	Includes materials and labor to install
	Install	S.Y.	1.35	6	.33	7.68	thin coat veneer plaster.
	Demolish and Install	S.Y.	1.35	13.40	.33	15.08	
	Clean	S.Y.		3.63		3.63	
	Paint	S.Y.	2.10	5.75		7.85	
	Minimum Charge	Job		254		254	

Walls / Ceilings

Wall Plaster

Wall Plaster	Unit	Material	Labor	Equip.	Total	Specification
Lath						
Metal						
Demolish	S.Y.		1.74		1.74	Includes material and labor to install
Install	S.Y.	16.45	6.95		23.40	wire lath on steel framed walls.
Demolish and Install	S.Y.	16.45	8.69		25.14	
Clean	S.Y.		3.63		3.63	
Minimum Charge	Job		123		123	
Gypsum						
Demolish	S.Y.		1.74		1.74	Includes material and labor to install
Install	S.Y.	5.35	6.50		11.85	gypsum lath on walls.
Demolish and Install	S.Y.	5.35	8.24		13.59	
Clean	S.Y.		3.63		3.63	
Minimum Charge	Job		254		254	
Repairs						
Fill in Small Cracks						
Install	L.F.	.11	4.06		4.17	Includes labor and material to fill in
Minimum Charge	Job		254		254	hairline cracks up to 1/8" with filler, sand and prep for paint.
Repair up to 1" Stock						
Install	L.F.	.44	5.05		5.49	Includes labor and material to fill in
Minimum Charge	Job		254		254	hairline cracks up to 1" with filler, sand and prep for paint.
Patch Small Hole (2")						
Install	Ea.	.11	5.35		5.46	Includes labor and material to patch
Minimum Charge	Job		254		254	up to a 2" diameter hole in plaster.
Patch Large Hole						
Install	Ea.	13.10	169		182.10	Includes labor and material to patch
Minimum Charge	Job		254		254	up to a 20" diameter hole in plaster.
Patch Plaster Section						
Install	Ea.	3.53	5.05		8.58	Includes labor and material to repair
Minimum Charge	Job		254		254	section of plaster wall, area less than 20 S.F.
Repair Inside Corner						
Install	L.F.	5.25	12.70		17.95	Includes material and labor to patch
Minimum Charge	Job		254		254	plaster corners.

Ceiling Plaster

Ceiling Plaster	Unit	Material	Labor	Equip.	Total	Specification
Interior						
Cement						
Demolish	S.Y.		7.40		7.40	Includes material and labor to install
Install	S.Y.	26	32.50	1.80	60.30	cement plaster on interior wood
Demolish and Install	S.Y.	26	39.90	1.80	67.70	framing with metal lath.
Clean	S.Y.		3.81		3.81	
Paint	S.Y.	2.10	5.75		7.85	
Minimum Charge	Job		254		254	
Acoustic						
Demolish	S.Y.		7.40		7.40	Includes material and labor to install
Install	S.Y.	15.45	39.50	1.80	56.75	acoustical cement plaster on interior
Demolish and Install	S.Y.	15.45	46.90	1.80	64.15	wood framing with metal lath.
Clean	S.Y.		3.81		3.81	
Paint	S.Y.	2.10	5.75		7.85	
Minimum Charge	Job		254		254	

For customer support on your Contractor's Pricing Guide: Residential Repair & Remodeling, call 800.448.8182.

Walls / Ceilings

Ceiling Plaster

Ceiling Plaster		Unit	Material	Labor	Equip.	Total	Specification
Gypsum							
	Demolish	S.Y.		7.40		7.40	Includes material and labor to install
	Install	S.Y.	16.25	33.50	1.53	51.28	gypsum wall or ceiling plaster on
	Demolish and Install	S.Y.	16.25	40.90	1.53	58.68	interior wood framing with metal lath.
	Clean	S.Y.		3.81		3.81	
	Paint	S.Y.	2.10	5.75		7.85	
	Minimum Charge	Job		254		254	
Stucco							
	Demolish	S.Y.		17.75		17.75	Includes material and labor to install
	Install	S.Y.	12.05	46.50	2.12	60.67	stucco on wire mesh over wood
	Demolish and Install	S.Y.	12.05	64.25	2.12	78.42	framing.
	Clean	S.Y.		3.81		3.81	
	Paint	S.Y.	2.10	5.75		7.85	
	Minimum Charge	Job		254		254	
Thin Coat							
Veneer							
	Demolish	S.Y.		7.40		7.40	Includes materials and labor to install
	Install	S.Y.	1.35	6	.33	7.68	thin coat veneer plaster.
	Demolish and Install	S.Y.	1.35	13.40	.33	15.08	
	Clean	S.Y.		3.81		3.81	
	Paint	S.Y.	2.10	5.75		7.85	
	Minimum Charge	Job		254		254	
Lath							
Metal							
	Demolish	S.Y.		1.74		1.74	Includes material and labor to install
	Install	S.Y.	16.45	8.70		25.15	wire lath on steel framed ceilings.
	Demolish and Install	S.Y.	16.45	10.44		26.89	
	Minimum Charge	Job		254		254	
Gypsum							
	Demolish	S.Y.		1.74		1.74	Includes material and labor to install
	Install	S.Y.	5.35	5.45		10.80	gypsum lath on ceilings.
	Demolish and Install	S.Y.	5.35	7.19		12.54	
	Minimum Charge	Job		254		254	
Repairs							
Fill in Small Cracks							
	Install	S.F.	.11	.92		1.03	Includes labor and material to fill in
	Minimum Charge	Job		505		505	hairline cracks up to 1/8" with filler, sand and prep for paint.

Gypsum Wallboard

Gypsum Wallboard		Unit	Material	Labor	Equip.	Total	Specification
Finished							
1/4"							
	Demolish	S.F.		.44		.44	Cost includes material and labor to
	Install	S.F.	.48	1.61		2.09	install 1/4" drywall, including joint
	Demolish and Install	S.F.	.48	2.05		2.53	taping, premixed compound and
	Clean	S.F.		.36		.36	screws.
	Paint	S.F.	.66	.38		1.04	
	Minimum Charge	Job		260		260	
3/8"							
	Demolish	S.F.		.44		.44	Cost includes material and labor to
	Install	S.F.	.54	1.63		2.17	install 3/8" drywall including joint
	Demolish and Install	S.F.	.54	2.07		2.61	tape, premixed compound and
	Clean	S.F.		.36		.36	screws.
	Paint	S.F.	.66	.38		1.04	
	Minimum Charge	Job		260		260	

For customer support on your Contractor's Pricing Guide: Residential Repair & Remodeling, call 800.448.8182.

Walls / Ceilings

Gypsum Wallboard		Unit	Material	Labor	Equip.	Total	Specification
1/2"							
	Demolish	S.F.		.44		.44	Cost includes material and labor to
	Install	S.F.	.59	1.65		2.24	install 1/2" drywall including joint
	Demolish and Install	S.F.	.59	2.09		2.68	tape, premixed compound and
	Clean	S.F.		.36		.36	screws.
	Paint	S.F.	.66	.38		1.04	
	Minimum Charge	Job		260		260	
1/2" Water Resistant							
	Demolish	S.F.		.44		.44	Cost includes material and labor to
	Install	S.F.	.48	1.65		2.13	install 1/2" WR drywall including joint
	Demolish and Install	S.F.	.48	2.09		2.57	tape, premixed compound and
	Clean	S.F.		.36		.36	screws.
	Paint	S.F.	.66	.38		1.04	
	Minimum Charge	Job		260		260	
1/2" Fire-rated							
	Demolish	S.F.		.44		.44	Cost includes material and labor to
	Install	S.F.	.58	1.65		2.23	install 1/2" FR drywall including
	Demolish and Install	S.F.	.58	2.09		2.67	screws. Taping and finishing not
	Clean	S.F.		.36		.36	included.
	Paint	S.F.	.66	.38		1.04	
	Minimum Charge	Job		260		260	
5/8"							
	Demolish	S.F.		.44		.44	Cost includes material and labor to
	Install	S.F.	.50	1.59		2.09	install 5/8" drywall including joint
	Demolish and Install	S.F.	.50	2.03		2.53	tape, premixed compound and
	Clean	S.F.		.36		.36	screws.
	Paint	S.F.	.66	.38		1.04	
	Minimum Charge	Job		260		260	
5/8" Fire-rated							
	Demolish	S.F.		.44		.44	Cost includes material and labor to
	Install	S.F.	.57	1.87		2.44	install 5/8" FR drywall, including joint
	Demolish and Install	S.F.	.57	2.31		2.88	taping, premixed compound and
	Clean	S.F.		.36		.36	screws.
	Paint	S.F.	.66	.38		1.04	
	Minimum Charge	Job		260		260	
2-hour Fire-rated							
	Demolish	S.F.		1.11		1.11	Cost includes material and labor to
	Install	S.F.	1.13	3.71		4.84	install 2 layers of 5/8" type X drywall
	Demolish and Install	S.F.	1.13	4.82		5.95	including joint taping, premixed
	Clean	S.F.		.36		.36	compound and screws.
	Paint	S.F.	.66	.38		1.04	
	Minimum Charge	Job		260		260	
Unfinished							
1/4"							
	Demolish	S.F.		.44		.44	Cost includes material and labor to
	Install	S.F.	.42	.56		.98	install 1/4" drywall including screws.
	Demolish and Install	S.F.	.42	1		1.42	Taping and finishing not included.
	Clean	S.F.		.36		.36	
	Minimum Charge	Job		260		260	
3/8"							
	Demolish	S.F.		.44		.44	Cost includes material and labor to
	Install	S.F.	.49	.56		1.05	install 3/8" drywall including screws.
	Demolish and Install	S.F.	.49	1		1.49	Tape, finish or texture is not included.
	Clean	S.F.		.36		.36	
	Minimum Charge	Job		260		260	

Walls / Ceilings

Gypsum Wallboard		Unit	Material	Labor	Equip.	Total	Specification
1/2"							Cost includes material and labor to
	Demolish	S.F.		.44		.44	install 1/2" drywall including screws.
	Install	S.F.	.53	.57		1.10	Taping and finishing not included.
	Demolish and Install	S.F.	.53	1.01		1.54	
	Clean	S.F.		.36		.36	
	Minimum Charge	Job		260		260	
1/2" Water Resistant							
	Demolish	S.F.		.44		.44	Cost includes material and labor to
	Install	S.F.	.42	.57		.99	install 1/2" WR drywall including
	Demolish and Install	S.F.	.42	1.01		1.43	screws. taping and finishing not
	Clean	S.F.		.36		.36	included.
	Minimum Charge	Job		260		260	
1/2" Fire-rated							
	Demolish	S.F.		.44		.44	Cost includes labor and material to
	Install	S.F.	.53	.38		.91	install 1/2" FR drywall including
	Demolish and Install	S.F.	.53	.82		1.35	screws. Tape, finish or texture is not
	Clean	S.F.		.36		.36	included.
	Minimum Charge	Job		260		260	
5/8"							
	Demolish	S.F.		.44		.44	Cost includes material and labor to
	Install	S.F.	.44	.64		1.08	install 5/8" drywall including screws.
	Demolish and Install	S.F.	.44	1.08		1.52	Taping and finishing not included.
	Clean	S.F.		.36		.36	
	Minimum Charge	Job		260		260	
5/8" Fire-rated							
	Demolish	S.F.		.44		.44	Cost includes material and labor to
	Install	S.F.	.50	.64		1.14	install 5/8" FR drywall including
	Demolish and Install	S.F.	.50	1.08		1.58	screws. Taping and finishing not
	Clean	S.F.		.36		.36	included.
	Minimum Charge	Job		260		260	
2-hour Fire-rated							
	Demolish	S.F.		1.11		1.11	Cost includes material and labor to
	Install	S.F.	1.13	3.71		4.84	install 2 layers of 5/8" type X drywall
	Demolish and Install	S.F.	1.13	4.82		5.95	including joint taping, premixed
	Clean	S.F.		.36		.36	compound and screws.
	Minimum Charge	Job		260		260	
Accessories							
Tape & Finish							
	Install	S.F.	.08	.52		.60	Includes labor and material for taping,
	Minimum Charge	Job		260		260	sanding and finish.
Texture							
Sprayed							
	Install	S.F.	.07	.65		.72	Includes labor and material to install
	Minimum Charge	Job		260		260	by spray, texture finish.
Trowel							
	Install	S.F.	.45	1.15		1.60	Includes labor and material to apply
	Minimum Charge	Job		260		260	hand troweled texture.
Patch Hole in Gypsum Wallboard							
	Install	S.F.	.96	71.50		72.46	Includes labor and material to patch a
	Minimum Charge	Job		260		260	one square foot area of damaged
							drywall.
Furring Strips							
	Demolish	S.F.		.11		.11	Cost includes material and labor to
	Install	S.F.	.13	1.16		1.29	install 1" x 2" furring strips, 16" O.C.
	Demolish and Install	S.F.	.13	1.27		1.40	per S.F. of surface area to be covered.
	Minimum Charge	Job		130		130	

For customer support on your Contractor's Pricing Guide: Residential Repair & Remodeling, call 800.448.8182.

Walls / Ceilings

Ceramic Wall Tile

	Unit	Material	Labor	Equip.	Total	Specification
Thin Set						
2″ x 2″						
Demolish	S.F.		1.48		1.48	Cost includes material and labor to
Install	S.F.	5.25	8.45		13.70	install 2″ x 2″ ceramic tile in organic
Demolish and Install	S.F.	5.25	9.93		15.18	adhesive including cutting and grout.
Clean	S.F.		.76		.76	
Minimum Charge	Job		254		254	
4-1/4″ x 4-1/4″						
Demolish	S.F.		1.48		1.48	Cost includes material and labor to
Install	S.F.	4.02	4.74		8.76	install 4-1/4″ x 4-1/4″ ceramic tile in
Demolish and Install	S.F.	4.02	6.22		10.24	organic adhesive including grout.
Clean	S.F.		.76		.76	
Minimum Charge	Job		254		254	
6″ x 6″						
Demolish	S.F.		1.48		1.48	Cost includes material and labor to
Install	S.F.	3.86	5.15		9.01	install 6″ x 6″ ceramic tile in organic
Demolish and Install	S.F.	3.86	6.63		10.49	adhesive including cutting and grout.
Clean	S.F.		.76		.76	
Minimum Charge	Job		254		254	
8″ x 8″						
Demolish	S.F.		1.48		1.48	Cost includes material and labor to
Install	S.F.	10.05	5.30		15.35	install 8″ x 8″ ceramic tile in organic
Demolish and Install	S.F.	10.05	6.78		16.83	adhesive including cutting and grout.
Clean	S.F.		.76		.76	
Minimum Charge	Job		254		254	
Thick Set						
2″ x 2″						
Demolish	S.F.		1.78		1.78	Cost includes material and labor to
Install	S.F.	11.10	10.15		21.25	install 2″ x 2″ ceramic tile in mortar
Demolish and Install	S.F.	11.10	11.93		23.03	bed including cutting and grout.
Clean	S.F.		.76		.76	
Minimum Charge	Job		254		254	
4-1/4″ x 4-1/4″						
Demolish	S.F.		1.78		1.78	Cost includes material and labor to
Install	S.F.	2.24	10.15		12.39	install 4-1/4″ x 4-1/4″ ceramic tile in
Demolish and Install	S.F.	2.24	11.93		14.17	mortar bed including cutting and
Clean	S.F.		.76		.76	grout.
Minimum Charge	Job		254		254	
6″ x 6″						
Demolish	S.F.		1.78		1.78	Cost includes material and labor to
Install	S.F.	2.34	9.25		11.59	install 6″ x 6″ ceramic tile in mortar
Demolish and Install	S.F.	2.34	11.03		13.37	bed including cutting and grout.
Clean	S.F.		.76		.76	
Minimum Charge	Job		254		254	
8″ x 8″						
Demolish	S.F.		1.78		1.78	Cost includes material and labor to
Install	S.F.	4.46	8.45		12.91	install 8″ x 8″ ceramic tile in mortar
Demolish and Install	S.F.	4.46	10.23		14.69	bed including cutting and grout.
Clean	S.F.		.76		.76	
Minimum Charge	Job		254		254	
Re-grout						
Install	S.F.	.23	5.10		5.33	Includes material and labor to regrout
Minimum Charge	Job		254		254	wall tiles.

Walls / Ceilings

Wall Tile

Wall Tile		Unit	Material	Labor	Equip.	Total	Specification
Mirror							
Custom Cut Wall							
	Demolish	S.F.		.85		.85	Includes material and labor to install
	Install	S.F.	3.51	5.20		8.71	distortion-free float glass mirror, 1/8"
	Demolish and Install	S.F.	3.51	6.05		9.56	to 3/16" thick with cut and polished
	Clean	S.F.	.02	.23		.25	edges.
	Minimum Charge	Job		252		252	
Beveled Wall							
	Demolish	S.F.		.85		.85	Includes material and labor to install
	Install	S.F.	36	9.20		45.20	distortion-free float glass mirror, 1/8"
	Demolish and Install	S.F.	36	10.05		46.05	to 3/16" thick with beveled edges.
	Clean	S.F.	.02	.23		.25	
	Minimum Charge	Job		252		252	
Smoked Wall							
	Demolish	S.F.		.85		.85	Includes material and labor to install
	Install	S.F.	3.66	9.20		12.86	distortion-free smoked float glass
	Demolish and Install	S.F.	3.66	10.05		13.71	mirror, 1/8" to 3/16" thick with cut
	Clean	S.F.	.02	.23		.25	and polished edges.
	Minimum Charge	Job		252		252	
Marble							
Good Grade							
	Demolish	S.F.		2.78		2.78	Cost includes material and labor to
	Install	S.F.	9.80	22		31.80	install 3/8" x 12" x 12" marble tile in
	Demolish and Install	S.F.	9.80	24.78		34.58	mortar bed with grout.
	Clean	S.F.		.76		.76	
	Minimum Charge	Job		254		254	
Better Grade							
	Demolish	S.F.		2.78		2.78	Cost includes material and labor to
	Install	S.F.	16.45	22		38.45	install 3/8" x 12" x 12" marble, in
	Demolish and Install	S.F.	16.45	24.78		41.23	mortar bed with grout.
	Clean	S.F.		.76		.76	
	Minimum Charge	Job		254		254	
Premium Grade							
	Demolish	S.F.		2.78		2.78	Cost includes material and labor to
	Install	S.F.	36.50	22		58.50	install 3/8" x 12" x 12" marble, in
	Demolish and Install	S.F.	36.50	24.78		61.28	mortar bed with grout.
	Clean	S.F.		.76		.76	
	Minimum Charge	Job		254		254	
Glass Block							
	Demolish	S.F.		2.22		2.22	Cost includes material and labor to
	Install	S.F.	29	21		50	install 3-7/8" thick, 6" x 6" smooth
	Demolish and Install	S.F.	29	23.22		52.22	face glass block including mortar and
	Clean	S.F.		.76		.76	cleaning.
	Minimum Charge	Job		475		475	
Cork							
	Demolish	S.F.		.34		.34	Cost includes material and labor to
	Install	S.F.	4.95	1.81		6.76	install 3/16" thick cork tile and
	Demolish and Install	S.F.	4.95	2.15		7.10	installation.
	Clean	S.F.		.76		.76	
	Minimum Charge	Job		217		217	
Re-grout							
	Install	S.F.	.23	5.10		5.33	Includes material and labor to regrout
	Minimum Charge	Job		254		254	wall tiles.

For customer support on your Contractor's Pricing Guide: Residential Repair & Remodeling, call 800.448.8182.

Walls / Ceilings

Wall Paneling

	Unit	Material	Labor	Equip.	Total	Specification
Plywood						
Good Grade						
Demolish	S.F.		.44		.44	Cost includes material and labor to
Install	S.F.	1.53	2.08		3.61	install 4' x 8' x 1/4" paneling,
Demolish and Install	S.F.	1.53	2.52		4.05	including nails and adhesive.
Clean	S.F.		.36		.36	
Paint	S.F.	.54	.82		1.36	
Minimum Charge	Job		260		260	
Better Grade						
Demolish	S.F.		.44		.44	Cost includes material and labor to
Install	S.F.	4.08	2.48		6.56	install 4' x 8' x 1/4" plywood wall
Demolish and Install	S.F.	4.08	2.92		7	panel, including nails and adhesive.
Clean	S.F.		.36		.36	
Paint	S.F.	.54	.82		1.36	
Minimum Charge	Job		260		260	
Premium Grade						
Demolish	S.F.		.44		.44	Cost includes material and labor to
Install	S.F.	1.62	2.98		4.60	install 4' x 8' x 1/4" plywood wall
Demolish and Install	S.F.	1.62	3.42		5.04	panel, including nails and adhesive.
Clean	S.F.		.36		.36	
Paint	S.F.	.54	.82		1.36	
Minimum Charge	Job		260		260	
Hardboard						
Embossed Paneling						
Demolish	S.F.		.44		.44	Includes material and labor to install
Install	S.F.	.68	2.08		2.76	hardboard embossed paneling
Demolish and Install	S.F.	.68	2.52		3.20	including adhesives.
Clean	S.F.		.36		.36	
Minimum Charge	Job		260		260	
Plastic Molding						
Demolish	L.F.		.44		.44	Includes material and labor to install
Install	L.F.	2.75	1.30		4.05	plastic trim for paneling.
Demolish and Install	L.F.	2.75	1.74		4.49	
Clean	L.F.	.02	.24		.26	
Minimum Charge	Job		260		260	
Hardboard						
Embossed Brick						
Demolish	S.F.		.44		.44	Cost includes material and labor to
Install	S.F.	.91	2.31		3.22	install 4' x 8' non-ceramic mineral
Demolish and Install	S.F.	.91	2.75		3.66	brick-like veneer wall panels including
Clean	S.F.		.36		.36	adhesive.
Paint	S.F.	.24	.65		.89	
Minimum Charge	Job		260		260	
Embossed Ceramic Tile						
Demolish	S.F.		.44		.44	Includes material and labor to install
Install	S.F.	.68	2.08		2.76	hardboard embossed paneling
Demolish and Install	S.F.	.68	2.52		3.20	including adhesives.
Clean	S.F.		.36		.36	
Paint	S.F.	.24	.65		.89	
Minimum Charge	Job		260		260	
Pegboard						
Demolish	S.F.		.44		.44	Cost includes material and labor to
Install	S.F.	1.17	2.08		3.25	install 4' x 8' x 1/4" pegboard panel,
Demolish and Install	S.F.	1.17	2.52		3.69	including adhesive.
Clean	S.F.		.36		.36	
Paint	S.F.	.24	.65		.89	
Minimum Charge	Job		260		260	

For customer support on your Contractor's Pricing Guide: Residential Repair & Remodeling, call 800.448.8182.

Walls / Ceilings

Wall Paneling

Wall Paneling		Unit	Material	Labor	Equip.	Total	Specification
Furring Strips							Cost includes material and labor to
	Demolish	S.F.		.11		.11	install 1" x 2" furring strips, 16" O.C.
	Install	S.F.	.13	1.16		1.29	per S.F. of surface area to be covered.
	Demolish and Install	S.F.	.13	1.27		1.40	
	Minimum Charge	Job		260		260	
Molding							
Unfinished Base							Cost includes material and labor to
	Demolish	L.F.		.74		.74	install 1/2" x 2-1/2" pine or birch
	Install	L.F.	1.38	1.57		2.95	base molding.
	Demolish and Install	L.F.	1.38	2.31		3.69	
	Reinstall	L.F.		1.26		1.26	
	Clean	L.F.	.02	.23		.25	
	Paint	L.F.	.10	.47		.57	
	Minimum Charge	Job		260		260	
Unfinished Corner							Includes material and labor to install
	Demolish	L.F.		.74		.74	unfinished pine corner molding.
	Install	L.F.	4.83	2.17		7	
	Demolish and Install	L.F.	4.83	2.91		7.74	
	Reinstall	L.F.		1.74		1.74	
	Clean	L.F.	.02	.23		.25	
	Paint	L.F.	.19	.69		.88	
	Minimum Charge	Job		260		260	
Unfinished Cove							Cost includes material and labor to
	Demolish	L.F.		.74		.74	install 11/16" x 2-3/4" pine cove
	Install	L.F.	2.92	1.97		4.89	molding.
	Demolish and Install	L.F.	2.92	2.71		5.63	
	Reinstall	L.F.		1.57		1.57	
	Clean	L.F.	.02	.23		.25	
	Paint	L.F.	.19	.69		.88	
	Minimum Charge	Job		260		260	
Unfinished Crown							Includes material and labor to install
	Demolish	L.F.		.89		.89	unfinished pine crown molding.
	Install	L.F.	1.72	2.08		3.80	
	Demolish and Install	L.F.	1.72	2.97		4.69	
	Reinstall	L.F.		1.67		1.67	
	Clean	L.F.	.03	.31		.34	
	Paint	L.F.	.18	.47		.65	
	Minimum Charge	Job		260		260	
Prefinished Base							Cost includes material and labor to
	Demolish	L.F.		.74		.74	install 9/16" x 3-5/16" prefinished
	Install	L.F.	1.97	1.67		3.64	base molding.
	Demolish and Install	L.F.	1.97	2.41		4.38	
	Reinstall	L.F.		1.34		1.34	
	Clean	L.F.	.02	.24		.26	
	Paint	L.F.	.10	.47		.57	
	Minimum Charge	Job		260		260	
Prefinished Corner							Includes material and labor to install
	Demolish	L.F.		.74		.74	inside or outside prefinished corner
	Install	L.F.	4.05	2.17		6.22	molding.
	Demolish and Install	L.F.	4.05	2.91		6.96	
	Reinstall	L.F.		1.74		1.74	
	Clean	L.F.	.02	.24		.26	
	Paint	L.F.	.19	.69		.88	
	Minimum Charge	Job		260		260	

Walls / Ceilings

Wall Paneling

Wall Paneling	Unit	Material	Labor	Equip.	Total	Specification
Prefinished Crown						Cost includes material and labor to install 3/4" x 3-13/16" prefinished crown molding.
Demolish	L.F.		.89		.89	
Install	L.F.	2.94	2.08		5.02	
Demolish and Install	L.F.	2.94	2.97		5.91	
Reinstall	L.F.		1.67		1.67	
Clean	L.F.	.03	.31		.34	
Paint	L.F.	.18	.47		.65	
Minimum Charge	Job		260		260	

Wallcovering

Wallcovering	Unit	Material	Labor	Equip.	Total	Specification
Surface Prep						
Patch Walls						Includes labor and material to repair section of plaster wall, area less than 20 S.F.
Install	S.F.	.18	5.05		5.23	
Minimum Charge	Job		217		217	
Sizing						Includes labor and materials to apply wall sizing.
Install	S.F.	.09	.22		.31	
Minimum Charge	Job		217		217	
Good Grade						
Walls						Includes material and labor to install pre-pasted, vinyl-coated wallcovering.
Demolish	S.F.		.80		.80	
Install	S.F.	1.20	.91		2.11	
Demolish and Install	S.F.	1.20	1.71		2.91	
Clean	S.F.	.01	.10		.11	
Minimum Charge	Job		217		217	
Ceiling						Includes material and labor to install pre-pasted, vinyl-coated wallcovering.
Demolish	S.F.		.80		.80	
Install	S.F.	1.20	.91		2.11	
Demolish and Install	S.F.	1.20	1.71		2.91	
Clean	S.F.	.02	.23		.25	
Minimum Charge	Job		217		217	
Better Grade						
Walls						Includes material and labor to install pre-pasted, vinyl-coated wallcovering.
Demolish	S.F.		.80		.80	
Install	S.F.	1.28	.91		2.19	
Demolish and Install	S.F.	1.28	1.71		2.99	
Clean	S.F.	.01	.10		.11	
Minimum Charge	Job		217		217	
Ceiling						Includes material and labor to install pre-pasted, vinyl-coated wallcovering.
Demolish	S.F.		.80		.80	
Install	S.F.	1.28	.91		2.19	
Demolish and Install	S.F.	1.28	1.71		2.99	
Clean	S.F.	.02	.23		.25	
Minimum Charge	Job		217		217	
Premium Grade						
Walls						Includes material and labor to install pre-pasted, vinyl-coated wallcovering.
Demolish	S.F.		1.01		1.01	
Install	S.F.	1.19	.91		2.10	
Demolish and Install	S.F.	1.19	1.92		3.11	
Clean	S.F.	.01	.10		.11	
Minimum Charge	Job		217		217	
Ceiling						Includes material and labor to install pre-pasted, vinyl-coated wallcovering.
Demolish	S.F.		1.01		1.01	
Install	S.F.	1.19	.91		2.10	
Demolish and Install	S.F.	1.19	1.92		3.11	
Clean	S.F.	.02	.23		.25	
Minimum Charge	Job		217		217	

For customer support on your Contractor's Pricing Guide: Residential Repair & Remodeling, call 800.448.8182.

Walls / Ceilings

Wallcovering

Wallcovering	Unit	Material	Labor	Equip.	Total	Specification
Mural						
Pre-printed						
Demolish	Ea.		139		139	Includes material and labor to install
Install	Ea.	320	325		645	wallcovering with a pre-printed mural.
Demolish and Install	Ea.	320	464		784	
Clean	Ea.		10		10	
Minimum Charge	Job		217		217	
Hand-painted						
Demolish	Ea.		139		139	Includes material and labor to install
Install	Ea.	620	660		1280	wallcovering with a hand-painted
Demolish and Install	Ea.	620	799		1419	mural.
Clean	Ea.		10		10	
Minimum Charge	Job		217		217	
Vinyl-coated						
Good Grade						
Demolish	Ea.		24.50		24.50	Includes material and labor to install
Install	Ea.	2.43	30.50		32.93	pre-pasted, vinyl-coated wallcovering.
Demolish and Install	Ea.	2.43	55		57.43	Based upon 30 S.F. rolls.
Clean	Ea.		10		10	
Minimum Charge	Job		217		217	
Better Grade						
Demolish	Ea.		24.50		24.50	Includes material and labor to install
Install	Ea.	29	30.50		59.50	pre-pasted, vinyl-coated wallcovering.
Demolish and Install	Ea.	29	55		84	Based upon 30 S.F. rolls.
Clean	Ea.		10		10	
Minimum Charge	Job		217		217	
Premium Grade						
Demolish	Ea.		24.50		24.50	Includes material and labor to install
Install	Ea.	42.50	30.50		73	pre-pasted, vinyl-coated wallcovering.
Demolish and Install	Ea.	42.50	55		97.50	Based upon 30 S.F. rolls.
Clean	Ea.		10		10	
Minimum Charge	Job		217		217	
Blank Underliner						
Demolish	Ea.		24.50		24.50	Includes material and labor to install
Install	Ea.	29.50	31		60.50	blank stock liner paper or bridging
Demolish and Install	Ea.	29.50	55.50		85	paper. Based upon 30 S.F. rolls.
Clean	Ea.		10		10	
Minimum Charge	Job		217		217	
Solid Vinyl						
Good Grade						
Demolish	Roll		27		27	Includes material and labor to install
Install	Ea.	28	30.50		58.50	pre-pasted, vinyl wallcovering. Based
Demolish and Install	Ea.	28	57.50		85.50	upon 30 S.F. rolls.
Clean	Ea.		10		10	
Minimum Charge	Job		217		217	
Better Grade						
Demolish	Roll		27		27	Includes material and labor to install
Install	Ea.	40	30.50		70.50	pre-pasted, vinyl wallcovering. Based
Demolish and Install	Ea.	40	57.50		97.50	upon 30 S.F. rolls.
Clean	Ea.		10		10	
Minimum Charge	Job		217		217	
Premium Grade						
Demolish	Roll		27		27	Includes material and labor to install
Install	Ea.	51.50	30.50		82	pre-pasted, vinyl wallcovering. Based
Demolish and Install	Ea.	51.50	57.50		109	upon 30 S.F. rolls.
Clean	Ea.		10		10	
Minimum Charge	Job		217		217	

For customer support on your Contractor's Pricing Guide: Residential Repair & Remodeling, call 800.448.8182.

Walls / Ceilings

Wallcovering

	Unit	Material	Labor	Equip.	Total	Specification
Blank Underliner						
Demolish	Ea.		24.50		24.50	Includes material and labor to install
Install	Ea.	29.50	31		60.50	blank stock liner paper or bridging
Demolish and Install	Ea.	29.50	55.50		85	paper. Based upon 30 S.F. rolls.
Clean	Ea.		10		10	
Minimum Charge	Job		217		217	
Grasscloth / String						
Good Grade						
Demolish	Ea.		24.50		24.50	Includes material and labor to install
Install	Ea.	55.50	39.50		95	grasscloth or string wallcovering.
Demolish and Install	Ea.	55.50	64		119.50	Based upon 30 S.F. rolls.
Clean	Ea.		13.30		13.30	
Minimum Charge	Job		217		217	
Better Grade						
Demolish	Ea.		24.50		24.50	Includes material and labor to install
Install	Ea.	68	39.50		107.50	grasscloth or string wallcovering.
Demolish and Install	Ea.	68	64		132	Based upon 30 S.F. rolls.
Clean	Ea.		13.30		13.30	
Minimum Charge	Job		217		217	
Premium Grade						
Demolish	Ea.		24.50		24.50	Includes material and labor to install
Install	Ea.	72.50	39.50		112	grasscloth or string wallcovering.
Demolish and Install	Ea.	72.50	64		136.50	Based upon 30 S.F. rolls.
Clean	Ea.		13.30		13.30	
Minimum Charge	Job		217		217	
Blank Underliner						
Demolish	Ea.		24.50		24.50	Includes material and labor to install
Install	Ea.	29.50	31		60.50	blank stock liner paper or bridging
Demolish and Install	Ea.	29.50	55.50		85	paper. Based upon 30 S.F. rolls.
Clean	Ea.		10		10	
Minimum Charge	Job		217		217	
Designer						
Premium Grade						
Demolish	S.F.		1.01		1.01	Includes material and labor to install
Install	Ea.	149	46		195	paperback vinyl, untrimmed and
Demolish and Install	Ea.	149	47.01		196.01	hand-painted. Based upon 30 S.F.
Clean	Ea.		10		10	rolls.
Minimum Charge	Job		217		217	
Blank Underliner						
Demolish	Ea.		24.50		24.50	Includes material and labor to install
Install	Ea.	29.50	31		60.50	blank stock liner paper or bridging
Demolish and Install	Ea.	29.50	55.50		85	paper. Based upon 30 S.F. rolls.
Clean	Ea.		10		10	
Minimum Charge	Job		217		217	
Foil						
Demolish	Ea.		24.50		24.50	Includes material and labor to install
Install	Ea.	93.50	39.50		133	foil coated wallcovering. Based upon
Demolish and Install	Ea.	93.50	64		157.50	30 S.F. rolls.
Clean	Ea.		10		10	
Minimum Charge	Job		217		217	
Wallcovering Border						
Good Grade						
Demolish	L.F.		1.61		1.61	Includes material and labor install
Install	L.F.	2.13	1.67		3.80	vinyl-coated border 4" to 6" wide.
Demolish and Install	L.F.	2.13	3.28		5.41	
Clean	L.F.		.15		.15	
Minimum Charge	Job		217		217	

For customer support on your Contractor's Pricing Guide: Residential Repair & Remodeling, call 800.448.8182.

Walls / Ceilings

Wallcovering

	Unit	Material	Labor	Equip.	Total	Specification
Better Grade						
Demolish	L.F.		1.61		1.61	Includes material and labor install
Install	L.F.	2.24	1.67		3.91	vinyl-coated border 4" to 6" wide.
Demolish and Install	L.F.	2.24	3.28		5.52	
Clean	L.F.		.15		.15	
Minimum Charge	Job		217		217	
Premium Grade						
Demolish	L.F.		1.61		1.61	Includes material and labor install
Install	L.F.	4.04	1.67		5.71	vinyl-coated border 4" to 6" wide.
Demolish and Install	L.F.	4.04	3.28		7.32	
Clean	L.F.		.15		.15	
Minimum Charge	Job		217		217	

Acoustical Ceiling

	Unit	Material	Labor	Equip.	Total	Specification
Complete System						
2' x 2' Square Edge						
Demolish	S.F.		.74		.74	Includes material and labor to install
Install	S.F.	4.26	1.60		5.86	T-bar grid acoustical ceiling system,
Demolish and Install	S.F.	4.26	2.34		6.60	random pinhole 5/8" thick square
Reinstall	S.F.		1.28		1.28	edge tile.
Clean	S.F.	.03	.30		.33	
Minimum Charge	Job		260		260	
2' x 2' Reveal Edge						
Demolish	S.F.		.74		.74	Includes material and labor to install
Install	S.F.	6.40	2.08		8.48	T-bar grid acoustical ceiling system,
Demolish and Install	S.F.	6.40	2.82		9.22	random pinhole 5/8" thick reveal
Reinstall	S.F.		1.67		1.67	edge tile.
Clean	S.F.	.03	.30		.33	
Minimum Charge	Job		260		260	
2' x 2' Fire-rated						
Demolish	S.F.		.74		.74	Includes material and labor to install
Install	S.F.	5.20	1.60		6.80	T-bar grid FR acoustical ceiling system,
Demolish and Install	S.F.	5.20	2.34		7.54	random pinhole 5/8" thick square
Reinstall	S.F.		1.28		1.28	edge tile.
Clean	S.F.	.03	.30		.33	
Minimum Charge	Job		260		260	
2' x 4' Square Edge						
Demolish	S.F.		.74		.74	Cost includes material and labor to
Install	S.F.	4.95	1.37		6.32	install 2' x 4' T-bar grid acoustical
Demolish and Install	S.F.	4.95	2.11		7.06	ceiling system, random pinhole 5/8"
Reinstall	S.F.		1.10		1.10	thick square edge tile.
Clean	S.F.	.03	.30		.33	
Minimum Charge	Job		260		260	
2' x 4' Reveal Edge						
Demolish	S.F.		.74		.74	Includes material and labor to install
Install	S.F.	5.60	1.89		7.49	T-bar grid acoustical ceiling system,
Demolish and Install	S.F.	5.60	2.63		8.23	random pinhole 5/8" thick reveal
Reinstall	S.F.		1.52		1.52	edge tile.
Clean	S.F.	.03	.30		.33	
Minimum Charge	Job		260		260	
2' x 4' Fire-rated						
Demolish	S.F.		.74		.74	Includes material and labor to install
Install	S.F.	5.20	1.60		6.80	T-bar grid FR acoustical ceiling system,
Demolish and Install	S.F.	5.20	2.34		7.54	random pinhole 5/8" thick square
Reinstall	S.F.		1.28		1.28	edge tile.
Clean	S.F.	.03	.30		.33	
Minimum Charge	Job		260		260	

For customer support on your Contractor's Pricing Guide: Residential Repair & Remodeling, call 800.448.8182.

Walls / Ceilings

Acoustical Ceiling

Acoustical Ceiling	Unit	Material	Labor	Equip.	Total	Specification
T-bar Grid						
2' x 2'						
Demolish	S.F.		.87		.87	Includes material and labor to install
Install	S.F.	2.35	.80		3.15	T-bar grid acoustical ceiling
Demolish and Install	S.F.	2.35	1.67		4.02	suspension system.
Reinstall	S.F.		.64		.64	
Clean	S.F.	.03	.30		.33	
Minimum Charge	Job		254		254	
2' x 4'						
Demolish	S.F.		.87		.87	Includes material and labor to install
Install	S.F.	1.89	.65		2.54	T-bar grid acoustical ceiling
Demolish and Install	S.F.	1.89	1.52		3.41	suspension system.
Reinstall	S.F.		.52		.52	
Clean	S.F.	.03	.30		.33	
Minimum Charge	Job		254		254	
Tile Panels						
2' x 2' Square Edge						
Demolish	S.F.		.25		.25	Includes material and labor to install
Install	S.F.	1.73	.83		2.56	square edge panels 5/8" thick with
Demolish and Install	S.F.	1.73	1.08		2.81	pinhole pattern. T-bar grid not
Reinstall	S.F.		.67		.67	included.
Clean	S.F.	.03	.30		.33	
Minimum Charge	Job		260		260	
2' x 2' Reveal Edge						
Demolish	S.F.		.25		.25	Includes material and labor to install
Install	S.F.	3.06	1.11		4.17	reveal edge panels 5/8" thick with
Demolish and Install	S.F.	3.06	1.36		4.42	pinhole pattern. T-bar grid not
Reinstall	S.F.		.89		.89	included.
Clean	S.F.	.03	.30		.33	
Minimum Charge	Job		260		260	
2' x 2' Fire-rated						
Demolish	S.F.		.25		.25	Includes material and labor to install
Install	S.F.	1.32	.77		2.09	FR square edge panels 5/8" thick with
Demolish and Install	S.F.	1.32	1.02		2.34	pinhole pattern. T-bar grid not
Reinstall	S.F.		.62		.62	included.
Clean	S.F.	.03	.30		.33	
Minimum Charge	Job		260		260	
2' x 4' Square Edge						
Demolish	S.F.		.25		.25	Includes material and labor to install
Install	S.F.	1.73	.83		2.56	square edge panels 5/8" thick with
Demolish and Install	S.F.	1.73	1.08		2.81	pinhole pattern. T-bar grid not
Reinstall	S.F.		.67		.67	included.
Clean	S.F.	.03	.30		.33	
Minimum Charge	Job		260		260	
2' x 4' Reveal Edge						
Demolish	S.F.		.25		.25	Includes material and labor to install
Install	S.F.	3.06	1.11		4.17	reveal edge panels 5/8" thick with
Demolish and Install	S.F.	3.06	1.36		4.42	pinhole pattern. T-bar grid not
Reinstall	S.F.		.89		.89	included.
Clean	S.F.	.03	.30		.33	
Minimum Charge	Job		260		260	
2' x 4' Fire-rated						
Demolish	S.F.		.25		.25	Includes material and labor to install
Install	S.F.	1.32	.77		2.09	FR square edge panels 5/8" thick with
Demolish and Install	S.F.	1.32	1.02		2.34	pinhole pattern. T-bar grid not
Reinstall	S.F.		.62		.62	included.
Clean	S.F.	.03	.30		.33	
Minimum Charge	Job		260		260	

For customer support on your Contractor's Pricing Guide: Residential Repair & Remodeling, call 800.448.8182.

Walls / Ceilings

Acoustical Ceiling

	Unit	Material	Labor	Equip.	Total	Specification
Bleach Out Stains						
Install	S.F.	.18	.40		.58	Includes labor and material to apply laundry type chlorine bleach to remove light staining of acoustic tile material.
Minimum Charge	Job		260		260	
Adhesive Tile						
Good Grade						
Demolish	S.F.		.99		.99	Cost includes material and labor to install 12" x 12" tile with embossed texture pattern. Furring strips not included.
Install	S.F.	2.89	1.74		4.63	
Demolish and Install	S.F.	2.89	2.73		5.62	
Clean	S.F.	.02	.23		.25	
Minimum Charge	Job		260		260	
Better Grade						
Demolish	S.F.		.99		.99	Cost includes material and labor to install 12" x 12" tile with embossed texture pattern. Furring strips not included.
Install	S.F.	3	1.74		4.74	
Demolish and Install	S.F.	3	2.73		5.73	
Clean	S.F.	.02	.23		.25	
Minimum Charge	Job		260		260	
Premium Grade						
Demolish	S.F.		.99		.99	Cost includes material and labor to install 12" x 12" tile with embossed texture pattern. Furring strips not included.
Install	S.F.	7.90	1.74		9.64	
Demolish and Install	S.F.	7.90	2.73		10.63	
Clean	S.F.	.02	.23		.25	
Minimum Charge	Job		260		260	
Bleach Out Stains						
Install	S.F.	.18	.40		.58	Includes labor and material to apply laundry type chlorine bleach to remove light staining of acoustic tile material.
Minimum Charge	Job		260		260	
Stapled Tile						
Good Grade						
Demolish	S.F.		.59		.59	Cost includes material and labor to install 12" x 12" tile with embossed texture pattern. Furring strips not included.
Install	S.F.	2.89	1.74		4.63	
Demolish and Install	S.F.	2.89	2.33		5.22	
Clean	S.F.	.02	.23		.25	
Minimum Charge	Job		260		260	
Better Grade						
Demolish	S.F.		.59		.59	Cost includes material and labor to install 12" x 12" tile with embossed texture pattern. Furring strips not included.
Install	S.F.	3	1.74		4.74	
Demolish and Install	S.F.	3	2.33		5.33	
Clean	S.F.	.02	.23		.25	
Minimum Charge	Job		260		260	
Premium Grade						
Demolish	S.F.		.59		.59	Cost includes material and labor to install 12" x 12" tile with embossed texture pattern. Furring strips not included.
Install	S.F.	7.90	1.74		9.64	
Demolish and Install	S.F.	7.90	2.33		10.23	
Clean	S.F.	.02	.23		.25	
Minimum Charge	Job		260		260	
Bleach Out Stains						
Install	S.F.	.18	.40		.58	Includes labor and material to apply laundry type chlorine bleach to remove light staining of acoustic tile material.
Minimum Charge	Job		260		260	

For customer support on your Contractor's Pricing Guide: Residential Repair & Remodeling, call 800.448.8182.

Walls / Ceilings

Acoustical Ceiling

	Unit	Material	Labor	Equip.	Total	Specification
Blown						
Scrape Off Existing						
Install	S.F.		.82		.82	Includes labor and material to remove existing blown acoustical ceiling material.
Minimum Charge	Job		260		260	
Seal Ceiling Gypsum Wallboard						
Install	S.F.	.12	.38		.50	Includes labor and material to seal existing drywall with shellac-based material.
Minimum Charge	Job		216		216	
Reblow Existing						
Install	S.F.	.07	.65		.72	Includes labor and material to install by spray, texture finish.
Minimum Charge	Job		675	95	770	
Bleach Out Stains						
Install	S.F.	.18	.40		.58	Includes labor and material to apply laundry type chlorine bleach to remove light staining of acoustic tile material.
Minimum Charge	Job		260		260	
New Blown						
Install	S.F.	.07	.65		.72	Includes labor and material to install by spray, texture finish.
Minimum Charge	Job		216		216	
Apply Glitter To						
Install	S.F.	.03	.20		.23	Includes labor and material to apply glitter material to existing or new drywall.
Minimum Charge	Job		216		216	
Furring Strips						
Demolish	S.F.		.11		.11	Cost includes material and labor to install 1" x 2" furring strips, 16" O.C. per S.F. of surface area to be covered.
Install	S.F.	.13	1.16		1.29	
Demolish and Install	S.F.	.13	1.27		1.40	
Minimum Charge	Job		130		130	

190

For customer support on your Contractor's Pricing Guide: Residential Repair & Remodeling, call 800.448.8182.

Finish Carpentry

Finished Stair

Bi-Fold Door

Raised Panel Door

Hollow Core Door	Unit	Material	Labor	Equip.	Total	Specification
Pre-hung Wood						
1' 6" x 6' 8"						
Demolish	Ea.		48		48	Cost includes material and labor to
Install	Ea.	320	29		349	install 1' 6" x 6' 8" pre-hung 1-3/8"
Demolish and Install	Ea.	320	77		397	lauan door with split pine jamb
Reinstall	Ea.		28.93		28.93	including casing, stops, and hinges.
Clean	Ea.	3.17	16.65		19.82	
Paint	Ea.	21	61.50		82.50	
Minimum Charge	Job		260		260	
2' x 6' 8"						
Demolish	Ea.		48		48	Cost includes material and labor to
Install	Ea.	330	30.50		360.50	install 2' x 6' 8" pre-hung 1-3/8"
Demolish and Install	Ea.	330	78.50		408.50	lauan door with split pine jamb
Reinstall	Ea.		30.64		30.64	including casing, stops, and hinges.
Clean	Ea.	3.17	16.65		19.82	
Paint	Ea.	21	61.50		82.50	
Minimum Charge	Job		260		260	
2' 4" x 6' 8"						
Demolish	Ea.		48		48	Cost includes material and labor to
Install	Ea.	345	30.50		375.50	install 2' 4" x 6' 8" pre-hung 1-3/8"
Demolish and Install	Ea.	345	78.50		423.50	lauan door with split pine jamb
Reinstall	Ea.		30.64		30.64	including casing, stops, and hinges.
Clean	Ea.	3.17	16.65		19.82	
Paint	Ea.	21	61.50		82.50	
Minimum Charge	Job		260		260	
2' 6" x 6' 8"						
Demolish	Ea.		48		48	Cost includes material and labor to
Install	Ea.	345	30.50		375.50	install 2' 6" x 6' 8" pre-hung 1-3/8"
Demolish and Install	Ea.	345	78.50		423.50	lauan door with split pine jamb
Reinstall	Ea.		30.64		30.64	including casing, stops, and hinges.
Clean	Ea.	3.17	16.65		19.82	
Paint	Ea.	21	61.50		82.50	
Minimum Charge	Job		260		260	
2' 8" x 6' 8"						
Demolish	Ea.		48		48	Cost includes material and labor to
Install	Ea.	350	30.50		380.50	install 2' 8" x 6' 8" pre-hung 1-3/8"
Demolish and Install	Ea.	350	78.50		428.50	lauan door with split pine jamb,
Reinstall	Ea.		30.64		30.64	including casing, stop and hinges.
Clean	Ea.	3.17	16.65		19.82	
Paint	Ea.	21	61.50		82.50	
Minimum Charge	Job		260		260	

For customer support on your Contractor's Pricing Guide: Residential Repair & Remodeling, call 800.448.8182.

Finish Carpentry

Hollow Core Door

	Unit	Material	Labor	Equip.	Total	Specification
3' x 6' 8"						
Demolish	Ea.		48		48	Cost includes material and labor to
Install	Ea.	360	32.50		392.50	install 3' x 6' 8" pre-hung 1-3/8"
Demolish and Install	Ea.	360	80.50		440.50	lauan door with split pine jamb
Reinstall	Ea.		32.55		32.55	including casing, stops, and hinges.
Clean	Ea.	3.17	16.65		19.82	
Paint	Ea.	21	61.50		82.50	
Minimum Charge	Job		260		260	
Pocket Type						
Demolish	Ea.		72.50		72.50	Cost includes material and labor to
Install	Ea.	120	58		178	install 2' 8" x 6' 8" lauan door.
Demolish and Install	Ea.	120	130.50		250.50	
Reinstall	Ea.		57.87		57.87	
Clean	Ea.	3.17	16.65		19.82	
Paint	Ea.	21	61.50		82.50	
Minimum Charge	Job		260		260	
Stain						
Install	Ea.	13.90	36		49.90	Includes labor and material to stain
Minimum Charge	Job		216		216	single door and trim on both sides.
Pre-hung Masonite						
1' 6" x 6' 8"						
Demolish	Ea.		48		48	Cost includes material and labor to
Install	Ea.	460	29		489	install 1' 6" x 6' 8" pre-hung 1-3/8"
Demolish and Install	Ea.	460	77		537	masonite door with split pine jamb
Reinstall	Ea.		28.93		28.93	including casing, stops, and hinges.
Clean	Ea.	3.17	16.65		19.82	
Paint	Ea.	21	61.50		82.50	
Minimum Charge	Job		260		260	
2' x 6' 8"						
Demolish	Ea.		48		48	Cost includes material and labor to
Install	Ea.	500	30.50		530.50	install 2' x 6' 8" pre-hung 1-3/8"
Demolish and Install	Ea.	500	78.50		578.50	masonite door with split pine jamb
Reinstall	Ea.		30.64		30.64	including casing, stops, and hinges.
Clean	Ea.	3.17	16.65		19.82	
Paint	Ea.	21	61.50		82.50	
Minimum Charge	Job		260		260	
2' 4" x 6' 8"						
Demolish	Ea.		48		48	Cost includes material and labor to
Install	Ea.	530	30.50		560.50	install 2' 4" x 6' 8" pre-hung 1-3/8"
Demolish and Install	Ea.	530	78.50		608.50	masonite door with split pine jamb
Reinstall	Ea.		30.64		30.64	including casing, stops, and hinges.
Clean	Ea.	3.17	16.65		19.82	
Paint	Ea.	21	61.50		82.50	
Minimum Charge	Job		260		260	
2' 8" x 6' 8"						
Demolish	Ea.		48		48	Cost includes material and labor to
Install	Ea.	555	30.50		585.50	install 2' 8" x 6' 8" pre-hung 1-3/8"
Demolish and Install	Ea.	555	78.50		633.50	masonite door with split pine jamb
Reinstall	Ea.		30.64		30.64	including casing, stops, and hinges.
Clean	Ea.	3.17	16.65		19.82	
Paint	Ea.	21	61.50		82.50	
Minimum Charge	Job		260		260	
3' x 6' 8"						
Demolish	Ea.		48		48	Cost includes material and labor to
Install	Ea.	585	32.50		617.50	install 3' x 6' 8" pre-hung 1-3/8"
Demolish and Install	Ea.	585	80.50		665.50	masonite door with split pine jamb
Reinstall	Ea.		32.55		32.55	including casing, stops, and hinges.
Clean	Ea.	3.17	16.65		19.82	
Paint	Ea.	21	61.50		82.50	
Minimum Charge	Job		260		260	

For customer support on your Contractor's Pricing Guide: Residential Repair & Remodeling, call 800.448.8182.

Finish Carpentry

Hollow Core Door

	Unit	Material	Labor	Equip.	Total	Specification
Pocket Type						
Demolish	Ea.		72.50		72.50	Cost includes material and labor to
Install	Ea.	120	58		178	install 2' 8" x 6' 8" lauan door.
Demolish and Install	Ea.	120	130.50		250.50	
Reinstall	Ea.		57.87		57.87	
Clean	Ea.	3.17	16.65		19.82	
Paint	Ea.	21	61.50		82.50	
Minimum Charge	Job		260		260	
Stain						
Install	Ea.	13.90	36		49.90	Includes labor and material to stain
Minimum Charge	Job		216		216	single door and trim on both sides.

Hollow Core Door Only

1' 6" x 6' 8"

	Unit	Material	Labor	Equip.	Total	Specification
Demolish	Ea.		10.35		10.35	Cost includes material and labor to
Install	Ea.	97.50	27.50		125	install 1' 6" x 6' 8" lauan hollow core
Demolish and Install	Ea.	97.50	37.85		135.35	door slab.
Reinstall	Ea.		27.41		27.41	
Clean	Ea.	1.32	12.50		13.82	
Paint	Ea.	8.70	54		62.70	
Minimum Charge	Job		260		260	

2' x 6' 8"

	Unit	Material	Labor	Equip.	Total	Specification
Demolish	Ea.		10.35		10.35	Cost includes material and labor to
Install	Ea.	111	29		140	install 2' x 6' 8" lauan hollow core
Demolish and Install	Ea.	111	39.35		150.35	door slab.
Reinstall	Ea.		28.93		28.93	
Clean	Ea.	1.32	12.50		13.82	
Paint	Ea.	8.70	54		62.70	
Minimum Charge	Job		260		260	

2' 4" x 6' 8"

	Unit	Material	Labor	Equip.	Total	Specification
Demolish	Ea.		10.35		10.35	Cost includes material and labor to
Install	Ea.	123	29		152	install 2' 4" x 6' 8" lauan hollow core
Demolish and Install	Ea.	123	39.35		162.35	door slab.
Reinstall	Ea.		28.93		28.93	
Clean	Ea.	1.32	12.50		13.82	
Paint	Ea.	8.70	54		62.70	
Minimum Charge	Job		260		260	

2' 8" x 6' 8"

	Unit	Material	Labor	Equip.	Total	Specification
Demolish	Ea.		10.35		10.35	Cost includes material and labor to
Install	Ea.	128	29		157	install 2' 8" x 6' 8" lauan hollow core
Demolish and Install	Ea.	128	39.35		167.35	door slab.
Reinstall	Ea.		28.93		28.93	
Clean	Ea.	1.32	12.50		13.82	
Paint	Ea.	8.70	54		62.70	
Minimum Charge	Job		260		260	

3' x 6' 8"

	Unit	Material	Labor	Equip.	Total	Specification
Demolish	Ea.		10.35		10.35	Cost includes material and labor to
Install	Ea.	135	30.50		165.50	install 3' x 6' 8" lauan hollow core
Demolish and Install	Ea.	135	40.85		175.85	door slab.
Reinstall	Ea.		30.64		30.64	
Clean	Ea.	1.32	12.50		13.82	
Paint	Ea.	8.70	54		62.70	
Minimum Charge	Job		260		260	

Pocket Type

	Unit	Material	Labor	Equip.	Total	Specification
Demolish	Ea.		10.35		10.35	Cost includes material and labor to
Install	Ea.	120	58		178	install 2' 8" x 6' 8" lauan door.
Demolish and Install	Ea.	120	68.35		188.35	
Reinstall	Ea.		57.87		57.87	
Clean	Ea.	1.32	12.50		13.82	
Paint	Ea.	8.70	54		62.70	
Minimum Charge	Job		260		260	

For customer support on your Contractor's Pricing Guide: Residential Repair & Remodeling, call 800.448.8182.

Finish Carpentry

Hollow Core Door

	Unit	Material	Labor	Equip.	Total	Specification
Stain						
Install	Ea.	7.70	48		55.70	Includes labor and materials to stain a
Minimum Charge	Job		216		216	flush style door on both sides by brush.
Casing Trim						
Single Width						
Demolish	Ea.		9.20		9.20	Cost includes material and labor to
Install	Opng.	31.50	43.50		75	install 11/16" x 2-1/2" pine ranch
Demolish and Install	Opng.	31.50	52.70		84.20	style casing for one side of a standard
Reinstall	Opng.		34.72		34.72	door opening.
Clean	Opng.	.45	4		4.45	
Paint	Ea.	10.20	8.65		18.85	
Minimum Charge	Job		260		260	
Double Width						
Demolish	Ea.		10.90		10.90	Cost includes material and labor to
Install	Opng.	37	52		89	install 11/16" x 2-1/2" pine ranch
Demolish and Install	Opng.	37	62.90		99.90	style door casing for one side of a
Reinstall	Opng.		41.66		41.66	double door opening.
Clean	Opng.	.53	8		8.53	
Paint	Ea.	10.20	9.80		20	
Minimum Charge	Job		260		260	
Hardware						
Doorknob w / Lock						
Demolish	Ea.		20.50		20.50	Includes material and labor to install
Install	Ea.	65.50	32.50		98	privacy lockset.
Demolish and Install	Ea.	65.50	53		118.50	
Reinstall	Ea.		26.04		26.04	
Clean	Ea.	.32	12.50		12.82	
Minimum Charge	Job		260		260	
Doorknob						
Demolish	Ea.		20.50		20.50	Includes material and labor to install
Install	Ea.	63.50	32.50		96	residential passage lockset, keyless
Demolish and Install	Ea.	63.50	53		116.50	bored type.
Reinstall	Ea.		26.04		26.04	
Clean	Ea.	.32	12.50		12.82	
Minimum Charge	Job		260		260	
Deadbolt						
Demolish	Ea.		18.10		18.10	Includes material and labor to install a
Install	Ea.	136	37		173	deadbolt.
Demolish and Install	Ea.	136	55.10		191.10	
Reinstall	Ea.		29.76		29.76	
Clean	Ea.	.32	12.50		12.82	
Minimum Charge	Job		260		260	
Lever Handle						
Demolish	Ea.		20.50		20.50	Includes material and labor to install
Install	Ea.	273	52		325	residential passage lockset, keyless
Demolish and Install	Ea.	273	72.50		345.50	bored type with lever handle.
Reinstall	Ea.		41.66		41.66	
Clean	Ea.	.32	12.50		12.82	
Minimum Charge	Job		260		260	
Closer						
Demolish	Ea.		10.85		10.85	Includes material and labor to install
Install	Ea.	261	80		341	pneumatic light duty closer for interior
Demolish and Install	Ea.	261	90.85		351.85	type doors.
Reinstall	Ea.		64.10		64.10	
Clean	Ea.	.15	16.65		16.80	
Minimum Charge	Job		260		260	

For customer support on your Contractor's Pricing Guide: Residential Repair & Remodeling, call 800.448.8182.

Finish Carpentry

Hollow Core Door

		Unit	Material	Labor	Equip.	Total	Specification
Push Plate							
	Demolish	Ea.		16.05		16.05	Includes material and labor to install
	Install	Ea.	38	43.50		81.50	bronze push-pull plate.
	Demolish and Install	Ea.	38	59.55		97.55	
	Reinstall	Ea.		34.72		34.72	
	Clean	Ea.	.32	12.50		12.82	
	Minimum Charge	Job		260		260	
Kickplate							
	Demolish	Ea.		16.05		16.05	Includes material and labor to install
	Install	Ea.	40	34.50		74.50	aluminum kickplate, 10″ x 28″.
	Demolish and Install	Ea.	40	50.55		90.55	
	Reinstall	Ea.		27.78		27.78	
	Clean	Ea.	.32	12.50		12.82	
	Minimum Charge	Job		260		260	
Exit Sign							
	Demolish	Ea.		31.50		31.50	Includes material and labor to install a
	Install	Ea.	121	77.50		198.50	wall mounted interior electric exit sign.
	Demolish and Install	Ea.	121	109		230	
	Reinstall	Ea.		62.08		62.08	
	Clean	Ea.	.02	10.10		10.12	
	Minimum Charge	Job		260		260	
Jamb							
	Demolish	Ea.		36		36	Includes material and labor to install
	Install	Ea.	161	14.10		175.10	flat pine jamb with square cut heads
	Demolish and Install	Ea.	161	50.10		211.10	and rabbeted sides for 6′ 8″ high and
	Clean	Ea.	.10	10		10.10	3-9/16″ door including trim sets for
	Paint	Ea.	20.50	29		49.50	both sides.
	Minimum Charge	Job		260		260	
Shave & Refit							
	Install	Ea.		43.50		43.50	Includes labor to shave and rework
	Minimum Charge	Job		260		260	door to fit opening at the job site.

Solid Core Door

		Unit	Material	Labor	Equip.	Total	Specification
Pre-hung Wood							
2′ 4″ x 6′ 8″							
	Demolish	Ea.		48		48	Cost includes material and labor to
	Install	Ea.	485	30.50		515.50	install 2′ 4″ x 6′ 8″ pre-hung 1-3/4″
	Demolish and Install	Ea.	485	78.50		563.50	door, 4-1/2″ split jamb, including
	Reinstall	Ea.		30.64		30.64	casing and stop, hinges, aluminum sill,
	Clean	Ea.	3.17	16.65		19.82	weatherstripping.
	Paint	Ea.	30.50	61.50		92	
	Minimum Charge	Job		260		260	
2′ 6″ x 6′ 8″							
	Demolish	Ea.		48		48	Cost includes material and labor to
	Install	Ea.	505	30.50		535.50	install 2′ 6″ x 6′ 8″ pre-hung 1-3/4″
	Demolish and Install	Ea.	505	78.50		583.50	door, 4-1/2″ split jamb, including
	Reinstall	Ea.		30.64		30.64	casing and stop, hinges, aluminum sill,
	Clean	Ea.	3.17	16.65		19.82	weatherstripping.
	Paint	Ea.	30.50	61.50		92	
	Minimum Charge	Job		260		260	

For customer support on your Contractor's Pricing Guide: Residential Repair & Remodeling, call 800.448.8182.

Finish Carpentry

Solid Core Door

	Unit	Material	Labor	Equip.	Total	Specification
2' 8" x 6' 8"						
Demolish	Ea.		48		48	Cost includes material and labor to
Install	Ea.	510	30.50		540.50	install 2' 8" x 6' 8" pre-hung 1-3/4"
Demolish and Install	Ea.	510	78.50		588.50	door, 4-1/2" split jamb, including
Reinstall	Ea.		30.64		30.64	casing and stop, hinges, aluminum sill,
Clean	Ea.	3.17	16.65		19.82	weatherstripping.
Paint	Ea.	30.50	61.50		92	
Minimum Charge	Job		260		260	
3' x 6' 8"						
Demolish	Ea.		48		48	Cost includes material and labor to
Install	Ea.	540	32.50		572.50	install 3' x 6' 8" pre-hung 1-3/4"
Demolish and Install	Ea.	540	80.50		620.50	door, including casing and stop,
Reinstall	Ea.		32.55		32.55	hinges, jamb, aluminum sill,
Clean	Ea.	3.17	16.65		19.82	weatherstripped.
Paint	Ea.	30.50	61.50		92	
Minimum Charge	Job		260		260	
3' 6" x 6' 8"						
Demolish	Ea.		48		48	Cost includes material and labor to
Install	Ea.	510	34.50		544.50	install 3' 6" x 6' 8" pre-hung 1-3/4"
Demolish and Install	Ea.	510	82.50		592.50	door, 4-1/2" split jamb, including
Reinstall	Ea.		34.72		34.72	casing and stop, hinges, aluminum sill,
Clean	Ea.	3.17	16.65		19.82	weatherstripping.
Paint	Ea.	30.50	61.50		92	
Minimum Charge	Job		260		260	
Raised Panel						
Demolish	Ea.		48		48	Cost includes material and labor to
Install	Ea.	330	29		359	install 3' x 6' 8" pre-hung 1-3/4"
Demolish and Install	Ea.	330	77		407	door, jamb, including casing and
Reinstall	Ea.		28.93		28.93	stop, hinges, aluminum sill,
Clean	Ea.	3.17	16.65		19.82	weatherstripping.
Paint	Ea.	30.50	144		174.50	
Minimum Charge	Job		260		260	
Stain						
Install	Ea.	13.90	36		49.90	Includes labor and material to stain
Minimum Charge	Job		216		216	single door and trim on both sides.

Bifold Door

	Unit	Material	Labor	Equip.	Total	Specification
Wood, Single						
3'						
Demolish	Ea.		36		36	Cost includes material and labor to
Install	Ea.	244	80		324	install 3' wide solid wood bi-fold door
Demolish and Install	Ea.	244	116		360	not including jamb, trim, track, and
Reinstall	Ea.		80.12		80.12	hardware.
Clean	Ea.	3.17	16.65		19.82	
Paint	Ea.	21	61.50		82.50	
Minimum Charge	Job		260		260	
5'						
Demolish	Ea.		36		36	Includes material and labor to install
Install	Ea.	430	94.50		524.50	pair of 2' 6" wide solid wood bi-fold
Demolish and Install	Ea.	430	130.50		560.50	doors not including jamb, trim, track,
Reinstall	Ea.		94.69		94.69	and hardware.
Clean	Ea.	1.39	28.50		29.89	
Paint	Ea.	60.50	108		168.50	
Minimum Charge	Job		260		260	

Finish Carpentry

Bifold Door

	Unit	Material	Labor	Equip.	Total	Specification
6'						
Demolish	Ea.		36		36	Includes material and labor to install
Install	Ea.	640	104		744	pair of 3' wide solid wood bi-fold
Demolish and Install	Ea.	640	140		780	doors not including jamb, trim, track,
Reinstall	Ea.		104.16		104.16	and hardware.
Clean	Ea.	1.58	33.50		35.08	
Paint	Ea.	60.50	108		168.50	
Minimum Charge	Job		260		260	
8'						
Demolish	Ea.		36		36	Includes material and labor to install
Install	Ea.	1125	116		1241	pair of 4' wide solid wood bi-fold
Demolish and Install	Ea.	1125	152		1277	doors not including jamb, trim, track,
Reinstall	Ea.		115.73		115.73	and hardware.
Clean	Ea.	2.18	40		42.18	
Paint	Ea.	31.50	108		139.50	
Minimum Charge	Job		260		260	

Mirrored

	Unit	Material	Labor	Equip.	Total	Specification
6'						
Demolish	Ea.		36		36	Includes material and labor to install
Install	Ea.	870	58		928	pair of 3' wide mirrored bi-fold doors
Demolish and Install	Ea.	870	94		964	including jamb, trim, track, and
Reinstall	Ea.		57.87		57.87	hardware.
Clean	Ea.	1.58	16.65		18.23	
Minimum Charge	Job		260		260	
8'						
Demolish	Ea.		36		36	Includes material and labor to install
Install	Ea.	1475	87		1562	pair of 4' wide mirrored bi-fold doors
Demolish and Install	Ea.	1475	123		1598	including jamb, trim, track, and
Reinstall	Ea.		86.80		86.80	hardware.
Clean	Ea.	2.18	20		22.18	
Minimum Charge	Job		260		260	

Louvered

	Unit	Material	Labor	Equip.	Total	Specification
3'						
Demolish	Ea.		36		36	Cost includes material and labor to
Install	Ea.	217	80		297	install 3' wide louvered bi-fold door
Demolish and Install	Ea.	217	116		333	not including jamb, trim, track, and
Reinstall	Ea.		80.12		80.12	hardware.
Clean	Ea.	3.17	16.65		19.82	
Paint	Ea.	21	61.50		82.50	
Minimum Charge	Job		260		260	
5'						
Demolish	Ea.		36		36	Includes material and labor to install
Install	Ea.	395	94.50		489.50	pair of 2' 6" wide louvered wood
Demolish and Install	Ea.	395	130.50		525.50	bi-fold doors not including jamb, trim,
Reinstall	Ea.		94.69		94.69	track, and hardware.
Clean	Ea.	1.39	28.50		29.89	
Paint	Ea.	60.50	108		168.50	
Minimum Charge	Job		260		260	
6'						
Demolish	Ea.		36		36	Includes material and labor to install
Install	Ea.	580	104		684	pair of 3' wide louvered wood bi-fold
Demolish and Install	Ea.	580	140		720	doors not including jamb, trim, track,
Reinstall	Ea.		104.16		104.16	and hardware.
Clean	Ea.	1.58	33.50		35.08	
Paint	Ea.	60.50	108		168.50	
Minimum Charge	Job		260		260	

For customer support on your Contractor's Pricing Guide: Residential Repair & Remodeling, call 800.448.8182.

Finish Carpentry

Bifold Door

	Unit	Material	Labor	Equip.	Total	Specification
8'						
Demolish	Ea.		36		36	Includes material and labor to install
Install	Ea.	1125	104		1229	pair of 4' wide louvered wood bi-fold
Demolish and Install	Ea.	1125	140		1265	doors not including jamb, trim, track,
Reinstall	Ea.		104.16		104.16	and hardware.
Clean	Ea.	7.55	28.50		36.05	
Paint	Ea.	31.50	108		139.50	
Minimum Charge	Job		260		260	
Accordion Type						
Demolish	S.F.		1.31		1.31	Includes material and labor to install
Install	Ea.	262	104		366	vinyl folding accordion closet doors,
Demolish and Install	Ea.	262	105.31		367.31	including track and frame.
Reinstall	Ea.		104.16		104.16	
Clean	S.F.		.54		.54	
Minimum Charge	Job		260		260	
Casing Trim						
Single Width						
Demolish	Ea.		9.20		9.20	Cost includes material and labor to
Install	Opng.	31.50	43.50		75	install 11/16" x 2-1/2" pine ranch
Demolish and Install	Opng.	31.50	52.70		84.20	style casing for one side of a standard
Reinstall	Opng.		34.72		34.72	door opening.
Clean	Opng.	.45	4		4.45	
Paint	Ea.	10.20	8.65		18.85	
Minimum Charge	Job		260		260	
Double Width						
Demolish	Ea.		10.90		10.90	Cost includes material and labor to
Install	Opng.	37	52		89	install 11/16" x 2-1/2" pine ranch
Demolish and Install	Opng.	37	62.90		99.90	style door casing for one side of a
Reinstall	Opng.		41.66		41.66	double door opening.
Clean	Opng.	.53	8		8.53	
Paint	Ea.	10.20	9.80		20	
Minimum Charge	Job		260		260	
Jamb						
Demolish	Ea.		36		36	Includes material and labor to install
Install	Ea.	161	14.10		175.10	flat pine jamb with square cut heads
Demolish and Install	Ea.	161	50.10		211.10	and rabbeted sides for 6' 8" high and
Clean	Ea.	.10	10		10.10	3-9/16" door including trim sets for
Paint	Ea.	20.50	29		49.50	both sides.
Minimum Charge	Job		130		130	
Stain						
Install	Ea.	13.90	36		49.90	Includes labor and material to stain
Minimum Charge	Job		216		216	single door and trim on both sides.
Shave & Refit						
Install	Ea.		43.50		43.50	Includes labor to shave and rework
Minimum Charge	Job		130		130	door to fit opening at the job site.

For customer support on your Contractor's Pricing Guide: Residential Repair & Remodeling, call 800.448.8182.

Finish Carpentry

Louver Door		Unit	Material	Labor	Equip.	Total	Specification
Full							
2' x 6' 8"							
	Demolish	Ea.		10.35		10.35	Cost includes material and labor to
	Install	Ea.	370	52		422	install 2' x 6' 8" pine full louver door,
	Demolish and Install	Ea.	370	62.35		432.35	jamb, hinges.
	Reinstall	Ea.		52.08		52.08	
	Clean	Ea.	5.85	50		55.85	
	Paint	Ea.	21	61.50		82.50	
	Minimum Charge	Job		260		260	
2' 8" x 6' 8"							
	Demolish	Ea.		10.35		10.35	Cost includes material and labor to
	Install	Ea.	430	52		482	install 2' 8" x 6' 8" pine full louver
	Demolish and Install	Ea.	430	62.35		492.35	door, jamb, hinges.
	Reinstall	Ea.		52.08		52.08	
	Clean	Ea.	5.85	50		55.85	
	Paint	Ea.	21	61.50		82.50	
	Minimum Charge	Job		260		260	
3' x 6' 8"							
	Demolish	Ea.		10.35		10.35	Cost includes material and labor to
	Install	Ea.	455	55		510	install 3' x 6' 8" pine full louver door,
	Demolish and Install	Ea.	455	65.35		520.35	jamb, hinges.
	Reinstall	Ea.		54.82		54.82	
	Clean	Ea.	5.85	50		55.85	
	Paint	Ea.	21	61.50		82.50	
	Minimum Charge	Job		260		260	
Half							
2' x 6' 8"							
	Demolish	Ea.		10.35		10.35	Cost includes material and labor to
	Install	Ea.	450	52		502	install 2' x 6' 8" pine half louver door,
	Demolish and Install	Ea.	450	62.35		512.35	jamb, hinges.
	Reinstall	Ea.		52.08		52.08	
	Clean	Ea.	5.85	50		55.85	
	Paint	Ea.	21	61.50		82.50	
	Minimum Charge	Job		260		260	
2' 8" x 6' 8"							
	Demolish	Ea.		10.35		10.35	Cost includes material and labor to
	Install	Ea.	470	52		522	install 2' 8" x 6' 8" pine half louver
	Demolish and Install	Ea.	470	62.35		532.35	door, jamb, hinges.
	Reinstall	Ea.		52.08		52.08	
	Clean	Ea.	5.85	50		55.85	
	Paint	Ea.	21	61.50		82.50	
	Minimum Charge	Job		260		260	
3' x 6' 8"							
	Demolish	Ea.		10.35		10.35	Cost includes material and labor to
	Install	Ea.	510	55		565	install 3' x 6' 8" pine half louver door,
	Demolish and Install	Ea.	510	65.35		575.35	jamb, hinges.
	Reinstall	Ea.		54.82		54.82	
	Clean	Ea.	5.85	50		55.85	
	Paint	Ea.	21	61.50		82.50	
	Minimum Charge	Job		260		260	
Jamb							
	Demolish	Ea.		36		36	Includes material and labor to install
	Install	Ea.	161	14.10		175.10	flat pine jamb with square cut heads
	Demolish and Install	Ea.	161	50.10		211.10	and rabbeted sides for 6' 8" high and
	Clean	Ea.	.10	10		10.10	3-9/16" door including trim sets for
	Paint	Ea.	20.50	29		49.50	both sides.
	Minimum Charge	Job		130		130	

199

For customer support on your Contractor's Pricing Guide: Residential Repair & Remodeling, call 800.448.8182.

Finish Carpentry

Louver Door

Louver Door		Unit	Material	Labor	Equip.	Total	Specification
Stain							
	Install	Ea.	13.90	36		49.90	Includes labor and material to stain
	Minimum Charge	Job		216		216	single door and trim on both sides.
Shave & Refit							
	Install	Ea.		43.50		43.50	Includes labor to shave and rework
	Minimum Charge	Job		216		216	door to fit opening at the job site.

Bypass Sliding Door

Bypass Sliding Door		Unit	Material	Labor	Equip.	Total	Specification
Wood							
5'							
	Demolish	Ea.		48		48	Includes material and labor to install
	Install	Opng.	475	94.50		569.50	pair of 2' 6" wide lauan hollow core
	Demolish and Install	Opng.	475	142.50		617.50	wood bi-pass doors with jamb, track
	Reinstall	Opng.		94.69		94.69	hardware.
	Clean	Ea.	1.39	28.50		29.89	
	Paint	Ea.	60.50	108		168.50	
	Minimum Charge	Job		260		260	
6'							
	Demolish	Ea.		48		48	Includes material and labor to install
	Install	Opng.	505	104		609	pair of 3' wide lauan hollow core
	Demolish and Install	Opng.	505	152		657	wood bi-pass doors with jamb, track
	Reinstall	Opng.		104.16		104.16	and hardware.
	Clean	Ea.	1.58	33.50		35.08	
	Paint	Ea.	60.50	108		168.50	
	Minimum Charge	Job		260		260	
8'							
	Demolish	Ea.		48		48	Includes material and labor to install
	Install	Opng.	1125	130		1255	pair of 4' wide lauan hollow core
	Demolish and Install	Opng.	1125	178		1303	wood bi-pass doors with jamb, track
	Reinstall	Opng.		130.20		130.20	and hardware.
	Clean	Ea.	2.18	40		42.18	
	Paint	Ea.	31.50	108		139.50	
	Minimum Charge	Job		260		260	
Mirrored							
6'							
	Demolish	Ea.		48		48	Includes material and labor to install
	Install	Opng.	460	104		564	pair of 3' wide mirrored bi-pass doors
	Demolish and Install	Opng.	460	152		612	with jamb, trim, track hardware.
	Reinstall	Opng.		104.16		104.16	
	Clean	Ea.	1.58	16.65		18.23	
	Minimum Charge	Job		260		260	
8'							
	Demolish	Ea.		48		48	Includes material and labor to install
	Install	Opng.	935	116		1051	pair of 4' wide mirrored bi-pass doors
	Demolish and Install	Opng.	935	164		1099	with jamb, trim, track hardware.
	Reinstall	Opng.		115.73		115.73	
	Clean	Ea.	2.18	20		22.18	
	Minimum Charge	Job		260		260	

For customer support on your **Contractor's Pricing Guide: Residential Repair & Remodeling, call 800.448.8182.**

Finish Carpentry

Bypass Sliding Door

	Unit	Material	Labor	Equip.	Total	Specification
Casing Trim						
Single Width						
Demolish	Ea.		9.20		9.20	Cost includes material and labor to
Install	Opng.	31.50	43.50		75	install 11/16" x 2-1/2" pine ranch
Demolish and Install	Opng.	31.50	52.70		84.20	style casing for one side of a standard
Reinstall	Opng.		34.72		34.72	door opening.
Clean	Opng.	.45	4		4.45	
Paint	Ea.	10.20	8.65		18.85	
Minimum Charge	Job		260		260	
Double Width						
Demolish	Ea.		10.90		10.90	Cost includes material and labor to
Install	Opng.	37	52		89	install 11/16" x 2-1/2" pine ranch
Demolish and Install	Opng.	37	62.90		99.90	style door casing for one side of a
Reinstall	Opng.		41.66		41.66	double door opening.
Clean	Opng.	.53	8		8.53	
Paint	Ea.	10.20	9.80		20	
Minimum Charge	Job		260		260	
Jamb						
Demolish	Ea.		36		36	Includes material and labor to install
Install	Ea.	161	14.10		175.10	flat pine jamb with square cut heads
Demolish and Install	Ea.	161	50.10		211.10	and rabbeted sides for 6' 8" high and
Clean	Ea.	.10	10		10.10	3-9/16" door including trim sets for
Paint	Ea.	20.50	29		49.50	both sides.
Minimum Charge	Job		130		130	
Stain						
Install	Ea.	13.90	36		49.90	Includes labor and material to stain
Minimum Charge	Job		216		216	single door and trim on both sides.
Shave & Refit						
Install	Ea.		43.50		43.50	Includes labor to shave and rework
Minimum Charge	Job		130		130	door to fit opening at the job site.

Base Molding

	Unit	Material	Labor	Equip.	Total	Specification
One Piece						
1-5/8"						
Demolish	L.F.		.74		.74	Includes material and labor to install
Install	L.F.	1.35	1.54		2.89	pine or birch base molding.
Demolish and Install	L.F.	1.35	2.28		3.63	
Reinstall	L.F.		1.23		1.23	
Clean	L.F.	.02	.23		.25	
Paint	L.F.	.10	.47		.57	
Minimum Charge	Job		260		260	
2-1/2"						
Demolish	L.F.		.74		.74	Cost includes material and labor to
Install	L.F.	1.38	1.57		2.95	install 1/2" x 2-1/2" pine or birch
Demolish and Install	L.F.	1.38	2.31		3.69	base molding.
Reinstall	L.F.		1.26		1.26	
Clean	L.F.	.02	.23		.25	
Paint	L.F.	.10	.47		.57	
Minimum Charge	Job		260		260	

Finish Carpentry

Base Molding

	Unit	Material	Labor	Equip.	Total	Specification
3-1/2″						
Demolish	L.F.		.74		.74	Cost includes material and labor to
Install	L.F.	.93	1.67		2.60	install 9/16″ x 3-1/2″ pine or birch
Demolish and Install	L.F.	.93	2.41		3.34	base molding.
Reinstall	L.F.		1.34		1.34	
Clean	L.F.	.02	.23		.25	
Paint	L.F.	.17	.51		.68	
Minimum Charge	Job		260		260	
6″						
Demolish	L.F.		.74		.74	Cost includes material and labor to
Install	L.F.	2.17	1.67		3.84	install 1″ x 6″ pine or birch base
Demolish and Install	L.F.	2.17	2.41		4.58	molding.
Reinstall	L.F.		1.34		1.34	
Clean	L.F.	.02	.33		.35	
Paint	L.F.	.18	.65		.83	
Minimum Charge	Job		260		260	
8″						
Demolish	L.F.		.89		.89	Cost includes material and labor to
Install	L.F.	2.91	1.67		4.58	install 1″ x 8″ pine or birch base
Demolish and Install	L.F.	2.91	2.56		5.47	molding.
Reinstall	L.F.		1.34		1.34	
Clean	L.F.	.03	.40		.43	
Paint	L.F.	.20	.72		.92	
Minimum Charge	Job		260		260	
Premium / Custom Grade						
Demolish	L.F.		.89		.89	Includes material and labor to install
Install	L.F.	3.70	5		8.70	custom 2 or 3 piece oak or other
Demolish and Install	L.F.	3.70	5.89		9.59	hardwood, stain grade trim.
Reinstall	L.F.		4.01		4.01	
Clean	L.F.	.03	.40		.43	
Paint	L.F.	.21	.72		.93	
Minimum Charge	Job		260		260	

One Piece Oak

	Unit	Material	Labor	Equip.	Total	Specification
1-5/8″						
Demolish	L.F.		.74		.74	Cost includes material and labor to
Install	L.F.	2.76	1.54		4.30	install 1/2″ x 1-5/8″ oak base
Demolish and Install	L.F.	2.76	2.28		5.04	molding.
Reinstall	L.F.		1.23		1.23	
Clean	L.F.	.02	.23		.25	
Paint	L.F.	.20	.64		.84	
Minimum Charge	Job		260		260	
2-1/2″						
Demolish	L.F.		.74		.74	Cost includes material and labor to
Install	L.F.	2.35	1.57		3.92	install 1/2″ x 2-1/2″ oak base
Demolish and Install	L.F.	2.35	2.31		4.66	molding.
Reinstall	L.F.		1.26		1.26	
Clean	L.F.	.02	.23		.25	
Paint	L.F.	.20	.65		.85	
Minimum Charge	Job		260		260	
3-1/2″						
Demolish	L.F.		.74		.74	Cost includes material and labor to
Install	L.F.	2.97	1.60		4.57	install 1/2″ x 3-1/2″ oak base
Demolish and Install	L.F.	2.97	2.34		5.31	molding.
Reinstall	L.F.		1.28		1.28	
Clean	L.F.	.02	.23		.25	
Paint	L.F.	.21	.66		.87	
Minimum Charge	Job		260		260	

Finish Carpentry

Base Molding

Base Molding	Unit	Material	Labor	Equip.	Total	Specification
6"						
Demolish	L.F.		.74		.74	Cost includes material and labor to
Install	L.F.	2.19	1.67		3.86	install 1" x 6" oak base molding.
Demolish and Install	L.F.	2.19	2.41		4.60	
Reinstall	L.F.		1.34		1.34	
Clean	L.F.	.03	.40		.43	
Paint	L.F.	.21	.69		.90	
Minimum Charge	Job		260		260	
8"						
Demolish	L.F.		.89		.89	Cost includes material and labor to
Install	L.F.	2.73	1.67		4.40	install 1" x 8" oak base molding.
Demolish and Install	L.F.	2.73	2.56		5.29	
Reinstall	L.F.		1.34		1.34	
Clean	L.F.	.03	.40		.43	
Paint	L.F.	.22	.72		.94	
Minimum Charge	Job		260		260	

Vinyl Base Molding

Vinyl Base Molding	Unit	Material	Labor	Equip.	Total	Specification
2-1/2"						
Demolish	L.F.		.74		.74	Cost includes material and labor to
Install	L.F.	2.31	1.61		3.92	install 2-1/2" vinyl or rubber cove
Demolish and Install	L.F.	2.31	2.35		4.66	base including adhesive.
Reinstall	L.F.		1.61		1.61	
Clean	L.F.		.36		.36	
Minimum Charge	Job		256		256	
4"						
Demolish	L.F.		.74		.74	Cost includes material and labor to
Install	Ea.	5.30	1.61		6.91	install 4" vinyl or rubber cove base
Demolish and Install	Ea.	5.30	2.35		7.65	including adhesive.
Reinstall	Ea.		1.61		1.61	
Clean	L.F.		.36		.36	
Minimum Charge	Job		256		256	
6"						
Demolish	L.F.		.74		.74	Cost includes material and labor to
Install	L.F.	2.95	1.61		4.56	install 6" vinyl or rubber cove base
Demolish and Install	L.F.	2.95	2.35		5.30	including adhesive.
Reinstall	L.F.		1.61		1.61	
Clean	L.F.		.36		.36	
Minimum Charge	Job		256		256	

Prefinished

Prefinished	Unit	Material	Labor	Equip.	Total	Specification
Demolish	L.F.		.74		.74	Cost includes material and labor to
Install	L.F.	1.97	1.67		3.64	install 9/16" x 3-5/16" prefinished
Demolish and Install	L.F.	1.97	2.41		4.38	base molding.
Reinstall	L.F.		1.34		1.34	
Clean	L.F.	.02	.24		.26	
Paint	L.F.	.17	.51		.68	
Minimum Charge	Job		260		260	

Shoe

Shoe	Unit	Material	Labor	Equip.	Total	Specification
Demolish	L.F.		.74		.74	Includes material and labor to install
Install	L.F.	1.35	1.54		2.89	pine or birch base molding.
Demolish and Install	L.F.	1.35	2.28		3.63	
Reinstall	L.F.		1.23		1.23	
Clean	L.F.	.02	.23		.25	
Paint	L.F.	.19	.69		.88	
Minimum Charge	Job		260		260	

For customer support on your Contractor's Pricing Guide: Residential Repair & Remodeling, call 800.448.8182.

Finish Carpentry

Base Molding

	Unit	Material	Labor	Equip.	Total	Specification
Carpeted						
Demolish	L.F.		.30		.30	Includes material and labor to install
Install	L.F.	1.24	3.08		4.32	carpet baseboard with adhesive.
Demolish and Install	L.F.	1.24	3.38		4.62	
Reinstall	L.F.		2.46		2.46	
Clean	L.F.		.15		.15	
Minimum Charge	Job		254		254	
Ceramic Tile						
Demolish	L.F.		1.61		1.61	Includes material and labor to install
Install	L.F.	3.97	7.05		11.02	thin set ceramic tile cove base
Demolish and Install	L.F.	3.97	8.66		12.63	including grout.
Clean	L.F.		.76		.76	
Minimum Charge	Job		254		254	
Quarry Tile						
Demolish	L.F.		1.61		1.61	Includes material and labor to install
Install	L.F.	4	8.20		12.20	quarry tile base including grout.
Demolish and Install	L.F.	4	9.81		13.81	
Clean	L.F.		.76		.76	
Minimum Charge	Job		254		254	
Caulk and Renail						
Install	L.F.	.03	.81		.84	Includes labor and material to renail
Minimum Charge	Job		260		260	base and caulk where necessary.

Molding

	Unit	Material	Labor	Equip.	Total	Specification
Oak Base						
Demolish	L.F.		.74		.74	Cost includes material and labor to
Install	L.F.	2.35	1.57		3.92	install 1/2" x 2-1/2" oak base
Demolish and Install	L.F.	2.35	2.31		4.66	molding.
Reinstall	L.F.		1.26		1.26	
Clean	L.F.	.02	.23		.25	
Paint	L.F.	.20	.65		.85	
Minimum Charge	Job		260		260	
Shoe						
Demolish	L.F.		.74		.74	Includes material and labor to install
Install	L.F.	1.35	1.54		2.89	pine or birch base molding.
Demolish and Install	L.F.	1.35	2.28		3.63	
Reinstall	L.F.		1.23		1.23	
Clean	L.F.	.02	.23		.25	
Paint	L.F.	.19	.69		.88	
Minimum Charge	Job		260		260	
Quarter Round						
Demolish	L.F.		.74		.74	Includes material and labor to install
Install	L.F.	1.62	2.04		3.66	pine base quarter round molding.
Demolish and Install	L.F.	1.62	2.78		4.40	
Reinstall	L.F.		1.63		1.63	
Clean	L.F.	.02	.23		.25	
Paint	L.F.	.21	1.08		1.29	
Minimum Charge	Job		260		260	
Chair Rail						
Demolish	L.F.		.74		.74	Cost includes material and labor to
Install	L.F.	1.16	1.93		3.09	install 5/8" x 2-1/2" oak chair rail
Demolish and Install	L.F.	1.16	2.67		3.83	molding.
Reinstall	L.F.		1.54		1.54	
Clean	L.F.	.02	.23		.25	
Paint	L.F.	.21	1.08		1.29	
Minimum Charge	Job		260		260	

Finish Carpentry

Molding

Molding		Unit	Material	Labor	Equip.	Total	Specification
Crown							
	Demolish	L.F.		.74		.74	Includes material and labor to install
	Install	L.F.	1.72	2.08		3.80	unfinished pine crown molding.
	Demolish and Install	L.F.	1.72	2.82		4.54	
	Reinstall	L.F.		1.67		1.67	
	Clean	L.F.	.03	.31		.34	
	Paint	L.F.	.21	1.08		1.29	
	Minimum Charge	Job		260		260	
Cove							
	Demolish	L.F.		.74		.74	Cost includes material and labor to
	Install	L.F.	2.92	1.97		4.89	install 11/16" x 2-3/4" pine cove
	Demolish and Install	L.F.	2.92	2.71		5.63	molding.
	Reinstall	L.F.		1.57		1.57	
	Clean	L.F.	.02	.23		.25	
	Paint	L.F.	.21	1.08		1.29	
	Minimum Charge	Job		260		260	
Corner							
	Demolish	L.F.		.74		.74	Includes material and labor to install
	Install	L.F.	2.76	.81		3.57	inside or outside pine corner molding.
	Demolish and Install	L.F.	2.76	1.55		4.31	
	Reinstall	L.F.		.65		.65	
	Clean	L.F.	.02	.23		.25	
	Paint	L.F.	.21	1.08		1.29	
	Minimum Charge	Job		260		260	
Picture							
	Demolish	L.F.		.74		.74	Cost includes material and labor to
	Install	L.F.	1.84	2.17		4.01	install 9/16" x 2-1/2" pine casing.
	Demolish and Install	L.F.	1.84	2.91		4.75	
	Reinstall	L.F.		1.74		1.74	
	Clean	L.F.	.02	.23		.25	
	Paint	L.F.	.21	1.08		1.29	
	Minimum Charge	Job		260		260	
Single Cased Opening							
	Demolish	Ea.		36		36	Includes material and labor to install
	Install	Ea.	194	32.50		226.50	flat pine jamb with square cut heads
	Demolish and Install	Ea.	194	68.50		262.50	and rabbeted sides for 6' 8" high
	Clean	Ea.	.22	8.35		8.57	opening including trim sets for both
	Paint	Ea.	3.77	10.80		14.57	sides.
	Minimum Charge	Job		260		260	
Double Cased Opening							
	Demolish	Ea.		36		36	Includes material and labor to install
	Install	Ea.	228	37		265	flat pine jamb with square cut heads
	Demolish and Install	Ea.	228	73		301	and rabbeted sides for 6' 8" high
	Clean	Ea.	.25	9.75		10	opening including trim sets for both
	Paint	Ea.	4.37	12.70		17.07	sides.
	Minimum Charge	Job		260		260	

Window Trim Set		Unit	Material	Labor	Equip.	Total	Specification
Single							
	Demolish	Ea.		9.85		9.85	Cost includes material and labor to
	Install	Opng.	71.50	52		123.50	install 11/16" x 2-1/2" pine ranch
	Demolish and Install	Opng.	71.50	61.85		133.35	style window casing.
	Reinstall	Opng.		41.66		41.66	
	Clean	Ea.	.32	5.70		6.02	
	Paint	Ea.	3.53	10.30		13.83	
	Minimum Charge	Job		260		260	

For customer support on your Contractor's Pricing Guide: Residential Repair & Remodeling, call 800.448.8182.

Finish Carpentry

Window Trim Set

	Unit	Material	Labor	Equip.	Total	Specification
Double						
Demolish	Ea.		11.10		11.10	Cost includes material and labor to
Install	Ea.	135	65		200	install 11/16″ x 2-1/2″ pine ranch
Demolish and Install	Ea.	135	76.10		211.10	style trim for casing one side of a
Reinstall	Ea.		52.08		52.08	double window opening.
Clean	Ea.	.56	6.65		7.21	
Paint	Ea.	5.95	17.25		23.20	
Minimum Charge	Job		260		260	
Triple						
Demolish	Ea.		12.70		12.70	Cost includes material and labor to
Install	Opng.	128	87		215	install 11/16″ x 2-1/2″ pine ranch
Demolish and Install	Opng.	128	99.70		227.70	style trim for casing one side of a
Reinstall	Opng.		69.44		69.44	triple window opening.
Clean	Ea.	.79	8		8.79	
Paint	Ea.	8.30	24		32.30	
Minimum Charge	Job		260		260	
Window Casing Per L.F.						
Demolish	L.F.		.74		.74	Cost includes material and labor to
Install	L.F.	1.86	2.08		3.94	install 11/16″ x 2-1/2″ pine ranch
Demolish and Install	L.F.	1.86	2.82		4.68	style trim for casing.
Reinstall	L.F.		1.67		1.67	
Clean	L.F.	.02	.23		.25	
Paint	L.F.	.21	1.08		1.29	
Minimum Charge	Job		260		260	

Window Sill

	Unit	Material	Labor	Equip.	Total	Specification
Wood						
Demolish	L.F.		.74		.74	Includes material and labor to install
Install	L.F.	4.22	2.60		6.82	flat, wood window stool.
Demolish and Install	L.F.	4.22	3.34		7.56	
Reinstall	L.F.		2.08		2.08	
Clean	L.F.	.02	.23		.25	
Paint	L.F.	.21	1.08		1.29	
Minimum Charge	Job		260		260	
Marble						
Demolish	L.F.		1.11		1.11	Includes material and labor to install
Install	L.F.	9.45	11.20		20.65	marble window sills, 6″ x 3/4″ thick.
Demolish and Install	L.F.	9.45	12.31		21.76	
Reinstall	L.F.		8.96		8.96	
Clean	L.F.	.02	.23		.25	
Minimum Charge	Job		254		254	

Shelving

	Unit	Material	Labor	Equip.	Total	Specification
Pine						
18″						
Demolish	L.F.		1.27		1.27	Cost includes material and labor to
Install	L.F.	5.50	5.50		11	install 1″ thick custom shelving board
Demolish and Install	L.F.	5.50	6.77		12.27	18″ wide.
Reinstall	L.F.		4.39		4.39	
Clean	L.F.	.04	.67		.71	
Paint	L.F.	2.66	.68		3.34	
Minimum Charge	Job		260		260	

For customer support on your Contractor's Pricing Guide: Residential Repair & Remodeling, call 800.448.8182.

Finish Carpentry

Shelving

Shelving		Unit	Material	Labor	Equip.	Total	Specification
24"							
	Demolish	L.F.		1.27		1.27	Cost includes material and labor to
	Install	L.F.	7.30	6.15		13.45	install 1" thick custom shelving board.
	Demolish and Install	L.F.	7.30	7.42		14.72	
	Reinstall	L.F.		4.90		4.90	
	Clean	L.F.	.07	1		1.07	
	Paint	L.F.	3.62	.91		4.53	
	Minimum Charge	Job		260		260	
Particle Board							
	Demolish	L.F.		1.17		1.17	Cost includes material and labor to
	Install	L.F.	1.78	2.74		4.52	install 3/4" thick particle shelving
	Demolish and Install	L.F.	1.78	3.91		5.69	board 18" wide.
	Reinstall	L.F.		2.19		2.19	
	Clean	L.F.	.04	.67		.71	
	Paint	S.F.	.29	1		1.29	
	Minimum Charge	Job		260		260	
Plywood							
	Demolish	L.F.		1.17		1.17	Cost includes material and labor to
	Install	L.F.	1.09	8		9.09	install 3/4" thick plywood shelving
	Demolish and Install	L.F.	1.09	9.17		10.26	board.
	Reinstall	L.F.		6.41		6.41	
	Clean	L.F.	.04	.67		.71	
	Paint	S.F.	.29	1		1.29	
	Minimum Charge	Job		260		260	
Hardwood (oak)							
	Demolish	L.F.		1.17		1.17	Includes material and labor to install
	Install	L.F.	12.75	3.59		16.34	oak shelving, 1" x 12", incl. cleats and
	Demolish and Install	L.F.	12.75	4.76		17.51	bracing.
	Reinstall	L.F.		2.87		2.87	
	Clean	L.F.	.04	.67		.71	
	Paint	S.F.	.29	1		1.29	
	Minimum Charge	Job		260		260	
Custom Bookcase							
	Demolish	L.F.		14.80		14.80	Includes material and labor to install
	Install	L.F.	28	52		80	custom modular bookcase unit with
	Demolish and Install	L.F.	28	66.80		94.80	clear pine faced frames, shelves 12"
	Reinstall	L.F.		41.66		41.66	O.C., 7' high, 8" deep.
	Clean	L.F.	1.27	14.25		15.52	
	Paint	L.F.	3.18	33		36.18	
	Minimum Charge	Job		260		260	
Glass							
	Demolish	L.F.		1.17		1.17	Includes material and labor to install
	Install	L.F.	58.50	3.59		62.09	glass shelving, 1" x 12", incl. cleats
	Demolish and Install	L.F.	58.50	4.76		63.26	and bracing.
	Reinstall	L.F.		2.87		2.87	
	Clean	L.F.	.06	.61		.67	
	Minimum Charge	Job		252		252	

Closet Shelving

Closet Shelving		Unit	Material	Labor	Equip.	Total	Specification
Closet Shelf and Rod							
	Demolish	L.F.		1.17		1.17	Cost includes material and labor to
	Install	L.F.	2.15	8.70		10.85	install 1" thick custom shelving board
	Demolish and Install	L.F.	2.15	9.87		12.02	18" wide and 1" diameter clothes pole
	Reinstall	L.F.		6.94		6.94	with brackets 3' O.C.
	Clean	L.F.	.03	.50		.53	
	Paint	L.F.	.42	1.08		1.50	
	Minimum Charge	Job		260		260	

For customer support on your Contractor's Pricing Guide: Residential Repair & Remodeling, call 800.448.8182.

Finish Carpentry

Closet Shelving

	Unit	Material	Labor	Equip.	Total	Specification
Clothing Rod						
Demolish	L.F.		.77		.77	Includes material and labor to install
Install	L.F.	3.48	2.60		6.08	fir closet pole, 1-5/8″ diameter.
Demolish and Install	L.F.	3.48	3.37		6.85	
Reinstall	L.F.		2.08		2.08	
Clean	L.F.	.03	.68		.71	
Paint	L.F.	2.98	1.23		4.21	
Minimum Charge	Job		260		260	

Stair Assemblies

	Unit	Material	Labor	Equip.	Total	Specification
Straight Hardwood						
Demolish	Ea.		515		515	Includes material and labor to install
Install	Flight	470	345		815	factory cut and assembled straight
Demolish and Install	Flight	470	860		1330	closed box stairs with oak treads and
Clean	Flight	3.17	34		37.17	prefinished stair rail with balusters.
Paint	Flight	37.50	57.50		95	
Minimum Charge	Job		260		260	
Clear Oak Tread						
Demolish	Ea.		13.15		13.15	Includes material and labor to install
Install	Ea.	94	29		123	1-1/4″ thick clear oak treads per riser.
Demolish and Install	Ea.	94	42.15		136.15	
Clean	Ea.	.10	3.39		3.49	
Paint	Ea.	1.24	3.20		4.44	
Minimum Charge	Job		260		260	
Landing						
Demolish	S.F.		4.62		4.62	Includes material and labor to install
Install	S.F.	4.30	29		33.30	clear oak landing.
Demolish and Install	S.F.	4.30	33.62		37.92	
Clean	S.F.		.50		.50	
Paint	S.F.	.88	1.07		1.95	
Minimum Charge	Job		260		260	
Refinish / Stain						
Install	S.F.	.34	3.07		3.41	Includes labor and material to refinish
Minimum Charge	Job		216		216	wood floor.
Spiral Hardwood						
Demolish	Ea.		515		515	Cost includes material and labor to
Install	Flight	3025	695		3720	install 4′ - 6′ diameter spiral stairs with
Demolish and Install	Flight	3025	1210		4235	oak treads, factory cut and assembled
Clean	Flight	3.17	34		37.17	with double handrails.
Paint	Flight	37.50	57.50		95	
Minimum Charge	Job		260		260	

Stairs

	Unit	Material	Labor	Equip.	Total	Specification
Disappearing						
Demolish	Ea.		28		28	Includes material and labor to install a
Install	Ea.	263	149		412	folding staircase.
Demolish and Install	Ea.	263	177		440	
Reinstall	Ea.		119.04		119.04	
Clean	Ea.	.47	25		25.47	
Minimum Charge	Job		260		260	

For customer support on your Contractor's Pricing Guide: Residential Repair & Remodeling, call 800.448.8182.

Finish Carpentry

Stair Components

Stair Components		Unit	Material	Labor	Equip.	Total	Specification
Handrail w / Balusters							
	Demolish	L.F.		3.70		3.70	Includes material and labor to install
	Install	L.F.	62.50	10.85		73.35	prefinished assembled stair rail with
	Demolish and Install	L.F.	62.50	14.55		77.05	brackets, turned balusters and newel.
	Clean	L.F.	.64	2.76		3.40	
	Paint	L.F.	4.54	5.10		9.64	
	Minimum Charge	Job		260		260	
Bannister							
	Demolish	L.F.		3.70		3.70	Includes material and labor to install
	Install	L.F.	23.50	8.70		32.20	built-up oak railings.
	Demolish and Install	L.F.	23.50	12.40		35.90	
	Clean	L.F.	.64	2.50		3.14	
	Paint	L.F.	4.54	3.92		8.46	
	Minimum Charge	Job		260		260	

For customer support on your Contractor's Pricing Guide: Residential Repair & Remodeling, call 800.448.8182.

Cabinets and Countertops

Wall and Base Cabinets

Countertop

Wall Cabinets	Unit	Material	Labor	Equip.	Total	Specification
Good Grade Laminated						
12" Wide, 1 Door						
Demolish	Ea.		11.10		11.10	Includes material and labor to install
Install	Ea.	180	47.50		227.50	good modular unit with melamine
Demolish and Install	Ea.	180	58.60		238.60	laminated to particle board, in
Reinstall	Ea.		37.88		37.88	textured colors or wood grain print
Clean	Ea.	.47	7.70		8.17	finish, hinges and pulls.
Paint	Ea.	1.36	15.40		16.76	
Minimum Charge	Job		260		260	
15" Wide, 1 Door						
Demolish	Ea.		13.90		13.90	Includes material and labor to install
Install	Ea.	213	48.50		261.50	good modular unit with melamine
Demolish and Install	Ea.	213	62.40		275.40	laminated to particle board, in
Reinstall	Ea.		38.94		38.94	textured colors or wood grain print
Clean	Ea.	.59	9.50		10.09	finish, hinges and pulls.
Paint	Ea.	1.71	19.60		21.31	
Minimum Charge	Job		260		260	
18" Wide, 1 Door						
Demolish	Ea.		16.75		16.75	Includes material and labor to install
Install	Ea.	212	50		262	good modular unit with melamine
Demolish and Install	Ea.	212	66.75		278.75	laminated to particle board, in
Reinstall	Ea.		39.87		39.87	textured colors or wood grain print
Clean	Ea.	.72	11.40		12.12	finish, hinges and pulls.
Paint	Ea.	2.04	22.50		24.54	
Minimum Charge	Job		260		260	
21" Wide, 1 Door						
Demolish	Ea.		19.30		19.30	Includes material and labor to install
Install	Ea.	160	50.50		210.50	good modular unit with melamine
Demolish and Install	Ea.	160	69.80		229.80	laminated to particle board, in
Reinstall	Ea.		40.45		40.45	textured colors or wood grain print
Clean	Ea.	.84	13.30		14.14	finish, hinges and pulls.
Paint	Ea.	2.39	27		29.39	
Minimum Charge	Job		260		260	
24" w / Blind Corner						
Demolish	Ea.		22		22	Includes material and labor to install
Install	Ea.	283	51.50		334.50	good modular unit with melamine
Demolish and Install	Ea.	283	73.50		356.50	laminated to particle board, in
Reinstall	Ea.		41.05		41.05	textured colors or wood grain print
Clean	Ea.	.95	15.35		16.30	finish, hinges and pulls.
Paint	Ea.	2.72	31		33.72	
Minimum Charge	Job		260		260	

Cabinets and Countertops

Wall Cabinets

	Unit	Material	Labor	Equip.	Total	Specification
27″ Wide, 2 Door						
Demolish	Ea.		24.50		24.50	Includes material and labor to install
Install	Ea.	98.50	52.50		151	good modular unit with melamine
Demolish and Install	Ea.	98.50	77		175.50	laminated to particle board, in
Reinstall	Ea.		42.08		42.08	textured colors or wood grain print
Clean	Ea.	1.07	17.35		18.42	finish, hinges and pulls.
Paint	Ea.	3.09	36		39.09	
Minimum Charge	Job		260		260	
30″ Wide, 2 Door						
Demolish	Ea.		28		28	Includes material and labor to install
Install	Ea.	325	54		379	good modular unit with melamine
Demolish and Install	Ea.	325	82		407	laminated to particle board, in
Reinstall	Ea.		43.18		43.18	textured colors or wood grain print
Clean	Ea.	1.19	19.05		20.24	finish, hinges and pulls.
Paint	Ea.	3.41	39		42.41	
Minimum Charge	Job		260		260	
36″ Wide, 2 Door						
Demolish	Ea.		33		33	Includes material and labor to install
Install	Ea.	360	55.50		415.50	good modular unit with melamine
Demolish and Install	Ea.	360	88.50		448.50	laminated to particle board, in
Reinstall	Ea.		44.32		44.32	textured colors or wood grain print
Clean	Ea.	1.42	23.50		24.92	finish, hinges and pulls.
Paint	Ea.	4.13	48		52.13	
Minimum Charge	Job		260		260	
48″ Wide, 2 Door						
Demolish	Ea.		44.50		44.50	Includes material and labor to install
Install	Ea.	370	56.50		426.50	good modular unit with melamine
Demolish and Install	Ea.	370	101		471	laminated to particle board, in
Reinstall	Ea.		45.29		45.29	textured colors or wood grain print
Clean	Ea.	1.90	30.50		32.40	finish, hinges and pulls.
Paint	Ea.	5.45	61.50		66.95	
Minimum Charge	Job		260		260	
Above-Appliance						
Demolish	Ea.		28		28	Includes material and labor to install
Install	Ea.	145	42		187	good modular unit with melamine
Demolish and Install	Ea.	145	70		215	laminated to particle board, in
Reinstall	Ea.		33.60		33.60	textured colors or wood grain print
Clean	Ea.	1.19	19.05		20.24	finish, hinges and pulls.
Paint	Ea.	3.41	39		42.41	
Minimum Charge	Job		260		260	
Better Grade Wood						
12″ Wide, 1 Door						
Demolish	Ea.		11.10		11.10	Includes material and labor to install
Install	Ea.	241	47.50		288.50	custom modular unit with solid
Demolish and Install	Ea.	241	58.60		299.60	hardwood faced frames, hardwood
Reinstall	Ea.		37.88		37.88	door frames and drawer fronts,
Clean	Ea.	.47	7.70		8.17	hardwood veneer on raised door
Paint	Ea.	1.36	16.60		17.96	panels, hinges and pulls.
Minimum Charge	Job		260		260	
15″ Wide, 1 Door						
Demolish	Ea.		13.90		13.90	Includes material and labor to install
Install	Ea.	284	48.50		332.50	custom modular unit with solid
Demolish and Install	Ea.	284	62.40		346.40	hardwood faced frames, hardwood
Reinstall	Ea.		38.94		38.94	door frames and drawer fronts,
Clean	Ea.	.59	9.50		10.09	hardwood veneer on raised door
Paint	Ea.	1.71	20.50		22.21	panels, hinges and pulls.
Minimum Charge	Job		260		260	

Cabinets and Countertops

Wall Cabinets

Wall Cabinets	Unit	Material	Labor	Equip.	Total	Specification
18" Wide, 1 Door						
Demolish	Ea.		16.75		16.75	Includes material and labor to install
Install	Ea.	283	50		333	custom modular unit with solid
Demolish and Install	Ea.	283	66.75		349.75	hardwood faced frames, hardwood
Reinstall	Ea.		39.87		39.87	door frames and drawer fronts,
Clean	Ea.	.72	11.40		12.12	hardwood veneer on raised door
Paint	Ea.	2.04	25.50		27.54	panels, hinges and pulls.
Minimum Charge	Job		260		260	
21" Wide, 1 Door						
Demolish	Ea.		19.30		19.30	Includes material and labor to install
Install	Ea.	213	50.50		263.50	custom modular unit with solid
Demolish and Install	Ea.	213	69.80		282.80	hardwood faced frames, hardwood
Reinstall	Ea.		40.45		40.45	door frames and drawer fronts,
Clean	Ea.	.84	13.30		14.14	hardwood veneer on raised door
Paint	Ea.	2.39	29		31.39	panels, hinges and pulls.
Minimum Charge	Job		260		260	
24" w / Blind Corner						
Demolish	Ea.		22		22	Includes material and labor to install
Install	Ea.	380	51.50		431.50	custom modular unit with solid
Demolish and Install	Ea.	380	73.50		453.50	hardwood faced frames, hardwood
Reinstall	Ea.		41.05		41.05	door frames and drawer fronts,
Clean	Ea.	.95	15.35		16.30	hardwood veneer on raised door
Paint	Ea.	2.72	33		35.72	panels, hinges and pulls.
Minimum Charge	Job		260		260	
27" Wide, 2 Door						
Demolish	Ea.		24.50		24.50	Includes material and labor to install
Install	Ea.	131	52.50		183.50	custom modular unit with solid
Demolish and Install	Ea.	131	77		208	hardwood faced frames, hardwood
Reinstall	Ea.		42.08		42.08	door frames and drawer fronts,
Clean	Ea.	1.07	17.35		18.42	hardwood veneer on raised door
Paint	Ea.	3.09	36		39.09	panels, hinges and pulls.
Minimum Charge	Job		260		260	
30" Wide, 2 Door						
Demolish	Ea.		28		28	Includes material and labor to install
Install	Ea.	430	54		484	custom modular unit with solid
Demolish and Install	Ea.	430	82		512	hardwood faced frames, hardwood
Reinstall	Ea.		43.18		43.18	door frames and drawer fronts,
Clean	Ea.	1.19	19.05		20.24	hardwood veneer on raised door
Paint	Ea.	3.41	43		46.41	panels, hinges and pulls.
Minimum Charge	Job		260		260	
36" Wide, 2 Door						
Demolish	Ea.		33		33	Includes material and labor to install
Install	Ea.	480	55.50		535.50	custom modular unit with solid
Demolish and Install	Ea.	480	88.50		568.50	hardwood faced frames, hardwood
Reinstall	Ea.		44.32		44.32	door frames and drawer fronts,
Clean	Ea.	1.42	23.50		24.92	hardwood veneer on raised door
Paint	Ea.	4.13	48		52.13	panels, hinges and pulls.
Minimum Charge	Job		260		260	
48" Wide, 2 Door						
Demolish	Ea.		44.50		44.50	Includes material and labor to install
Install	Ea.	490	56.50		546.50	custom modular unit with solid
Demolish and Install	Ea.	490	101		591	hardwood faced frames, hardwood
Reinstall	Ea.		45.29		45.29	door frames and drawer fronts,
Clean	Ea.	1.90	30.50		32.40	hardwood veneer on raised door
Paint	Ea.	5.45	61.50		66.95	panels, hinges and pulls.
Minimum Charge	Job		260		260	

For customer support on your Contractor's Pricing Guide: Residential Repair & Remodeling, call 800.448.8182.

Cabinets and Countertops

Wall Cabinets	Unit	Material	Labor	Equip.	Total	Specification
Above-appliance						
Demolish	Ea.		28		28	Includes material and labor to install
Install	Ea.	194	42		236	custom modular unit with solid
Demolish and Install	Ea.	194	70		264	hardwood faced frames, hardwood
Reinstall	Ea.		33.60		33.60	door frames and drawer fronts,
Clean	Ea.	1.19	19.05		20.24	hardwood veneer on raised door
Paint	Ea.	3.41	43		46.41	panels, hinges and pulls.
Minimum Charge	Job		260		260	
Premium Grade Wood						
12″ Wide, 1 Door						
Demolish	Ea.		11.10		11.10	Includes material and labor to install
Install	Ea.	288	47.50		335.50	premium modular unit with solid
Demolish and Install	Ea.	288	58.60		346.60	hardwood faced frames, hardwood
Reinstall	Ea.		37.88		37.88	door frames and drawer fronts,
Clean	Ea.	.47	7.70		8.17	hardwood veneer on raised door
Paint	Ea.	1.36	18		19.36	panels, hinges and pulls.
Minimum Charge	Job		260		260	
15″ Wide, 1 Door						
Demolish	Ea.		13.90		13.90	Includes material and labor to install
Install	Ea.	380	48.50		428.50	premium modular unit with solid
Demolish and Install	Ea.	380	62.40		442.40	hardwood faced frames, hardwood
Reinstall	Ea.		38.94		38.94	door frames and drawer fronts,
Clean	Ea.	.59	9.50		10.09	hardwood veneer on raised door
Paint	Ea.	1.71	22.50		24.21	panels, hinges and pulls.
Minimum Charge	Job		260		260	
18″ Wide, 1 Door						
Demolish	Ea.		16.75		16.75	Includes material and labor to install
Install	Ea.	340	50		390	premium modular unit with solid
Demolish and Install	Ea.	340	66.75		406.75	hardwood faced frames, hardwood
Reinstall	Ea.		39.87		39.87	door frames and drawer fronts,
Clean	Ea.	.72	11.40		12.12	hardwood veneer on raised door
Paint	Ea.	2.04	27		29.04	panels, hinges and pulls.
Minimum Charge	Job		260		260	
21″ Wide, 1 Door						
Demolish	Ea.		19.30		19.30	Includes material and labor to install
Install	Ea.	255	50.50		305.50	premium modular unit with solid
Demolish and Install	Ea.	255	69.80		324.80	hardwood faced frames, hardwood
Reinstall	Ea.		40.45		40.45	door frames and drawer fronts,
Clean	Ea.	.84	13.30		14.14	hardwood veneer on raised door
Paint	Ea.	2.39	31		33.39	panels, hinges and pulls.
Minimum Charge	Job		260		260	
24″ w / Blind Corner						
Demolish	Ea.		22		22	Includes material and labor to install
Install	Ea.	450	51.50		501.50	premium modular unit with solid
Demolish and Install	Ea.	450	73.50		523.50	hardwood faced frames, hardwood
Reinstall	Ea.		41.05		41.05	door frames and drawer fronts,
Clean	Ea.	.95	15.35		16.30	hardwood veneer on raised door
Paint	Ea.	2.72	36		38.72	panels, hinges and pulls.
Minimum Charge	Job		260		260	
27″ Wide, 2 Door						
Demolish	Ea.		24.50		24.50	Includes material and labor to install
Install	Ea.	360	52.50		412.50	premium modular unit with solid
Demolish and Install	Ea.	360	77		437	hardwood faced frames, hardwood
Reinstall	Ea.		42.08		42.08	door frames and drawer fronts,
Clean	Ea.	1.07	17.35		18.42	hardwood veneer on raised door
Paint	Ea.	3.09	39		42.09	panels, hinges and pulls.
Minimum Charge	Job		260		260	

For customer support on your Contractor's Pricing Guide: Residential Repair & Remodeling, call 800.448.8182.

Cabinets and Countertops

Wall Cabinets

Wall Cabinets		Unit	Material	Labor	Equip.	Total	Specification
30″ Wide, 2 Door							
	Demolish	Ea.		28		28	Includes material and labor to install
	Install	Ea.	515	54		569	premium modular unit with solid
	Demolish and Install	Ea.	515	82		597	hardwood faced frames, hardwood
	Reinstall	Ea.		43.18		43.18	door frames and drawer fronts,
	Clean	Ea.	1.19	19.05		20.24	hardwood veneer on raised door
	Paint	Ea.	3.41	43		46.41	panels, hinges and pulls.
	Minimum Charge	Job		260		260	
36″ Wide, 2 Door							
	Demolish	Ea.		33		33	Includes material and labor to install
	Install	Ea.	570	55.50		625.50	premium modular unit with solid
	Demolish and Install	Ea.	570	88.50		658.50	hardwood faced frames, hardwood
	Reinstall	Ea.		44.32		44.32	door frames and drawer fronts,
	Clean	Ea.	1.42	23.50		24.92	hardwood veneer on raised door
	Paint	Ea.	4.13	54		58.13	panels, hinges and pulls.
	Minimum Charge	Job		260		260	
48″ Wide, 2 Door							
	Demolish	Ea.		44.50		44.50	Includes material and labor to install
	Install	Ea.	590	56.50		646.50	premium modular unit with solid
	Demolish and Install	Ea.	590	101		691	hardwood faced frames, hardwood
	Reinstall	Ea.		45.29		45.29	door frames and drawer fronts,
	Clean	Ea.	1.90	30.50		32.40	hardwood veneer on raised door
	Paint	Ea.	5.45	72		77.45	panels, hinges and pulls.
	Minimum Charge	Job		260		260	
Above-appliance							
	Demolish	Ea.		28		28	Includes material and labor to install
	Install	Ea.	233	42		275	premium modular unit with solid
	Demolish and Install	Ea.	233	70		303	hardwood faced frames, hardwood
	Reinstall	Ea.		33.60		33.60	door frames and drawer fronts,
	Clean	Ea.	1.19	19.05		20.24	hardwood veneer on raised door
	Paint	Ea.	3.41	43		46.41	panels, hinges, pulls.
	Minimum Charge	Job		260		260	

Premium Hardwood

		Unit	Material	Labor	Equip.	Total	Specification
12″ Wide, 1 Door							
	Demolish	Ea.		11.10		11.10	Includes material and labor to install
	Install	Ea.	315	47.50		362.50	premium hardwood modular unit with
	Demolish and Install	Ea.	315	58.60		373.60	solid hardwood faced frames, drawer
	Reinstall	Ea.		37.88		37.88	fronts, and door panels, steel drawer
	Clean	Ea.	.47	7.70		8.17	guides, hinges, pulls, laminated
	Paint	Ea.	1.36	19.60		20.96	interior.
	Minimum Charge	Job		260		260	
15″ Wide, 1 Door							
	Demolish	Ea.		13.90		13.90	Includes material and labor to install
	Install	Ea.	495	48.50		543.50	premium hardwood modular unit with
	Demolish and Install	Ea.	495	62.40		557.40	solid hardwood faced frames, drawer
	Reinstall	Ea.		38.94		38.94	fronts, and door panels, steel drawer
	Clean	Ea.	.59	9.50		10.09	guides, hinges, pulls, laminated
	Paint	Ea.	1.71	24		25.71	interior.
	Minimum Charge	Job		260		260	
18″ Wide, 1 Door							
	Demolish	Ea.		16.75		16.75	Includes material and labor to install
	Install	Ea.	370	50		420	premium hardwood modular unit with
	Demolish and Install	Ea.	370	66.75		436.75	solid hardwood faced frames, drawer
	Reinstall	Ea.		39.87		39.87	fronts, and door panels, steel drawer
	Clean	Ea.	.72	11.40		12.12	guides, hinges, pulls, laminated
	Paint	Ea.	2.04	29		31.04	interior.
	Minimum Charge	Job		260		260	

For customer support on your Contractor's Pricing Guide: Residential Repair & Remodeling, call 800.448.8182.

Cabinets and Countertops

Wall Cabinets

	Unit	Material	Labor	Equip.	Total	Specification
21" Wide, 1 Door						
Demolish	Ea.		19.30		19.30	Includes material and labor to install
Install	Ea.	279	50.50		329.50	premium hardwood modular unit with
Demolish and Install	Ea.	279	69.80		348.80	solid hardwood faced frames, drawer
Reinstall	Ea.		40.45		40.45	fronts, and door panels, steel drawer
Clean	Ea.	.84	13.30		14.14	guides, hinges, pulls, laminated
Paint	Ea.	2.39	33		35.39	interior.
Minimum Charge	Job		260		260	
24" w / Blind Corner						
Demolish	Ea.		22		22	Includes material and labor to install
Install	Ea.	495	51.50		546.50	premium hardwood modular unit with
Demolish and Install	Ea.	495	73.50		568.50	solid hardwood faced frames, drawer
Reinstall	Ea.		41.05		41.05	fronts, and door panels, steel drawer
Clean	Ea.	.95	15.35		16.30	guides, hinges, pulls, laminated
Paint	Ea.	2.72	39		41.72	interior.
Minimum Charge	Job		260		260	
27" Wide, 2 Door						
Demolish	Ea.		24.50		24.50	Includes material and labor to install
Install	Ea.	172	52.50		224.50	premium hardwood modular unit with
Demolish and Install	Ea.	172	77		249	solid hardwood faced frames, drawer
Reinstall	Ea.		42.08		42.08	fronts, and door panels, steel drawer
Clean	Ea.	1.07	17.35		18.42	guides, hinges, pulls, laminated
Paint	Ea.	3.09	43		46.09	interior.
Minimum Charge	Job		260		260	
30" Wide, 2 Door						
Demolish	Ea.		28		28	Includes material and labor to install
Install	Ea.	565	54		619	premium hardwood modular unit with
Demolish and Install	Ea.	565	82		647	solid hardwood faced frames, drawer
Reinstall	Ea.		43.18		43.18	fronts, and door panels, steel drawer
Clean	Ea.	1.19	19.05		20.24	guides, hinges, pulls, laminated
Paint	Ea.	3.41	48		51.41	interior.
Minimum Charge	Job		260		260	
36" Wide, 2 Door						
Demolish	Ea.		33		33	Includes material and labor to install
Install	Ea.	625	55.50		680.50	premium hardwood modular unit with
Demolish and Install	Ea.	625	88.50		713.50	solid hardwood faced frames, drawer
Reinstall	Ea.		44.32		44.32	fronts, and door panels, steel drawer
Clean	Ea.	1.42	23.50		24.92	guides, hinges, pulls, laminated
Paint	Ea.	4.13	61.50		65.63	interior.
Minimum Charge	Job		260		260	
48" Wide, 2 Door						
Demolish	Ea.		44.50		44.50	Includes material and labor to install
Install	Ea.	645	56.50		701.50	premium hardwood modular unit with
Demolish and Install	Ea.	645	101		746	solid hardwood faced frames, drawer
Reinstall	Ea.		45.29		45.29	fronts, and door panels, steel drawer
Clean	Ea.	1.90	30.50		32.40	guides, hinges, pulls, laminated
Paint	Ea.	5.45	72		77.45	interior.
Minimum Charge	Job		260		260	
Above-appliance						
Demolish	Ea.		28		28	Includes material and labor to install
Install	Ea.	254	42		296	premium hardwood modular unit with
Demolish and Install	Ea.	254	70		324	solid hardwood faced frames, drawer
Reinstall	Ea.		33.60		33.60	fronts, and door panels, steel drawer
Clean	Ea.	1.19	19.05		20.24	guides, hinges, pulls, laminated
Paint	Ea.	3.41	48		51.41	interior.
Minimum Charge	Job		260		260	

For customer support on your Contractor's Pricing Guide: Residential Repair & Remodeling, call 800.448.8182.

Cabinets and Countertops

Wall Cabinets

	Unit	Material	Labor	Equip.	Total	Specification
Medicine Cabinet						
Good Grade						
Demolish	Ea.		18.50		18.50	Cost includes material and labor to
Install	Ea.	292	37		329	install 14″ x 18″ recessed medicine
Demolish and Install	Ea.	292	55.50		347.50	cabinet with stainless steel frame,
Reinstall	Ea.		29.76		29.76	mirror. Light not included.
Clean	Ea.	.02	20		20.02	
Paint	Ea.	3.62	27		30.62	
Minimum Charge	Job		260		260	
Good Grade Lighted						
Demolish	Ea.		18.50		18.50	Cost includes material and labor to
Install	Ea.	340	43.50		383.50	install 14″ x 18″ recessed medicine
Demolish and Install	Ea.	340	62		402	cabinet with stainless steel frame,
Reinstall	Ea.		34.72		34.72	mirror and light.
Clean	Ea.	.02	20		20.02	
Paint	Ea.	3.62	27		30.62	
Minimum Charge	Job		260		260	
Better Grade						
Demolish	Ea.		18.50		18.50	Includes material and labor to install
Install	Ea.	445	37		482	recessed medicine cabinet with
Demolish and Install	Ea.	445	55.50		500.50	polished chrome frame and mirror.
Reinstall	Ea.		29.76		29.76	Light not included.
Clean	Ea.	.02	20		20.02	
Paint	Ea.	3.62	27		30.62	
Minimum Charge	Job		260		260	
Better Grade Lighted						
Demolish	Ea.		18.50		18.50	Includes material and labor to install
Install	Ea.	490	43.50		533.50	recessed medicine cabinet with
Demolish and Install	Ea.	490	62		552	polished chrome frame, mirror and
Reinstall	Ea.		34.72		34.72	light.
Clean	Ea.	.02	20		20.02	
Paint	Ea.	3.62	27		30.62	
Minimum Charge	Job		260		260	
Premium Grade						
Demolish	Ea.		18.50		18.50	Cost includes material and labor to
Install	Ea.	239	37		276	install 14″ x 24″ recessed medicine
Demolish and Install	Ea.	239	55.50		294.50	cabinet with beveled mirror, frameless
Reinstall	Ea.		29.76		29.76	swing door. Light not included.
Clean	Ea.	.02	20		20.02	
Paint	Ea.	3.62	27		30.62	
Minimum Charge	Job		260		260	
Premium Grade Lighted						
Demolish	Ea.		18.50		18.50	Cost includes material and labor to
Install	Ea.	286	43.50		329.50	install 36″ x 30″ surface mounted
Demolish and Install	Ea.	286	62		348	medicine cabinet with beveled tri-view
Reinstall	Ea.		34.72		34.72	mirror, frameless swing door and light.
Clean	Ea.	.02	20		20.02	
Paint	Ea.	3.62	27		30.62	
Minimum Charge	Job		260		260	
Bath Mirror						
Demolish	Ea.		13.90		13.90	Cost includes material and labor to
Install	Ea.	68	26		94	install 18″ x 24″ surface mount mirror
Demolish and Install	Ea.	68	39.90		107.90	with stainless steel frame.
Reinstall	Ea.		20.83		20.83	
Clean	Ea.	.02	1.35		1.37	
Minimum Charge	Job		260		260	

For customer support on your Contractor's Pricing Guide: Residential Repair & Remodeling, call 800.448.8182.

Cabinets and Countertops

Wall Cabinets

	Unit	Material	Labor	Equip.	Total	Specification
Garage						Includes material and labor to install
Demolish	L.F.		11.10		11.10	good modular unit with melamine
Install	L.F.	103	26		129	laminated to particle board, in
Demolish and Install	L.F.	103	37.10		140.10	textured colors or wood grain print
Reinstall	L.F.		20.83		20.83	finish, hinges and pulls.
Clean	L.F.	.47	7.70		8.17	
Paint	L.F.	1.36	15.40		16.76	
Minimum Charge	Job		260		260	
Strip and Refinish						Includes labor and material to strip,
Install	L.F.	7.05	54		61.05	prep and refinish exterior of cabinets.
Minimum Charge	Job		216		216	

Tall Cabinets

	Unit	Material	Labor	Equip.	Total	Specification
Good Grade Laminated						Includes material and labor to install
Demolish	L.F.		19.75		19.75	custom modular unit with solid
Install	L.F.	370	58		428	hardwood faced frames, hardwood
Demolish and Install	L.F.	370	77.75		447.75	door frames and drawer fronts,
Reinstall	L.F.		46.29		46.29	hardwood veneer on raised door
Clean	L.F.	1.27	14.25		15.52	panels, hinges and pulls.
Paint	L.F.	3.18	31		34.18	
Minimum Charge	Job		260		260	
Better Grade Wood						Includes material and labor to install
Demolish	L.F.		19.75		19.75	custom modular unit with solid
Install	L.F.	530	58		588	hardwood faced frames, hardwood
Demolish and Install	L.F.	530	77.75		607.75	door frames and drawer fronts,
Reinstall	L.F.		46.29		46.29	hardwood veneer on raised door
Clean	L.F.	1.27	14.25		15.52	panels, hinges and pulls.
Paint	L.F.	3.18	33		36.18	
Minimum Charge	Job		260		260	
Premium Grade Wood						Includes material and labor to install
Demolish	L.F.		19.75		19.75	premium modular unit with solid
Install	L.F.	755	58		813	hardwood faced frames, hardwood
Demolish and Install	L.F.	755	77.75		832.75	door frames and drawer fronts,
Reinstall	L.F.		46.29		46.29	hardwood veneer on raised door
Clean	L.F.	1.27	14.25		15.52	panels, hinges and pulls.
Paint	L.F.	3.18	36		39.18	
Minimum Charge	Job		260		260	
Premium Hardwood						Includes material and labor to install
Demolish	L.F.		19.75		19.75	premium hardwood modular unit with
Install	L.F.	1050	58		1108	solid hardwood faced frames, drawer
Demolish and Install	L.F.	1050	77.75		1127.75	fronts, and door panels, steel drawer
Reinstall	L.F.		46.29		46.29	guides, hinges and pulls, laminated
Clean	L.F.	1.27	14.25		15.52	interior.
Paint	L.F.	3.18	39		42.18	
Minimum Charge	Job		260		260	

Cabinets and Countertops

Tall Cabinets	Unit	Material	Labor	Equip.	Total	Specification
Single Oven, 27″ Wide						
Good Grade Laminated						
Demolish	Ea.		44.50		44.50	Includes material and labor to install
Install	Ea.	1275	130		1405	good modular unit with melamine
Demolish and Install	Ea.	1275	174.50		1449.50	laminated to particle board, in
Reinstall	Ea.		104.16		104.16	textured colors or wood grain print
Clean	Ea.	3.96	25		28.96	finish, hinges and pulls.
Paint	Ea.	5.35	69		74.35	
Minimum Charge	Job		260		260	
Better Grade Veneer						
Demolish	Ea.		44.50		44.50	Includes material and labor to install
Install	Ea.	2250	130		2380	custom modular unit with solid
Demolish and Install	Ea.	2250	174.50		2424.50	hardwood faced frames, hardwood
Reinstall	Ea.		104.16		104.16	door frames and drawer fronts,
Clean	Ea.	3.96	25		28.96	hardwood veneer on raised door
Paint	Ea.	5.35	74.50		79.85	panels, hinges and pulls.
Minimum Charge	Job		260		260	
Premium Grade Veneer						
Demolish	Ea.		44.50		44.50	Includes material and labor to install
Install	Ea.	2025	130		2155	premium modular unit with solid
Demolish and Install	Ea.	2025	174.50		2199.50	hardwood faced frames, hardwood
Reinstall	Ea.		104.16		104.16	door frames and drawer fronts,
Clean	Ea.	3.96	25		28.96	hardwood veneer on raised door
Paint	Ea.	5.35	81.50		86.85	panels, hinges and pulls.
Minimum Charge	Job		260		260	
Premium Hardwood						
Demolish	Ea.		44.50		44.50	Includes material and labor to install
Install	Ea.	2225	130		2355	premium hardwood modular unit with
Demolish and Install	Ea.	2225	174.50		2399.50	solid hardwood faced frames, drawer
Reinstall	Ea.		104.16		104.16	fronts, and door panels, steel drawer
Clean	Ea.	3.96	25		28.96	guides, hinges and pulls, laminated
Paint	Ea.	5.35	88		93.35	interior.
Minimum Charge	Job		260		260	
Single Oven, 30″ Wide						
Good Grade Laminated						
Demolish	Ea.		44.50		44.50	Includes material and labor to install
Install	Ea.	1275	130		1405	good modular unit with melamine
Demolish and Install	Ea.	1275	174.50		1449.50	laminated to particle board, in
Reinstall	Ea.		104.16		104.16	textured colors or wood grain print
Clean	Ea.	3.96	25		28.96	finish, hinges and pulls.
Paint	Ea.	5.35	69		74.35	
Minimum Charge	Job		260		260	
Better Grade Veneer						
Demolish	Ea.		44.50		44.50	Includes material and labor to install
Install	Ea.	1350	130		1480	custom modular unit with solid
Demolish and Install	Ea.	1350	174.50		1524.50	hardwood faced frames, hardwood
Reinstall	Ea.		104.16		104.16	door frames and drawer fronts,
Clean	Ea.	3.96	25		28.96	hardwood veneer on raised door
Paint	Ea.	5.35	69		74.35	panels, hinges and pulls.
Minimum Charge	Job		260		260	

For customer support on your Contractor's Pricing Guide: Residential Repair & Remodeling, call 800.448.8182.

Cabinets and Countertops

Tall Cabinets

	Unit	Material	Labor	Equip.	Total	Specification
Premium Grade Veneer						
Demolish	Ea.		44.50		44.50	Includes material and labor to install
Install	Ea.	1475	130		1605	premium modular unit with solid
Demolish and Install	Ea.	1475	174.50		1649.50	hardwood faced frames, hardwood
Reinstall	Ea.		104.16		104.16	door frames and drawer fronts,
Clean	Ea.	3.96	25		28.96	hardwood veneer on raised door
Paint	Ea.	5.35	69		74.35	panels, hinges and pulls.
Minimum Charge	Job		260		260	
Premium Hardwood						
Demolish	Ea.		44.50		44.50	Includes material and labor to install
Install	Ea.	1900	130		2030	premium hardwood modular unit with
Demolish and Install	Ea.	1900	174.50		2074.50	solid hardwood faced frames, drawer
Reinstall	Ea.		104.16		104.16	fronts, and door panels, steel drawer
Clean	Ea.	3.96	25		28.96	guides, hinges and pulls, laminated
Paint	Ea.	5.35	69		74.35	interior.
Minimum Charge	Job		260		260	

Utility 18″ W x 12″ D

	Unit	Material	Labor	Equip.	Total	Specification
Good Grade Laminated						
Demolish	Ea.		44.50		44.50	Includes material and labor to install
Install	Ea.	325	104		429	good modular unit with melamine
Demolish and Install	Ea.	325	148.50		473.50	laminated to particle board, in
Reinstall	Ea.		83.33		83.33	textured colors or wood grain print
Clean	Ea.	.94	12.10		13.04	finish, hinges and pulls.
Paint	Ea.	2.04	27		29.04	
Minimum Charge	Job		260		260	
Better Grade Veneer						
Demolish	Ea.		44.50		44.50	Includes material and labor to install
Install	Ea.	435	104		539	custom modular unit with solid
Demolish and Install	Ea.	435	148.50		583.50	hardwood faced frames, hardwood
Reinstall	Ea.		83.33		83.33	door frames and drawer fronts,
Clean	Ea.	.94	12.10		13.04	hardwood veneer on raised door
Paint	Ea.	2.04	29		31.04	panels, hinges and pulls.
Minimum Charge	Job		260		260	
Premium Grade Veneer						
Demolish	Ea.		44.50		44.50	Includes material and labor to install
Install	Ea.	515	104		619	premium modular unit with solid
Demolish and Install	Ea.	515	148.50		663.50	hardwood faced frames, hardwood
Reinstall	Ea.		83.33		83.33	door frames and drawer fronts,
Clean	Ea.	.94	12.10		13.04	hardwood veneer on raised door
Paint	Ea.	2.04	33		35.04	panels, hinges and pulls.
Minimum Charge	Job		260		260	
Premium Hardwood						
Demolish	Ea.		44.50		44.50	Includes material and labor to install
Install	Ea.	565	104		669	premium hardwood modular unit with
Demolish and Install	Ea.	565	148.50		713.50	solid hardwood faced frames, drawer
Reinstall	Ea.		83.33		83.33	fronts, and door panels, steel drawer
Clean	Ea.	.94	12.10		13.04	guides, hinges and pulls, laminated
Paint	Ea.	2.04	36		38.04	interior.
Minimum Charge	Job		260		260	

For customer support on your Contractor's Pricing Guide: Residential Repair & Remodeling, call 800.448.8182.

Cabinets and Countertops

Tall Cabinets	Unit	Material	Labor	Equip.	Total	Specification
Utility 24″ W x 24″ D						
Good Grade Laminated						
Demolish	Ea.		44.50		44.50	Includes material and labor to install
Install	Ea.	1100	130		1230	good modular unit with melamine
Demolish and Install	Ea.	1100	174.50		1274.50	laminated to particle board, in
Reinstall	Ea.		104.16		104.16	textured colors or wood grain print
Clean	Ea.	1.22	16		17.22	finish, hinges and pulls.
Paint	Ea.	2.72	36		38.72	
Minimum Charge	Job		260		260	
Better Grade Veneer						
Demolish	Ea.		44.50		44.50	Includes material and labor to install
Install	Ea.	1450	130		1580	custom modular unit with solid
Demolish and Install	Ea.	1450	174.50		1624.50	hardwood faced frames, hardwood
Reinstall	Ea.		104.16		104.16	door frames and drawer fronts,
Clean	Ea.	1.22	16		17.22	hardwood veneer on raised door
Paint	Ea.	2.72	39		41.72	panels, hinges and pulls.
Minimum Charge	Job		260		260	
Premium Grade Veneer						
Demolish	Ea.		44.50		44.50	Includes material and labor to install
Install	Ea.	1750	130		1880	premium modular unit with solid
Demolish and Install	Ea.	1750	174.50		1924.50	hardwood faced frames, hardwood
Reinstall	Ea.		104.16		104.16	door frames and drawer fronts,
Clean	Ea.	1.22	16		17.22	hardwood veneer on raised door
Paint	Ea.	2.72	43		45.72	panels, hinges and pulls.
Minimum Charge	Job		260		260	
Premium Hardwood						
Demolish	Ea.		44.50		44.50	Includes material and labor to install
Install	Ea.	1925	130		2055	premium hardwood modular unit with
Demolish and Install	Ea.	1925	174.50		2099.50	solid hardwood faced frames, drawer
Reinstall	Ea.		104.16		104.16	fronts, and door panels, steel drawer
Clean	Ea.	1.22	16		17.22	guides, hinges and pulls, laminated
Paint	Ea.	2.72	48		50.72	interior.
Minimum Charge	Job		260		260	
Plain Shelves						
Install	Ea.	35	29		64	Includes labor and material to install
Rotating Shelves						tall cabinet shelving, per set.
Install	Ea.	243	26		269	Includes labor and material to install
Strip and Refinish						lazy susan shelving for wall cabinet.
Install	L.F.	14.10	86.50		100.60	Includes labor and material to strip,
Minimum Charge	Job		216		216	prep and refinish exterior of cabinets.
Stain						
Install	L.F.	.54	3.92		4.46	Includes labor and material to replace
Minimum Charge	Job		216		216	normal prep and stain on doors and exterior cabinets.

Cabinets and Countertops

Base Cabinets

	Unit	Material	Labor	Equip.	Total	Specification
Good Grade Laminated						
12″ w, 1 Door, 1 Drawer						
Demolish	Ea.		11.10		11.10	Includes material and labor to install
Install	Ea.	350	42		392	good modular unit with melamine
Demolish and Install	Ea.	350	53.10		403.10	laminate, textured colors or wood
Reinstall	Ea.		33.60		33.60	grain print finish including hinges and
Clean	Ea.	.64	8		8.64	pulls.
Paint	Ea.	1.36	18		19.36	
Minimum Charge	Job		260		260	
15″ w, 1 Door, 1 Drawer						
Demolish	Ea.		13.90		13.90	Includes material and labor to install
Install	Ea.	370	43.50		413.50	good modular unit with melamine
Demolish and Install	Ea.	370	57.40		427.40	laminate, textured colors or wood
Reinstall	Ea.		34.72		34.72	grain print finish including hinges and
Clean	Ea.	.79	10		10.79	pulls.
Paint	Ea.	1.71	22.50		24.21	
Minimum Charge	Job		260		260	
18″ w, 1 Door, 1 Drawer						
Demolish	Ea.		16.45		16.45	Includes material and labor to install
Install	Ea.	385	44.50		429.50	good modular unit with melamine
Demolish and Install	Ea.	385	60.95		445.95	laminate, textured colors or wood
Reinstall	Ea.		35.76		35.76	grain print finish including hinges and
Clean	Ea.	.94	12.10		13.04	pulls.
Paint	Ea.	2.04	27		29.04	
Minimum Charge	Job		260		260	
21″ w, 1 Door, 1 Drawer						
Demolish	Ea.		19.30		19.30	Includes material and labor to install
Install	Ea.	405	46		451	good modular unit with melamine
Demolish and Install	Ea.	405	65.30		470.30	laminate, textured colors or wood
Reinstall	Ea.		36.71		36.71	grain print finish including hinges and
Clean	Ea.	1.13	14.25		15.38	pulls.
Paint	Ea.	2.39	31		33.39	
Minimum Charge	Job		260		260	
24″ w, 1 Door, 1 Drawer						
Demolish	Ea.		22		22	Includes material and labor to install
Install	Ea.	450	46.50		496.50	good modular unit with melamine
Demolish and Install	Ea.	450	68.50		518.50	laminate, textured colors or wood
Reinstall	Ea.		37.37		37.37	grain print finish including hinges and
Clean	Ea.	1.22	16		17.22	pulls.
Paint	Ea.	2.72	36		38.72	
Minimum Charge	Job		260		260	
30″ Wide Sink Base						
Demolish	Ea.		28		28	Includes material and labor to install
Install	Ea.	375	48.50		423.50	good modular unit with melamine
Demolish and Install	Ea.	375	76.50		451.50	laminate, textured colors or wood
Reinstall	Ea.		38.94		38.94	grain print finish including hinges and
Clean	Ea.	1.58	20		21.58	pulls.
Paint	Ea.	3.41	43		46.41	
Minimum Charge	Job		260		260	
36″ Wide Sink Base						
Demolish	Ea.		34		34	Includes material and labor to install
Install	Ea.	395	51.50		446.50	good modular unit with melamine
Demolish and Install	Ea.	395	85.50		480.50	laminate, textured colors or wood
Reinstall	Ea.		41.05		41.05	grain print finish including hinges and
Clean	Ea.	1.76	23.50		25.26	pulls.
Paint	Ea.	4.13	54		58.13	
Minimum Charge	Job		260		260	

222

For customer support on your Contractor's Pricing Guide: Residential Repair & Remodeling, call 800.448.8182.

Cabinets and Countertops

Base Cabinets		Unit	Material	Labor	Equip.	Total	Specification
36″ Blind Corner							
	Demolish	Ea.		34		34	Includes material and labor to install
	Install	Ea.	560	58		618	good modular unit with melamine
	Demolish and Install	Ea.	560	92		652	laminate, textured colors or wood
	Reinstall	Ea.		46.29		46.29	grain print finish including hinges and
	Clean	Ea.	1.76	23.50		25.26	pulls.
	Paint	Ea.	4.13	54		58.13	
	Minimum Charge	Job		260		260	
42″ w, 2 Door, 2 Drawer							
	Demolish	Ea.		40.50		40.50	Includes material and labor to install
	Install	Ea.	585	52.50		637.50	good modular unit with melamine
	Demolish and Install	Ea.	585	93		678	laminate, textured colors or wood
	Reinstall	Ea.		42.08		42.08	grain print finish including hinges and
	Clean	Ea.	2.26	28.50		30.76	pulls.
	Paint	Ea.	4.73	61.50		66.23	
	Minimum Charge	Job		260		260	
Clean Interior							
	Clean	L.F.	.64	6.05		6.69	Includes labor and materials to clean
	Minimum Charge	Job		200		200	the interior of a 27″ cabinet.

Better Grade Wood

		Unit	Material	Labor	Equip.	Total	Specification
12″ w, 1 Door, 1 Drawer							
	Demolish	Ea.		11.10		11.10	Includes material and labor to install
	Install	Ea.	470	42		512	custom modular unit including solid
	Demolish and Install	Ea.	470	53.10		523.10	hardwood faced frames, hardwood
	Reinstall	Ea.		33.60		33.60	door frames and drawer fronts,
	Clean	Ea.	.64	8		8.64	hardwood veneer on raised door
	Paint	Ea.	1.36	19.60		20.96	panels, hinges and pulls.
	Minimum Charge	Job		260		260	
15″ w, 1 Door, 1 Drawer							
	Demolish	Ea.		13.90		13.90	Includes material and labor to install
	Install	Ea.	490	43.50		533.50	custom modular unit including solid
	Demolish and Install	Ea.	490	57.40		547.40	hardwood faced frames, hardwood
	Reinstall	Ea.		34.72		34.72	door frames and drawer fronts,
	Clean	Ea.	.79	10		10.79	hardwood veneer on raised door
	Paint	Ea.	1.71	25.50		27.21	panels, hinges and pulls.
	Minimum Charge	Job		260		260	
18″ w, 1 Door, 1 Drawer							
	Demolish	Ea.		16.45		16.45	Includes material and labor to install
	Install	Ea.	510	44.50		554.50	custom modular unit including solid
	Demolish and Install	Ea.	510	60.95		570.95	hardwood faced frames, hardwood
	Reinstall	Ea.		35.76		35.76	door frames and drawer fronts,
	Clean	Ea.	.94	12.10		13.04	hardwood veneer on raised door
	Paint	Ea.	2.04	29		31.04	panels, hinges and pulls.
	Minimum Charge	Job		260		260	
21″ w, 1 Door, 1 Drawer							
	Demolish	Ea.		19.30		19.30	Includes material and labor to install
	Install	Ea.	540	46		586	custom modular unit including solid
	Demolish and Install	Ea.	540	65.30		605.30	hardwood faced frames, hardwood
	Reinstall	Ea.		36.71		36.71	door frames and drawer fronts,
	Clean	Ea.	1.13	14.25		15.38	hardwood veneer on raised door
	Paint	Ea.	2.39	33		35.39	panels, hinges and pulls.
	Minimum Charge	Job		260		260	

For customer support on your Contractor's Pricing Guide: Residential Repair & Remodeling, call 800.448.8182.

Cabinets and Countertops

Base Cabinets	Unit	Material	Labor	Equip.	Total	Specification
24″ w, 1 Door, 1 Drawer						
Demolish	Ea.		22		22	Includes material and labor to install
Install	Ea.	600	46.50		646.50	custom modular unit including solid
Demolish and Install	Ea.	600	68.50		668.50	hardwood faced frames, hardwood
Reinstall	Ea.		37.37		37.37	door frames and drawer fronts,
Clean	Ea.	1.22	16		17.22	hardwood veneer on raised door
Paint	Ea.	2.72	39		41.72	panels, hinges and pulls.
Minimum Charge	Job		260		260	
30″ Wide Sink Base						
Demolish	Ea.		28		28	Includes material and labor to install
Install	Ea.	500	48.50		548.50	custom modular unit including solid
Demolish and Install	Ea.	500	76.50		576.50	hardwood faced frames, hardwood
Reinstall	Ea.		38.94		38.94	door frames and drawer fronts,
Clean	Ea.	1.58	20		21.58	hardwood veneer on raised door
Paint	Ea.	3.41	48		51.41	panels, hinges and pulls.
Minimum Charge	Job		260		260	
36″ Wide Sink Base						
Demolish	Ea.		34		34	Includes material and labor to install
Install	Ea.	535	51.50		586.50	custom modular unit including solid
Demolish and Install	Ea.	535	85.50		620.50	hardwood faced frames, hardwood
Reinstall	Ea.		41.05		41.05	door frames and drawer fronts,
Clean	Ea.	1.76	23.50		25.26	hardwood veneer on raised door
Paint	Ea.	4.13	61.50		65.63	panels, hinges and pulls.
Minimum Charge	Job		260		260	
36″ Blind Corner						
Demolish	Ea.		34		34	Includes material and labor to install
Install	Ea.	750	58		808	custom modular unit including solid
Demolish and Install	Ea.	750	92		842	hardwood faced frames, hardwood
Reinstall	Ea.		46.29		46.29	door frames and drawer fronts,
Clean	Ea.	1.76	23.50		25.26	hardwood veneer on raised door
Paint	Ea.	4.13	61.50		65.63	panels, hinges and pulls.
Minimum Charge	Job		260		260	
42″ w, 2 Door, 2 Drawer						
Demolish	Ea.		40.50		40.50	Includes material and labor to install
Install	Ea.	780	52.50		832.50	modular unit including solid hardwood
Demolish and Install	Ea.	780	93		873	faced frames, hardwood door frames
Reinstall	Ea.		42.08		42.08	and drawer fronts, hardwood veneer
Clean	Ea.	2.26	28.50		30.76	on raised door panels, hinges and
Paint	Ea.	4.73	72		76.73	pulls.
Minimum Charge	Job		260		260	
Clean Interior						
Clean	L.F.	.64	6.05		6.69	Includes labor and materials to clean
Minimum Charge	Job		200		200	the interior of a 27″ cabinet.

Premium Grade Wood

	Unit	Material	Labor	Equip.	Total	Specification
12″ w, 1 Door, 1 Drawer						
Demolish	Ea.		11.10		11.10	Includes material and labor to install
Install	Ea.	565	21		586	premium hardwood modular unit with
Demolish and Install	Ea.	565	32.10		597.10	solid hardwood faced frames, drawer
Reinstall	Ea.		16.80		16.80	fronts, and door panels, steel drawer
Clean	Ea.	.64	8		8.64	guides, hinges, pulls and laminated
Paint	Ea.	1.36	21.50		22.86	interior.
Minimum Charge	Job		260		260	

224

For customer support on your Contractor's Pricing Guide: Residential Repair & Remodeling, call 800.448.8182.

Cabinets and Countertops

Base Cabinets	Unit	Material	Labor	Equip.	Total	Specification
15" w, 1 Door, 1 Drawer						
Demolish	Ea.		13.90		13.90	Includes material and labor to install
Install	Ea.	590	43.50		633.50	premium modular unit including solid
Demolish and Install	Ea.	590	57.40		647.40	hardwood faced frames, hardwood
Reinstall	Ea.		34.72		34.72	door frames and drawer fronts,
Clean	Ea.	.79	10		10.79	hardwood veneer on raised door
Paint	Ea.	1.71	27		28.71	panels, hinges and pulls.
Minimum Charge	Job		260		260	
18" w, 1 Door, 1 Drawer						
Demolish	Ea.		16.45		16.45	Includes material and labor to install
Install	Ea.	615	44.50		659.50	premium modular unit including solid
Demolish and Install	Ea.	615	60.95		675.95	hardwood faced frames, hardwood
Reinstall	Ea.		35.76		35.76	door frames and drawer fronts,
Clean	Ea.	.94	12.10		13.04	hardwood veneer on raised door
Paint	Ea.	2.04	33		35.04	panels, hinges and pulls.
Minimum Charge	Job		260		260	
21" w, 1 Door, 1 Drawer						
Demolish	Ea.		19.30		19.30	Includes material and labor to install
Install	Ea.	655	46		701	premium modular unit including solid
Demolish and Install	Ea.	655	65.30		720.30	hardwood faced frames, hardwood
Reinstall	Ea.		36.71		36.71	door frames and drawer fronts,
Clean	Ea.	1.13	14.25		15.38	hardwood veneer on raised door
Paint	Ea.	2.39	36		38.39	panels, hinges and pulls.
Minimum Charge	Job		260		260	
24" w, 1 Door, 1 Drawer						
Demolish	Ea.		22		22	Includes material and labor to install
Install	Ea.	715	46.50		761.50	premium modular unit including solid
Demolish and Install	Ea.	715	68.50		783.50	hardwood faced frames, hardwood
Reinstall	Ea.		37.37		37.37	door frames and drawer fronts,
Clean	Ea.	1.22	16		17.22	hardwood veneer on raised door
Paint	Ea.	2.72	43		45.72	panels, hinges and pulls.
Minimum Charge	Job		260		260	
30" Wide Sink Base						
Demolish	Ea.		28		28	Includes material and labor to install
Install	Ea.	600	48.50		648.50	premium modular unit including solid
Demolish and Install	Ea.	600	76.50		676.50	hardwood faced frames, hardwood
Reinstall	Ea.		38.94		38.94	door frames and drawer fronts,
Clean	Ea.	1.58	20		21.58	hardwood veneer on raised door
Paint	Ea.	3.41	54		57.41	panels, hinges and pulls.
Minimum Charge	Job		260		260	
36" Wide Sink Base						
Demolish	Ea.		34		34	Includes material and labor to install
Install	Ea.	640	51.50		691.50	premium modular unit including solid
Demolish and Install	Ea.	640	85.50		725.50	hardwood faced frames, hardwood
Reinstall	Ea.		41.05		41.05	door frames and drawer fronts,
Clean	Ea.	1.76	23.50		25.26	hardwood veneer on raised door
Paint	Ea.	4.13	61.50		65.63	panels, hinges and pulls.
Minimum Charge	Job		260		260	
36" Blind Corner						
Demolish	Ea.		34		34	Includes material and labor to install
Install	Ea.	895	58		953	premium modular unit including solid
Demolish and Install	Ea.	895	92		987	hardwood faced frames, hardwood
Reinstall	Ea.		46.29		46.29	door frames and drawer fronts,
Clean	Ea.	1.76	23.50		25.26	hardwood veneer on raised door
Paint	Ea.	4.13	61.50		65.63	panels, hinges and pulls.
Minimum Charge	Job		260		260	

For customer support on your Contractor's Pricing Guide: Residential Repair & Remodeling, call 800.448.8182.

Cabinets and Countertops

Base Cabinets		Unit	Material	Labor	Equip.	Total	Specification
42" w, 2 Door, 2 Drawer							
	Demolish	Ea.		40.50		40.50	Includes material and labor to install
	Install	Ea.	1050	52.50		1102.50	premium modular unit including solid
	Demolish and Install	Ea.	1050	93		1143	hardwood faced frames, hardwood
	Reinstall	Ea.		42.08		42.08	door frames and drawer fronts,
	Clean	Ea.	2.26	28.50		30.76	hardwood veneer on raised door
	Paint	Ea.	4.73	72		76.73	panels, hinges and pulls.
	Minimum Charge	Job		260		260	
Clean Interior							
	Clean	L.F.	.64	6.05		6.69	Includes labor and materials to clean
	Minimum Charge	Job		200		200	the interior of a 27" cabinet.

Premium Hardwood

		Unit	Material	Labor	Equip.	Total	Specification
12" w, 1 Door, 1 Drawer							
	Demolish	Ea.		11.10		11.10	Includes material and labor to install
	Install	Ea.	615	42		657	premium hardwood modular unit with
	Demolish and Install	Ea.	615	53.10		668.10	solid hardwood faced frames, drawer
	Reinstall	Ea.		33.60		33.60	fronts, and door panels, steel drawer
	Clean	Ea.	.64	8		8.64	guides, hinges, pulls and laminated
	Paint	Ea.	1.36	24		25.36	interior.
	Minimum Charge	Job		260		260	
15" w, 1 Door, 1 Drawer							
	Demolish	Ea.		13.90		13.90	Includes material and labor to install
	Install	Ea.	645	43.50		688.50	premium hardwood modular unit with
	Demolish and Install	Ea.	645	57.40		702.40	solid hardwood faced frames, drawer
	Reinstall	Ea.		34.72		34.72	fronts, and door panels, steel drawer
	Clean	Ea.	.79	10		10.79	guides, hinges, pulls and laminated
	Paint	Ea.	1.71	31		32.71	interior.
	Minimum Charge	Job		260		260	
18" w, 1 Door, 1 Drawer							
	Demolish	Ea.		16.45		16.45	Includes material and labor to install
	Install	Ea.	670	44.50		714.50	premium hardwood modular unit with
	Demolish and Install	Ea.	670	60.95		730.95	solid hardwood faced frames, drawer
	Reinstall	Ea.		35.76		35.76	fronts, and door panels, steel drawer
	Clean	Ea.	.94	12.10		13.04	guides, hinges, pulls and laminated
	Paint	Ea.	2.04	36		38.04	interior.
	Minimum Charge	Job		260		260	
21" w, 1 Door, 1 Drawer							
	Demolish	Ea.		19.30		19.30	Includes material and labor to install
	Install	Ea.	710	46		756	premium hardwood modular unit with
	Demolish and Install	Ea.	710	65.30		775.30	solid hardwood faced frames, drawer
	Reinstall	Ea.		36.71		36.71	fronts, and door panels, steel drawer
	Clean	Ea.	1.13	14.25		15.38	guides, hinges, pulls and laminated
	Paint	Ea.	2.39	39		41.39	interior.
	Minimum Charge	Job		260		260	
24" w, 1 Door, 1 Drawer							
	Demolish	Ea.		22		22	Includes material and labor to install
	Install	Ea.	780	46.50		826.50	premium hardwood modular unit with
	Demolish and Install	Ea.	780	68.50		848.50	solid hardwood faced frames, drawer
	Reinstall	Ea.		37.37		37.37	fronts, and door panels, steel drawer
	Clean	Ea.	1.22	16		17.22	guides, hinges and pulls, laminated
	Paint	Ea.	2.72	48		50.72	interior.
	Minimum Charge	Job		260		260	

Cabinets and Countertops

Base Cabinets	Unit	Material	Labor	Equip.	Total	Specification
30″ Wide Sink Base						
Demolish	Ea.		28		28	Includes material and labor to install
Install	Ea.	655	48.50		703.50	premium hardwood modular unit with
Demolish and Install	Ea.	655	76.50		731.50	solid hardwood faced frames, drawer
Reinstall	Ea.		38.94		38.94	fronts, and door panels, steel drawer
Clean	Ea.	1.58	20		21.58	guides, hinges, pulls and laminated
Paint	Ea.	3.41	54		57.41	interior.
Minimum Charge	Job		260		260	
36″ Wide Sink Base						
Demolish	Ea.		34		34	Includes material and labor to install
Install	Ea.	695	51.50		746.50	premium hardwood modular unit with
Demolish and Install	Ea.	695	85.50		780.50	solid hardwood faced frames, drawer
Reinstall	Ea.		41.05		41.05	fronts, and door panels, steel drawer
Clean	Ea.	1.76	23.50		25.26	guides, hinges, pulls and laminated
Paint	Ea.	4.13	72		76.13	interior.
Minimum Charge	Job		260		260	
36″ Blind Corner						
Demolish	Ea.		34		34	Includes material and labor to install
Install	Ea.	980	58		1038	premium hardwood modular unit with
Demolish and Install	Ea.	980	92		1072	solid hardwood faced frames, drawer
Reinstall	Ea.		46.29		46.29	fronts, and door panels, steel drawer
Clean	Ea.	1.76	23.50		25.26	guides, hinges, pulls and laminated
Paint	Ea.	4.13	72		76.13	interior.
Minimum Charge	Job		260		260	
42″ w, 2 Door, 2 Drawer						
Demolish	Ea.		40.50		40.50	Includes material and labor to install
Install	Ea.	1375	52.50		1427.50	premium hardwood modular unit with
Demolish and Install	Ea.	1375	93		1468	solid hardwood faced frames, drawer
Reinstall	Ea.		42.08		42.08	fronts, and door panels, steel drawer
Clean	Ea.	2.26	28.50		30.76	guides, hinges, pulls and laminated
Paint	Ea.	4.73	86.50		91.23	interior.
Minimum Charge	Job		260		260	
Clean Interior						
Clean	L.F.	.64	6.05		6.69	Includes labor and materials to clean
Minimum Charge	Job		200		200	the interior of a 27″ cabinet.
Island						
24″ Wide						
Demolish	Ea.		22		22	Includes material and labor to install
Install	Ea.	805	43.50		848.50	custom modular unit with solid
Demolish and Install	Ea.	805	65.50		870.50	hardwood faced frames, hardwood
Reinstall	Ea.		34.72		34.72	door frames and drawer fronts,
Clean	Ea.	1.76	23.50		25.26	hardwood veneer on raised door
Paint	Ea.	4.08	61.50		65.58	panels, hinges and pulls.
Minimum Charge	Job		260		260	
30″ Wide						
Demolish	Ea.		28		28	Includes material and labor to install
Install	Ea.	830	54.50		884.50	custom modular unit with solid
Demolish and Install	Ea.	830	82.50		912.50	hardwood faced frames, hardwood
Reinstall	Ea.		43.40		43.40	door frames and drawer fronts,
Clean	Ea.	2.26	28.50		30.76	hardwood veneer on raised door
Paint	Ea.	5.15	72		77.15	panels, hinges and pulls.
Minimum Charge	Job		260		260	

For customer support on your Contractor's Pricing Guide: Residential Repair & Remodeling, call 800.448.8182.

Cabinets and Countertops

Base Cabinets

	Unit	Material	Labor	Equip.	Total	Specification
36″ Wide						
Demolish	Ea.		34		34	Includes material and labor to install
Install	Ea.	940	65		1005	custom modular unit with solid
Demolish and Install	Ea.	940	99		1039	hardwood faced frames, hardwood
Reinstall	Ea.		52.08		52.08	door frames and drawer fronts,
Clean	Ea.	2.64	36.50		39.14	hardwood veneer on raised door
Paint	Ea.	6.20	86.50		92.70	panels, hinges and pulls.
Minimum Charge	Job		260		260	
48″ Wide						
Demolish	Ea.		44.50		44.50	Includes material and labor to install
Install	Ea.	1600	87		1687	custom modular unit with solid
Demolish and Install	Ea.	1600	131.50		1731.50	hardwood faced frames, hardwood
Reinstall	Ea.		69.44		69.44	door frames and drawer fronts,
Clean	Ea.	3.17	44.50		47.67	hardwood veneer on raised door
Paint	Ea.	7.10	108		115.10	panels, hinges and pulls.
Minimum Charge	Job		260		260	
60″ Wide						
Demolish	Ea.		49.50		49.50	Includes material and labor to install
Install	Ea.	1650	109		1759	custom modular unit with solid
Demolish and Install	Ea.	1650	158.50		1808.50	hardwood faced frames, hardwood
Reinstall	Ea.		86.80		86.80	door frames and drawer fronts,
Clean	Ea.	3.96	57		60.96	hardwood veneer on raised door
Paint	Ea.	7.10	144		151.10	panels, hinges and pulls.
Minimum Charge	Job		260		260	
Clean Interior						
Clean	L.F.	.64	6.05		6.69	Includes labor and materials to clean
Minimum Charge	Job		200		200	the interior of a 27″ cabinet.
Liquor Bar						
Demolish	L.F.		11.10		11.10	Includes material and labor to install
Install	L.F.	289	52		341	custom modular unit with solid
Demolish and Install	L.F.	289	63.10		352.10	hardwood faced frames, hardwood
Reinstall	L.F.		41.66		41.66	door frames and drawer fronts,
Clean	L.F.	.64	8.35		8.99	hardwood veneer on raised door
Paint	L.F.	10.90	19.60		30.50	panels, hinges and pulls.
Minimum Charge	Job		260		260	
Clean Interior						
Clean	L.F.	.64	6.05		6.69	Includes labor and materials to clean
Minimum Charge	Job		200		200	the interior of a 27″ cabinet.
Seal Interior						
Install	L.F.	1.03	7.10		8.13	Includes labor and materials to seal a
Minimum Charge	Job		216		216	cabinet interior.
Back Bar						
Demolish	L.F.		11.10		11.10	Includes material and labor to install
Install	L.F.	289	52		341	custom modular unit with solid
Demolish and Install	L.F.	289	63.10		352.10	hardwood faced frames, hardwood
Reinstall	L.F.		41.66		41.66	door frames and drawer fronts,
Clean	L.F.	.64	8.35		8.99	hardwood veneer on raised door
Paint	L.F.	10.90	19.60		30.50	panels, hinges and pulls.
Minimum Charge	Job		260		260	
Clean Interior						
Clean	L.F.	.64	6.05		6.69	Includes labor and materials to clean
Minimum Charge	Job		200		200	the interior of a 27″ cabinet.

For customer support on your Contractor's Pricing Guide: Residential Repair & Remodeling, call 800.448.8182.

Cabinets and Countertops

Base Cabinets		Unit	Material	Labor	Equip.	Total	Specification
Seal Interior							
	Install	L.F.	1.03	7.10		8.13	Includes labor and materials to seal a
	Minimum Charge	Job		216		216	cabinet interior.
Storage (garage)							
	Demolish	L.F.		11.10		11.10	Includes material and labor to install
	Install	L.F.	217	52		269	good modular unit with melamine
	Demolish and Install	L.F.	217	63.10		280.10	laminate, textured colors or wood
	Reinstall	L.F.		41.66		41.66	grain print finish including hinges and
	Clean	L.F.	.64	8.35		8.99	pulls.
	Paint	L.F.	10.90	19.60		30.50	
	Minimum Charge	Job		260		260	
Clean Interior							
	Clean	L.F.	.64	6.05		6.69	Includes labor and materials to clean
	Minimum Charge	Job		200		200	the interior of a 27" cabinet.
Seal Interior							
	Install	L.F.	1.03	7.10		8.13	Includes labor and materials to seal a
	Minimum Charge	Job		216		216	cabinet interior.
Built-in Desk							
	Demolish	L.F.		11.10		11.10	Includes material and labor to install
	Install	L.F.	289	52		341	custom modular unit with solid
	Demolish and Install	L.F.	289	63.10		352.10	hardwood faced frames, hardwood
	Reinstall	L.F.		41.66		41.66	door frames and drawer fronts,
	Clean	L.F.	.64	8.35		8.99	hardwood veneer on raised door
	Paint	L.F.	10.90	19.60		30.50	panels, hinges and pulls.
	Minimum Charge	Job		260		260	
Built-in Bookcase							
	Demolish	L.F.		14.80		14.80	Includes material and labor to install
	Install	L.F.	28	52		80	custom modular bookcase unit with
	Demolish and Install	L.F.	28	66.80		94.80	clear pine faced frames, shelves 12"
	Reinstall	L.F.		41.66		41.66	O.C., 7' high, 8" deep.
	Clean	L.F.	1.27	14.25		15.52	
	Paint	L.F.	3.18	33		36.18	
	Minimum Charge	Job		260		260	
Strip and Refinish							
	Install	L.F.	7.05	54		61.05	Includes labor and material to strip,
	Minimum Charge	Job		216		216	prep and refinish exterior of cabinets.
Stain							
	Install	L.F.	1.22	2.01		3.23	Includes labor and material to stain
	Minimum Charge	Job		260		260	door and exterior of cabinets.

Countertop		Unit	Material	Labor	Equip.	Total	Specification
Laminated							
With Splash							
	Demolish	L.F.		7.40		7.40	Includes material and labor to install
	Install	L.F.	49	18.60		67.60	one piece laminated top with 4"
	Demolish and Install	L.F.	49	26		75	backsplash.
	Reinstall	L.F.		14.88		14.88	
	Clean	L.F.	.08	1		1.08	
	Minimum Charge	Job		260		260	

229

For customer support on your Contractor's Pricing Guide: Residential Repair & Remodeling, call 800.448.8182.

Cabinets and Countertops

Countertop		Unit	Material	Labor	Equip.	Total	Specification
Roll Top							
	Demolish	L.F.		7.40		7.40	Includes material and labor to install
	Install	L.F.	13.65	17.35		31	one piece laminated top with rolled
	Demolish and Install	L.F.	13.65	24.75		38.40	drip edge (post formed) and
	Reinstall	L.F.		13.89		13.89	backsplash.
	Clean	L.F.	.11	1.51		1.62	
	Minimum Charge	Job		260		260	
Ceramic Tile							
	Demolish	S.F.		3.70		3.70	Cost includes material and labor to
	Install	S.F.	24	10.40		34.40	install 4-1/4" x 4-1/4" to 6" x 6"
	Demolish and Install	S.F.	24	14.10		38.10	glazed tile set in mortar bed and grout
	Clean	S.F.	.06	.76		.82	on particle board substrate.
	Minimum Charge	Job		254		254	
Cultured Marble							
	Demolish	L.F.		7.40		7.40	Includes material and labor to install
	Install	L.F.	248	51.50		299.50	marble countertop, 24" wide, no
	Demolish and Install	L.F.	248	58.90		306.90	backsplash.
	Reinstall	L.F.		41.38		41.38	
	Clean	L.F.	.11	1.51		1.62	
	Minimum Charge	Job		254		254	
With Splash							
	Demolish	L.F.		7.40		7.40	Includes material and labor to install
	Install	L.F.	282	51.50		333.50	marble countertop, 24" wide, with
	Demolish and Install	L.F.	282	58.90		340.90	backsplash.
	Reinstall	L.F.		41.38		41.38	
	Clean	L.F.	.11	1.51		1.62	
	Minimum Charge	Job		254		254	
Quarry Tile							
	Demolish	S.F.		3.70		3.70	Includes material and labor to install
	Install	S.F.	11.90	10.40		22.30	quarry tile countertop with no
	Demolish and Install	S.F.	11.90	14.10		26	backsplash.
	Clean	S.F.	.06	.76		.82	
	Minimum Charge	Job		254		254	
With Splash							
	Demolish	S.F.		3.70		3.70	Includes material and labor to install
	Install	S.F.	13.40	11.55		24.95	quarry tile countertop with backsplash.
	Demolish and Install	S.F.	13.40	15.25		28.65	
	Clean	S.F.	.06	.76		.82	
	Minimum Charge	Job		254		254	
Butcher Block							
	Demolish	S.F.		3.70		3.70	Includes material and labor to install
	Install	S.F.	48	9.30		57.30	solid laminated maple countertop with
	Demolish and Install	S.F.	48	13		61	no backsplash.
	Reinstall	S.F.		7.44		7.44	
	Clean	S.F.	.03	.31		.34	
	Minimum Charge	Job		260		260	
Solid Surface							
	Demolish	L.F.		7.40		7.40	Includes material and labor to install
	Install	L.F.	60	37		97	solid surface counter top 22" deep.
	Demolish and Install	L.F.	60	44.40		104.40	
	Reinstall	L.F.		29.76		29.76	
	Clean	L.F.	.11	1.51		1.62	
	Minimum Charge	Job		260		260	

For customer support on your Contractor's Pricing Guide: Residential Repair & Remodeling, call 800.448.8182.

Cabinets and Countertops

Vanity Cabinets

Vanity Cabinets	Unit	Material	Labor	Equip.	Total	Specification
Good Grade Laminated						
24″ Wide, 2 Door						
Demolish	Ea.		22		22	Includes material and labor to install
Install	Ea.	380	52		432	modular unit with melamine laminated
Demolish and Install	Ea.	380	74		454	to particle board, textured colors or
Reinstall	Ea.		41.66		41.66	wood grain print finish, hinges and
Clean	Ea.	1.22	16		17.22	pulls.
Paint	Ea.	2.72	36		38.72	
Minimum Charge	Job		260		260	
30″ Wide, 2 Door						
Demolish	Ea.		28		28	Includes material and labor to install
Install	Ea.	470	65		535	modular unit with melamine laminated
Demolish and Install	Ea.	470	93		563	to particle board, textured colors or
Reinstall	Ea.		52.08		52.08	wood grain print finish, hinges and
Clean	Ea.	1.58	20		21.58	pulls.
Paint	Ea.	3.41	43		46.41	
Minimum Charge	Job		260		260	
36″ Wide, 2 Door						
Demolish	Ea.		34		34	Includes material and labor to install
Install	Ea.	490	78		568	modular unit with melamine laminated
Demolish and Install	Ea.	490	112		602	to particle board, textured colors or
Reinstall	Ea.		62.51		62.51	wood grain print finish, hinges and
Clean	Ea.	1.76	23.50		25.26	pulls.
Paint	Ea.	4.13	54		58.13	
Minimum Charge	Job		260		260	
42″ w, 3 Doors / Drawer						
Demolish	Ea.		40.50		40.50	Includes material and labor to install
Install	Ea.	980	84		1064	modular unit with melamine laminated
Demolish and Install	Ea.	980	124.50		1104.50	to particle board, textured colors or
Reinstall	Ea.		67.31		67.31	wood grain print finish, hinges and
Clean	Ea.	2.26	28.50		30.76	pulls.
Paint	Ea.	4.73	61.50		66.23	
Minimum Charge	Job		260		260	
48″ w, 3 Doors / Drawer						
Demolish	Ea.		44.50		44.50	Includes material and labor to install
Install	Ea.	640	91		731	modular unit with melamine laminated
Demolish and Install	Ea.	640	135.50		775.50	to particle board, textured colors or
Reinstall	Ea.		72.90		72.90	wood grain print finish, hinges and
Clean	Ea.	3.17	44.50		47.67	pulls.
Paint	Ea.	7.10	108		115.10	
Minimum Charge	Job		260		260	
Clean Interior						
Install	L.F.	.64	8.35		8.99	Includes labor and materials to clean
Minimum Charge	Job		200		200	base cabinetry.
Better Grade Wood						
24″ Wide, 2 Door						
Demolish	Ea.		22		22	Includes material and labor to install
Install	Ea.	505	52		557	custom modular unit with solid
Demolish and Install	Ea.	505	74		579	hardwood faced frames, hardwood
Reinstall	Ea.		41.66		41.66	door frames and drawer fronts,
Clean	Ea.	1.22	16		17.22	hardwood veneer on raised door
Paint	Ea.	2.72	39		41.72	panels, hinges and pulls.
Minimum Charge	Job		260		260	

For customer support on your Contractor's Pricing Guide: Residential Repair & Remodeling, call 800.448.8182.

Cabinets and Countertops

Vanity Cabinets		Unit	Material	Labor	Equip.	Total	Specification
30" Wide, 2 Door							
	Demolish	Ea.		28		28	Includes material and labor to install
	Install	Ea.	625	65		690	custom modular unit with solid
	Demolish and Install	Ea.	625	93		718	hardwood faced frames, hardwood
	Reinstall	Ea.		52.08		52.08	door frames and drawer fronts,
	Clean	Ea.	1.58	20		21.58	hardwood veneer on raised door
	Paint	Ea.	3.41	48		51.41	panels, hinges and pulls.
	Minimum Charge	Job		260		260	
36" Wide, 2 Door							
	Demolish	Ea.		34		34	Includes material and labor to install
	Install	Ea.	650	78		728	custom modular unit with solid
	Demolish and Install	Ea.	650	112		762	hardwood faced frames, hardwood
	Reinstall	Ea.		62.51		62.51	door frames and drawer fronts,
	Clean	Ea.	1.76	23.50		25.26	hardwood veneer on raised door
	Paint	Ea.	4.13	61.50		65.63	panels, hinges and pulls.
	Minimum Charge	Job		260		260	
42" w, 3 Doors / Drawer							
	Demolish	Ea.		40.50		40.50	Includes material and labor to install
	Install	Ea.	980	84		1064	custom modular unit with solid
	Demolish and Install	Ea.	980	124.50		1104.50	hardwood faced frames, hardwood
	Reinstall	Ea.		67.31		67.31	door frames and drawer fronts,
	Clean	Ea.	2.26	28.50		30.76	hardwood veneer on raised door
	Paint	Ea.	4.73	72		76.73	panels, hinges and pulls.
	Minimum Charge	Job		260		260	
48" w, 3 Doors / Drawer							
	Demolish	Ea.		44.50		44.50	Includes material and labor to install
	Install	Ea.	845	91		936	custom modular unit with solid
	Demolish and Install	Ea.	845	135.50		980.50	hardwood faced frames, hardwood
	Reinstall	Ea.		72.90		72.90	door frames and drawer fronts,
	Clean	Ea.	3.17	44.50		47.67	hardwood veneer on raised door
	Paint	Ea.	7.10	108		115.10	panels, hinges and pulls.
	Minimum Charge	Job		260		260	
Clean Interior							
	Install	L.F.	.64	8.35		8.99	Includes labor and materials to clean
	Minimum Charge	Job		200		200	base cabinetry.
Premium Grade Wood							
24" Wide, 2 Door							
	Demolish	Ea.		22		22	Includes material and labor to install
	Install	Ea.	585	52		637	premium modular unit with solid
	Demolish and Install	Ea.	585	74		659	hardwood faced frames, hardwood
	Reinstall	Ea.		41.66		41.66	door frames and drawer fronts,
	Clean	Ea.	1.22	16		17.22	hardwood veneer on raised door
	Paint	Ea.	2.72	43		45.72	panels, hinges and pulls.
	Minimum Charge	Job		260		260	
30" Wide, 2 Door							
	Demolish	Ea.		28		28	Includes material and labor to install
	Install	Ea.	715	65		780	premium modular unit with solid
	Demolish and Install	Ea.	715	93		808	hardwood faced frames, hardwood
	Reinstall	Ea.		52.08		52.08	door frames and drawer fronts,
	Clean	Ea.	1.58	20		21.58	hardwood veneer on raised door
	Paint	Ea.	3.41	54		57.41	panels, hinges and pulls.
	Minimum Charge	Job		260		260	

Cabinets and Countertops

Vanity Cabinets

	Unit	Material	Labor	Equip.	Total	Specification
36" Wide, 2 Door						
Demolish	Ea.		34		34	Includes material and labor to install
Install	Ea.	770	78		848	premium modular unit with solid
Demolish and Install	Ea.	770	112		882	hardwood faced frames, hardwood
Reinstall	Ea.		62.51		62.51	door frames and drawer fronts,
Clean	Ea.	1.76	23.50		25.26	hardwood veneer on raised door
Paint	Ea.	4.13	61.50		65.63	panels, hinges and pulls.
Minimum Charge	Job		260		260	
42" w, 3 Doors / Drawer						
Demolish	Ea.		40.50		40.50	Includes material and labor to install
Install	Ea.	870	84		954	premium modular unit with solid
Demolish and Install	Ea.	870	124.50		994.50	hardwood faced frames, hardwood
Reinstall	Ea.		67.31		67.31	door frames and drawer fronts,
Clean	Ea.	2.26	28.50		30.76	hardwood veneer on raised door
Paint	Ea.	4.73	72		76.73	panels, hinges and pulls.
Minimum Charge	Job		260		260	
48" w, 3 Doors / Drawer						
Demolish	Ea.		44.50		44.50	Includes material and labor to install
Install	Ea.	1125	91		1216	premium modular unit with solid
Demolish and Install	Ea.	1125	135.50		1260.50	hardwood faced frames, hardwood
Reinstall	Ea.		72.90		72.90	door frames and drawer fronts,
Clean	Ea.	3.17	44.50		47.67	hardwood veneer on raised door
Paint	Ea.	7.10	108		115.10	panels, hinges and pulls.
Minimum Charge	Job		260		260	
Clean Interior						
Install	L.F.	.64	8.35		8.99	Includes labor and materials to clean
Minimum Charge	Job		200		200	base cabinetry.

Premium Hardwood

	Unit	Material	Labor	Equip.	Total	Specification
24" Wide, 2 Door						
Demolish	Ea.		22		22	Includes material and labor to install
Install	Ea.	715	52		767	premium modular unit with solid
Demolish and Install	Ea.	715	74		789	hardwood faced frames, hardwood
Reinstall	Ea.		41.66		41.66	door frames and drawer fronts,
Clean	Ea.	1.22	16		17.22	hardwood veneer on raised door
Paint	Ea.	2.72	48		50.72	panels, hinges and pulls.
Minimum Charge	Job		260		260	
30" Wide, 2 Door						
Demolish	Ea.		28		28	Includes material and labor to install
Install	Ea.	840	65		905	premium modular unit with solid
Demolish and Install	Ea.	840	93		933	hardwood faced frames, hardwood
Reinstall	Ea.		52.08		52.08	door frames and drawer fronts,
Clean	Ea.	1.58	20		21.58	hardwood veneer on raised door
Paint	Ea.	3.41	54		57.41	panels, hinges and pulls.
Minimum Charge	Job		260		260	
36" Wide, 2 Door						
Demolish	Ea.		34		34	Includes material and labor to install
Install	Ea.	1375	78		1453	premium modular unit with solid
Demolish and Install	Ea.	1375	112		1487	hardwood faced frames, hardwood
Reinstall	Ea.		62.51		62.51	door frames and drawer fronts,
Clean	Ea.	1.76	23.50		25.26	hardwood veneer on raised door
Paint	Ea.	4.13	72		76.13	panels, hinges and pulls.
Minimum Charge	Job		260		260	

For customer support on your Contractor's Pricing Guide: Residential Repair & Remodeling, call 800.448.8182.

Cabinets and Countertops

Vanity Cabinets

	Unit	Material	Labor	Equip.	Total	Specification
42″ w, 3 Doors / Drawer						
Demolish	Ea.		40.50		40.50	Includes material and labor to install
Install	Ea.	1000	84		1084	premium modular unit with solid
Demolish and Install	Ea.	1000	124.50		1124.50	hardwood faced frames, hardwood
Reinstall	Ea.		67.31		67.31	door frames and drawer fronts,
Clean	Ea.	2.26	28.50		30.76	hardwood veneer on raised door
Paint	Ea.	4.73	86.50		91.23	panels, hinges and pulls.
Minimum Charge	Job		260		260	
48″ w, 3 Doors / Drawer						
Demolish	Ea.		44.50		44.50	Includes material and labor to install
Install	Ea.	1125	91		1216	premium modular unit with solid
Demolish and Install	Ea.	1125	135.50		1260.50	hardwood faced frames, hardwood
Reinstall	Ea.		72.90		72.90	door frames and drawer fronts,
Clean	Ea.	3.17	44.50		47.67	hardwood veneer on raised door
Paint	Ea.	7.10	108		115.10	panels, hinges and pulls.
Minimum Charge	Job		260		260	
Clean Interior						
Install	L.F.	.64	8.35		8.99	Includes labor and materials to clean
Minimum Charge	Job		200		200	base cabinetry.
Strip and Refinish						
Install	L.F.	7.05	54		61.05	Includes labor and material to strip,
Minimum Charge	Job		216		216	prep and refinish exterior of cabinets.

Vanity Cabinet Tops

	Unit	Material	Labor	Equip.	Total	Specification
Cultured Marble Top						
Demolish	L.F.		7.40		7.40	Includes material and labor to install
Install	L.F.	58	30.50		88.50	cultured marble vanity top with
Demolish and Install	L.F.	58	37.90		95.90	integral sink, 22″ deep.
Reinstall	L.F.		24.34		24.34	
Clean	L.F.	.11	1.51		1.62	
Minimum Charge	Job		254		254	
Laminated Top						
Demolish	L.F.		4.44		4.44	Includes material and labor to install
Install	L.F.	49	18.60		67.60	one piece laminated top with 4″
Demolish and Install	L.F.	49	23.04		72.04	backsplash.
Reinstall	L.F.		14.88		14.88	
Clean	L.F.	.08	1		1.08	
Minimum Charge	Job		260		260	
Solid Surface						
Demolish	L.F.		7.40		7.40	Includes material and labor to install
Install	L.F.	89	38.50		127.50	solid surface vanity top with integral
Demolish and Install	L.F.	89	45.90		134.90	sink installed, 22″ deep.
Reinstall	L.F.		30.86		30.86	
Clean	L.F.	.11	1.51		1.62	
Minimum Charge	Job		260		260	

For customer support on your Contractor's Pricing Guide: Residential Repair & Remodeling, call 800.448.8182.

Painting

Roller Handle Paint Brush

Paint Preparation, Walls

		Unit	Material	Labor	Equip.	Total	Specification
Cover / Protect Floors							
	Install	S.F.	.08	.17		.25	Includes material and labor to install plastic masking sheet on large uninterrupted areas.
	Minimum Charge	Job		217		217	
Cover / Protect Walls							
	Install	S.F.	.08	.17		.25	Includes material and labor to install plastic masking sheet on walls.
	Minimum Charge	Job		217		217	
Clean Walls							
Light							
	Clean	S.F.		.13		.13	Includes labor to wash gypsum drywall or plaster wall surfaces.
	Minimum Charge	Job		217		217	
Heavy							
	Clean	S.F.		.17		.17	Includes labor and material for heavy cleaning (multiple applications) with detergent and solvent.
	Minimum Charge	Job		217		217	
Prep Walls							
	Paint	S.F.	.07	.22		.29	Includes labor and material to prepare for painting including scraping, patching and puttying.
	Minimum Charge	Job		217		217	
Sand Walls							
	Paint	S.F.		.14		.14	Includes labor to sand gypsum drywall wall surfaces.
	Minimum Charge	Job		217		217	
Seal Walls							
Large Area							
	Paint	S.F.	.09	.32		.41	Includes labor and material to paint with primer / sealer by roller.
	Minimum Charge	Job		217		217	
Spot							
	Paint	S.F.	.12	.38		.50	Includes labor and material to seal existing drywall with shellac-based material.
	Minimum Charge	Job		217		217	

Painting

Paint Preparation, Ceilings

		Unit	Material	Labor	Equip.	Total	Specification
Cover / Protect Floors							Includes material and labor to install plastic masking sheet on large uninterrupted areas.
	Install	S.F.	.08	.17		.25	
	Minimum Charge	Job		217		217	
Cover / Protect Walls							Includes material and labor to install plastic masking sheet on walls.
	Install	S.F.	.08	.17		.25	
	Minimum Charge	Job		217		217	
Prep Ceiling							Includes labor and material to fill in hairline cracks up to 1/8" with filler, sand and prep for paint.
	Paint	S.F.	.11	.92		1.03	
	Minimum Charge	Job		217		217	
Sand Ceiling							Includes labor to sand gypsum drywall ceiling surfaces.
	Paint	S.F.		.20		.20	
	Minimum Charge	Job		217		217	
Seal Ceiling							
Large Area							Includes labor and material to paint ceiling, one coat flat latex.
	Paint	S.F.	.24	.33		.57	
	Minimum Charge	Job		216		216	
Spot							Includes labor and material to seal existing drywall with shellac-based material.
	Paint	S.F.	.12	.38		.50	
	Minimum Charge	Job		216		216	

Paint / Texture, Walls

		Unit	Material	Labor	Equip.	Total	Specification
Texture Walls							
Spray							Includes labor and material to install by spray, texture finish.
	Install	S.F.	.07	.65		.72	
	Minimum Charge	Job		216		216	
Trowel							Includes labor and material to apply hand troweled texture.
	Install	S.F.	.45	1.15		1.60	
	Minimum Charge	Job		216		216	
Paint Walls 1 Coat							Includes labor and material to paint, one coat flat latex.
	Paint	S.F.	.24	.33		.57	
	Minimum Charge	Job		216		216	
Paint Walls 2 Coats							Includes labor and material to paint, two coats flat latex.
	Paint	S.F.	.48	.54		1.02	
	Minimum Charge	Job		216		216	
Paint Walls 3 Coats							Includes labor and material to paint, three coats flat latex.
	Paint	S.F.	.73	.66		1.39	
	Minimum Charge	Job		216		216	
Clean and Seal							Includes labor and material to do light cleaning, sealing with 2 coats of latex paint.
	Paint	S.F.	.44	1.15		1.59	
	Minimum Charge	Job		216		216	

Paint / Texture, Ceilings

		Unit	Material	Labor	Equip.	Total	Specification
Texture Ceiling							
Spray							Includes labor and material to install by spray, texture finish.
	Install	S.F.	.07	.65		.72	
	Minimum Charge	Job		216		216	

Painting

Paint / Texture, Ceilings		Unit	Material	Labor	Equip.	Total	Specification
Trowel							
	Install	S.F.	.45	1.15		1.60	Includes labor and material to apply
	Minimum Charge	Job		216		216	hand troweled texture.
Paint Ceiling, 1 Coat							
	Paint	S.F.	.24	.33		.57	Includes labor and material to paint
	Minimum Charge	Job		216		216	ceiling, one coat flat latex.
Paint Ceiling, 2 Coats							
	Paint	S.F.	.25	.54		.79	Includes labor and material to paint
	Minimum Charge	Job		216		216	two coats flat latex paint.
Paint Ceiling, 3 Coats							
	Paint	S.F.	.37	.63		1	Includes labor and material to paint
	Minimum Charge	Job		216		216	ceiling, three coats flat latex.
Clean and Seal							
	Paint	S.F.	.37	.91		1.28	Includes labor and material for light
	Minimum Charge	Job		216		216	cleaning, sealing and paint 3 coats of flat latex paint.

Flooring

Plank Flooring Wood Strip Flooring Tile Flooring

Wood Plank Flooring		Unit	Material	Labor	Equip.	Total	Specification
Maple							
2-1/4"							
	Demolish	S.F.		1.78		1.78	Includes material and labor to install
	Install	S.F.	6.40	2.60		9	unfinished T&G maple flooring,
	Demolish and Install	S.F.	6.40	4.38		10.78	25/32" thick, 2-1/4" wide, random
	Reinstall	S.F.		2.08		2.08	3' to 16' lengths including felt
	Clean	S.F.		.50		.50	underlayment, nailed in place over
	Paint	S.F.	.34	3.07		3.41	prepared subfloor.
	Minimum Charge	Job		260		260	
3-1/4"							
	Demolish	S.F.		1.78		1.78	Includes material and labor to install
	Install	S.F.	10.70	2.60		13.30	unfinished T&G maple flooring,
	Demolish and Install	S.F.	10.70	4.38		15.08	33/32" thick, 3-1/4" wide, random
	Reinstall	S.F.		2.08		2.08	3' to 16' lengths including felt
	Clean	S.F.		.50		.50	underlayment, nailed in place over
	Paint	S.F.	.34	3.07		3.41	prepared subfloor.
	Minimum Charge	Job		260		260	
Sand and Finish							
	Install	S.F.	.34	1		1.34	Includes labor and material to refinish
	Minimum Charge	Job		260		260	wood floor.
Oak							
2-1/4"							
	Demolish	S.F.		1.78		1.78	Includes material and labor to install
	Install	S.F.	7.20	2.60		9.80	unfinished T&G oak flooring, 25/32"
	Demolish and Install	S.F.	7.20	4.38		11.58	thick, 2-1/4" wide, random 3' to 16'
	Reinstall	S.F.		2.08		2.08	lengths including felt underlayment,
	Clean	S.F.		.50		.50	nailed in place over prepared
	Paint	S.F.	.34	3.07		3.41	subfloor.
	Minimum Charge	Job		260		260	
3-1/4"							
	Demolish	S.F.		1.78		1.78	Includes material and labor to install
	Install	S.F.	5.15	1.84		6.99	unfinished T&G oak flooring, 25/32"
	Demolish and Install	S.F.	5.15	3.62		8.77	thick, 3-1/4" wide, random 3' to 16'
	Reinstall	S.F.		1.47		1.47	lengths including felt underlayment,
	Clean	S.F.		.50		.50	nailed in place over prepared
	Paint	S.F.	.34	3.07		3.41	subfloor.
	Minimum Charge	Job		260		260	
Sand and Finish							
	Install	S.F.	.34	3.07		3.41	Includes labor and material to refinish
	Minimum Charge	Job		260		260	wood floor.

For customer support on your Contractor's Pricing Guide: Residential Repair & Remodeling, call 800.448.8182.

Flooring

Wood Plank Flooring

Wood Plank Flooring	Unit	Material	Labor	Equip.	Total	Specification
Walnut						
3″ to 7″						
Demolish	S.F.		1.78		1.78	Includes material and labor to install
Install	S.F.	26.50	1.84		28.34	unfinished T&G walnut flooring,
Demolish and Install	S.F.	26.50	3.62		30.12	25/32″ thick, random 3′ to 16′
Reinstall	S.F.		1.47		1.47	lengths including felt underlayment,
Clean	S.F.		.50		.50	nailed in place over prepared
Paint	S.F.	.34	3.07		3.41	subfloor.
Minimum Charge	Job		260		260	
Sand and Finish						
Install	S.F.	.34	3.07		3.41	Includes labor and material to refinish
Minimum Charge	Job		260		260	wood floor.
Pine						
2-1/4″						
Demolish	S.F.		1.78		1.78	Includes material and labor to install
Install	S.F.	1.40	1.96		3.36	unfinished T&G pine flooring, 25/32″
Demolish and Install	S.F.	1.40	3.74		5.14	thick, 2-1/4″ wide, random 3′ to 16′
Reinstall	S.F.		1.57		1.57	lengths including felt underlayment,
Clean	S.F.		.50		.50	nailed in place over prepared
Paint	S.F.	.34	1		1.34	subfloor.
Minimum Charge	Job		260		260	
Sand and Finish						
Install	S.F.	.34	3.07		3.41	Includes labor and material to refinish
Minimum Charge	Job		260		260	wood floor.
Prefinished Oak						
2-1/4″						
Demolish	S.F.		1.78		1.78	Includes material and labor to install
Install	S.F.	6.45	2.60		9.05	prefinished T&G oak flooring, 2-1/4″
Demolish and Install	S.F.	6.45	4.38		10.83	wide, random 3′ to 16′ lengths
Reinstall	S.F.		2.08		2.08	including felt underlayment, nailed in
Clean	S.F.		.50		.50	place over prepared subfloor.
Minimum Charge	Job		260		260	
3-1/4″						
Demolish	S.F.		1.78		1.78	Includes material and labor to install
Install	S.F.	6.65	2.39		9.04	prefinished T&G oak flooring, 25/32″
Demolish and Install	S.F.	6.65	4.17		10.82	thick, 3″ to 7″ wide, random 3′ to 16′
Reinstall	S.F.		1.91		1.91	lengths including felt underlayment,
Clean	S.F.		.50		.50	nailed in place over prepared
Minimum Charge	Job		260		260	subfloor.
Sand and Finish New						
Install	S.F.	.34	1.35		1.69	Includes labor and material to refinish
Minimum Charge	Job		260		260	wood floor.
Refinish Existing						
Install	S.F.	.34	1		1.34	Includes labor and material to refinish
Minimum Charge	Job		260		260	wood floor.

Flooring

Wood Plank Flooring

Wood Plank Flooring	Unit	Material	Labor	Equip.	Total	Specification
Urethane Coat						
Install	S.F.	.19	.23		.42	Includes labor and material to install
Minimum Charge	Job		260		260	two coats of urethane, on existing floor.
Clean and Wax						
Clean	S.F.	.03	.31		.34	Includes labor and material to wash
Minimum Charge	Job		260		260	and wax hardwood.
Underlayment Per S.F.						
Demolish	S.F.		.76		.76	Cost includes material and labor to
Install	S.F.	1.01	.69		1.70	install 1/4" lauan subfloor, standard
Demolish and Install	S.F.	1.01	1.45		2.46	interior grade, nailed every 6".
Clean	S.F.		.39		.39	
Minimum Charge	Job		260		260	

Wood Parquet Tile Floor

Wood Parquet Tile Floor	Unit	Material	Labor	Equip.	Total	Specification
Oak						
9" x 9"						
Demolish	S.F.		1.29		1.29	Includes material and labor to install
Install	S.F.	22	4.17		26.17	prefinished oak 9" x 9" parquet block
Demolish and Install	S.F.	22	5.46		27.46	flooring, 5/16" thick, installed in
Reinstall	S.F.		3.33		3.33	mastic.
Clean	S.F.		.50		.50	
Paint	S.F.	.34	3.07		3.41	
Minimum Charge	Job		260		260	
13" x 13"						
Demolish	S.F.		1.29		1.29	Includes material and labor to install
Install	S.F.	21	3.26		24.26	prefinished oak 13" x 13" parquet
Demolish and Install	S.F.	21	4.55		25.55	block flooring, 5/16" thick, installed in
Reinstall	S.F.		2.60		2.60	mastic.
Clean	S.F.		.50		.50	
Paint	S.F.	.34	3.07		3.41	
Minimum Charge	Job		260		260	
Cherry						
9" x 9"						
Demolish	S.F.		1.29		1.29	Includes material and labor to install
Install	S.F.	35	4.17		39.17	prefinished cherry 9" x 9" parquet
Demolish and Install	S.F.	35	5.46		40.46	block flooring, 5/16" thick, installed in
Reinstall	S.F.		4.17		4.17	mastic.
Clean	S.F.		.50		.50	
Paint	S.F.	.34	3.07		3.41	
Minimum Charge	Job		260		260	
13" x 13"						
Demolish	S.F.		1.29		1.29	Includes material and labor to install
Install	S.F.	36	3.26		39.26	prefinished cherry 13" x 13" parquet
Demolish and Install	S.F.	36	4.55		40.55	block flooring, 5/16" thick, installed in
Reinstall	S.F.		3.26		3.26	mastic.
Clean	S.F.		.50		.50	
Paint	S.F.	.34	3.07		3.41	
Minimum Charge	Job		260		260	

For customer support on your Contractor's Pricing Guide: Residential Repair & Remodeling, call 800.448.8182.

Flooring

Wood Parquet Tile Floor

Wood Parquet Tile Floor	Unit	Material	Labor	Equip.	Total	Specification
Walnut						
9″ x 9″						
Demolish	S.F.		1.29		1.29	Includes material and labor to install
Install	S.F.	24	4.17		28.17	prefinished walnut 9″ x 9″ parquet
Demolish and Install	S.F.	24	5.46		29.46	flooring in mastic.
Reinstall	S.F.		3.33		3.33	
Clean	S.F.		.50		.50	
Paint	S.F.	.34	3.07		3.41	
Minimum Charge	Job		260		260	
13″ x 13″						
Demolish	S.F.		1.29		1.29	Includes material and labor to install
Install	S.F.	24	3.26		27.26	prefinished walnut 13″ x 13″ parquet
Demolish and Install	S.F.	24	4.55		28.55	block flooring, 5/16″ thick, installed in
Reinstall	S.F.		3.26		3.26	mastic.
Clean	S.F.		.50		.50	
Paint	S.F.	.34	3.07		3.41	
Minimum Charge	Job		260		260	
Acrylic Impregnated						
Oak 12″ x 12″						
Demolish	S.F.		1.29		1.29	Includes material and labor to install
Install	S.F.	23	3.72		26.72	red oak 12″ x 12″ acrylic
Demolish and Install	S.F.	23	5.01		28.01	impregnated parquet block flooring,
Reinstall	S.F.		3.72		3.72	5/16″ thick, installed in mastic.
Clean	S.F.		.50		.50	
Minimum Charge	Job		260		260	
Cherry 12″ x 12″						
Demolish	S.F.		1.29		1.29	Includes material and labor to install
Install	S.F.	43	3.72		46.72	prefinished cherry 12″ x 12″ acrylic
Demolish and Install	S.F.	43	5.01		48.01	impregnated parquet block flooring,
Reinstall	S.F.		3.72		3.72	5/16″ thick, installed in mastic.
Clean	S.F.		.50		.50	
Minimum Charge	Job		260		260	
Ash 12″ x 12″						
Demolish	S.F.		1.29		1.29	Includes material and labor to install
Install	S.F.	44.50	3.72		48.22	ash 12″ x 12″ acrylic impregnated
Demolish and Install	S.F.	44.50	5.01		49.51	parquet block flooring, 5/16″ thick in
Reinstall	S.F.		3.72		3.72	mastic.
Clean	S.F.		.50		.50	
Minimum Charge	Job		260		260	
Teak						
Demolish	S.F.		1.29		1.29	Includes material and labor to install
Install	S.F.	23	3.26		26.26	prefinished teak parquet block
Demolish and Install	S.F.	23	4.55		27.55	flooring, 5/16″ thick, and installed in
Reinstall	S.F.		3.26		3.26	mastic.
Clean	S.F.		.50		.50	
Paint	S.F.	.34	3.07		3.41	
Minimum Charge	Job		260		260	
Clean and Wax						
Clean	S.F.	.03	.31		.34	Includes labor and material to wash
Minimum Charge	Job		260		260	and wax hardwood.

Flooring

Ceramic Tile Flooring

Ceramic Tile Flooring	Unit	Material	Labor	Equip.	Total	Specification
Economy Grade						
Demolish	S.F.		1.32		1.32	Cost includes material and labor to
Install	S.F.	6.25	9.95		16.20	install 4-1/4" x 4-1/4" ceramic tile in
Demolish and Install	S.F.	6.25	11.27		17.52	mortar bed including grout.
Clean	S.F.		.76		.76	
Minimum Charge	Job		254		254	
Average Grade						
Demolish	S.F.		1.32		1.32	Cost includes material and labor to
Install	S.F.	6.15	10.15		16.30	install 4-1/4" x 4-1/4" ceramic tile in
Demolish and Install	S.F.	6.15	11.47		17.62	mortar bed including grout.
Clean	S.F.		.76		.76	
Minimum Charge	Job		254		254	
Premium Grade						
Demolish	S.F.		1.32		1.32	Cost includes material and labor to
Install	S.F.	13.40	10.15		23.55	install 4-1/4" x 4-1/4" ceramic tile in
Demolish and Install	S.F.	13.40	11.47		24.87	mortar bed including grout.
Clean	S.F.		.76		.76	
Minimum Charge	Job		254		254	
Re-grout						
Install	S.F.	.07	4.07		4.14	Includes material and labor to regrout
Minimum Charge	Job		254		254	tile floors.

Hard Tile Flooring

Hard Tile Flooring	Unit	Material	Labor	Equip.	Total	Specification
Thick Set Paver						
Brick						
Demolish	S.F.		2.78		2.78	Cost includes material and labor to
Install	S.F.	3.72	22.50		26.22	install 6" x 12" adobe brick paver with
Demolish and Install	S.F.	3.72	25.28		29	1/2" mortar joints.
Clean	S.F.		.76		.76	
Minimum Charge	Job		254		254	
Mexican Red						
Demolish	S.F.		2.78		2.78	Cost includes material and labor to
Install	S.F.	2.45	10.60		13.05	install 12" x 12" red Mexican paver
Demolish and Install	S.F.	2.45	13.38		15.83	tile in mortar bed with grout.
Clean	S.F.		.76		.76	
Minimum Charge	Job		254		254	
Saltillo						
Demolish	S.F.		1.32		1.32	Cost includes material and labor to
Install	S.F.	4.18	10.60		14.78	install 12" x 12" saltillo tile in mortar
Demolish and Install	S.F.	4.18	11.92		16.10	bed with grout.
Clean	S.F.		.76		.76	
Minimum Charge	Job		254		254	
Marble						
Premium Grade						
Demolish	S.F.		1.32		1.32	Cost includes material and labor to
Install	S.F.	36.50	22		58.50	install 3/8" x 12" x 12" marble, in
Demolish and Install	S.F.	36.50	23.32		59.82	mortar bed with grout.
Clean	S.F.		.76		.76	
Minimum Charge	Job		254		254	

For customer support on your Contractor's Pricing Guide: Residential Repair & Remodeling, call 800.448.8182.

Flooring

Hard Tile Flooring

Hard Tile Flooring		Unit	Material	Labor	Equip.	Total	Specification
Terrazzo							
Gray Cement							
	Demolish	S.F.		1.76		1.76	Cost includes material and labor to
	Install	S.F.	5.10	12.30	5.40	22.80	install 1-3/4" terrazzo , #1 and #2
	Demolish and Install	S.F.	5.10	14.06	5.40	24.56	chips in gray Portland cement.
	Clean	S.F.		.76		.76	
	Minimum Charge	Job		460	203	663	
White Cement							
	Demolish	S.F.		1.76		1.76	Cost includes material and labor to
	Install	S.F.	6	12.30	5.40	23.70	install 1-3/4" terrazzo, #1 and #2
	Demolish and Install	S.F.	6	14.06	5.40	25.46	chips in white Portland cement.
	Clean	S.F.		.76		.76	
	Minimum Charge	Job		460	203	663	
Non-skid Gray							
	Demolish	S.F.		1.76		1.76	Cost includes material and labor to
	Install	S.F.	9.45	15.35	6.75	31.55	install 1-3/4" terrazzo, #1 and #2
	Demolish and Install	S.F.	9.45	17.11	6.75	33.31	chips in gray Portland cement with
	Clean	S.F.		.76		.76	light non-skid abrasive.
	Minimum Charge	Job		460	203	663	
Non-skid White							
	Demolish	S.F.		1.76		1.76	Cost includes material and labor to
	Install	S.F.	10.45	15.35	6.75	32.55	install 1-3/4" terrazzo, #1 and #2
	Demolish and Install	S.F.	10.45	17.11	6.75	34.31	chips in white Portland cement with
	Clean	S.F.		.76		.76	light non-skid abrasive.
	Minimum Charge	Job		460	203	663	
Gray w / Brass Divider							
	Demolish	S.F.		1.78		1.78	Cost includes material and labor to
	Install	S.F.	14.80	16.75	7.35	38.90	install 1-3/4" terrazzo, #1 and #2
	Demolish and Install	S.F.	14.80	18.53	7.35	40.68	chips in gray Portland cement with
	Clean	S.F.		.76		.76	brass strips 2' O.C. each way.
	Minimum Charge	Job		460	203	663	
Clean							
	Clean	S.F.	.13	.51		.64	Includes labor and material to clean
	Minimum Charge	Job		460	203	663	terrazzo flooring.
Slate							
	Demolish	S.F.		1.59		1.59	Includes material and labor to install
	Install	S.F.	12.95	5		17.95	slate tile flooring in thin set with grout.
	Demolish and Install	S.F.	12.95	6.59		19.54	
	Clean	S.F.		.76		.76	
	Minimum Charge	Job		254		254	
Granite							
	Demolish	S.F.		1.31		1.31	Includes material and labor to install
	Install	S.F.	30.50	5.10		35.60	granite tile flooring in thin set with
	Demolish and Install	S.F.	30.50	6.41		36.91	grout.
	Reinstall	S.F.		5.08		5.08	
	Clean	S.F.		.76		.76	
	Minimum Charge	Job		254		254	
Re-grout							
	Install	S.F.	.07	4.07		4.14	Includes material and labor to regrout
	Minimum Charge	Job		254		254	tile floors.

For customer support on your Contractor's Pricing Guide: Residential Repair & Remodeling, call 800.448.8182.

Flooring

Sheet Vinyl Per S.F.

		Unit	Material	Labor	Equip.	Total	Specification
Economy Grade							
	Demolish	S.F.		.63		.63	Includes material and labor to install
	Install	S.F.	3.13	1.59		4.72	resilient sheet vinyl flooring.
	Demolish and Install	S.F.	3.13	2.22		5.35	
	Clean	S.F.		.50		.50	
	Minimum Charge	Job		254		254	
Average Grade							
	Demolish	S.F.		.63		.63	Includes material and labor to install
	Install	S.F.	4.48	1.59		6.07	resilient sheet vinyl flooring.
	Demolish and Install	S.F.	4.48	2.22		6.70	
	Clean	S.F.		.50		.50	
	Minimum Charge	Job		254		254	
Premium Grade							
	Demolish	S.F.		.63		.63	Includes material and labor to install
	Install	S.F.	4.36	1.59		5.95	resilient sheet vinyl flooring.
	Demolish and Install	S.F.	4.36	2.22		6.58	
	Clean	S.F.		.50		.50	
	Minimum Charge	Job		254		254	
Clean and Wax							
	Clean	S.F.	.03	.55		.58	Includes labor and material to wash
	Minimum Charge	Job		254		254	and wax a vinyl tile floor.

Sheet Vinyl

		Unit	Material	Labor	Equip.	Total	Specification
Economy Grade							
	Demolish	S.Y.		5.75		5.75	Includes material and labor to install
	Install	S.Y.	28.50	14.30		42.80	no wax sheet vinyl flooring.
	Demolish and Install	S.Y.	28.50	20.05		48.55	
	Clean	S.Y.		4.49		4.49	
	Minimum Charge	Job		254		254	
Average Grade							
	Demolish	S.Y.		5.75		5.75	Includes material and labor to install
	Install	S.Y.	40	14.30		54.30	no wax sheet vinyl flooring.
	Demolish and Install	S.Y.	40	20.05		60.05	
	Clean	S.Y.		4.49		4.49	
	Minimum Charge	Job		254		254	
Premium Grade							
	Demolish	S.Y.		5.75		5.75	Includes material and labor to install
	Install	S.Y.	39	14.30		53.30	no wax sheet vinyl flooring.
	Demolish and Install	S.Y.	39	20.05		59.05	
	Clean	S.Y.		4.49		4.49	
	Minimum Charge	Job		254		254	
Clean and Wax							
	Clean	S.F.	.03	.55		.58	Includes labor and material to wash
	Minimum Charge	Job		254		254	and wax a vinyl tile floor.

Vinyl Tile

		Unit	Material	Labor	Equip.	Total	Specification
Economy Grade							
	Demolish	S.F.		.89		.89	Includes material and labor to install
	Install	S.F.	1.96	1.02		2.98	no wax 12" x 12" vinyl tile.
	Demolish and Install	S.F.	1.96	1.91		3.87	
	Clean	S.F.		.50		.50	
	Minimum Charge	Job		254		254	

For customer support on your Contractor's Pricing Guide: Residential Repair & Remodeling, call 800.448.8182.

Flooring

Vinyl Tile

Vinyl Tile		Unit	Material	Labor	Equip.	Total	Specification
Average Grade							
	Demolish	S.F.		.89		.89	Includes material and labor to install
	Install	S.F.	2.02	1.02		3.04	no wax 12″ x 12″ vinyl tile.
	Demolish and Install	S.F.	2.02	1.91		3.93	
	Clean	S.F.		.50		.50	
	Minimum Charge	Job		254		254	
Premium Grade							
	Demolish	S.F.		.89		.89	Includes material and labor to install
	Install	S.F.	1.94	1.02		2.96	no wax 12″ x 12″ vinyl tile.
	Demolish and Install	S.F.	1.94	1.91		3.85	
	Clean	S.F.		.50		.50	
	Minimum Charge	Job		254		254	
Clean and Wax							
	Clean	S.F.	.03	.55		.58	Includes labor and material to wash
	Minimum Charge	Job		254		254	and wax a vinyl tile floor.

Carpeting Per S.F.

Carpeting Per S.F.		Unit	Material	Labor	Equip.	Total	Specification
Economy Grade							
	Demolish	S.F.		.10		.10	Includes material and labor to install
	Install	S.F.	1.38	.75		2.13	carpet including tack strips and hot
	Demolish and Install	S.F.	1.38	.85		2.23	melt tape on seams.
	Clean	S.F.		.69		.69	
	Minimum Charge	Job		254		254	
Average Grade							
	Demolish	S.F.		.10		.10	Includes material and labor to install
	Install	S.F.	1.38	.75		2.13	carpet including tack strips and hot
	Demolish and Install	S.F.	1.38	.85		2.23	melt tape on seams.
	Clean	S.F.		.69		.69	
	Minimum Charge	Job		254		254	
Premium Grade							
	Demolish	S.F.		.10		.10	Includes material and labor to install
	Install	S.F.	2.18	.75		2.93	carpet including tack strips and hot
	Demolish and Install	S.F.	2.18	.85		3.03	melt tape on seams.
	Clean	S.F.		.69		.69	
	Minimum Charge	Job		254		254	
Indoor/Outdoor							
	Demolish	S.F.		.10		.10	Includes material and labor to install
	Install	S.F.	1.12	.75		1.87	indoor-outdoor carpet.
	Demolish and Install	S.F.	1.12	.85		1.97	
	Clean	S.F.		.69		.69	
	Minimum Charge	Job		254		254	
100% Wool							
Average Grade							
	Demolish	S.F.		.10		.10	Includes material and labor to install
	Install	S.F.	8.60	1.01		9.61	carpet including tack strips and hot
	Demolish and Install	S.F.	8.60	1.11		9.71	melt tape on seams.
	Clean	S.F.		.69		.69	
	Minimum Charge	Job		254		254	
Premium Grade							
	Demolish	S.F.		.10		.10	Includes material and labor to install
	Install	S.F.	12.40	1.01		13.41	carpet including tack strips and hot
	Demolish and Install	S.F.	12.40	1.11		13.51	melt tape on seams.
	Clean	S.F.		.69		.69	
	Minimum Charge	Job		254		254	

For customer support on your Contractor's Pricing Guide: Residential Repair & Remodeling, call 800.448.8182.

Flooring

Carpeting Per S.F.

Carpeting Per S.F.		Unit	Material	Labor	Equip.	Total	Specification
Luxury Grade							
	Demolish	S.F.		.10		.10	Includes material and labor to install
	Install	S.F.	12.75	1.01		13.76	carpet including tack strips and hot
	Demolish and Install	S.F.	12.75	1.11		13.86	melt tape on seams.
	Clean	S.F.		.69		.69	
	Minimum Charge	Job		254		254	
Berber							
Average Grade							
	Demolish	S.F.		.10		.10	Includes material and labor to install
	Install	S.F.	7.60	1.20		8.80	carpet including tack strips and hot
	Demolish and Install	S.F.	7.60	1.30		8.90	melt tape on seams.
	Clean	S.F.		.69		.69	
	Minimum Charge	Job		254		254	
Premium Grade							
	Demolish	S.F.		.10		.10	Includes material and labor to install
	Install	S.F.	8.55	1.20		9.75	carpet including tack strips and hot
	Demolish and Install	S.F.	8.55	1.30		9.85	melt tape on seams.
	Clean	S.F.		.69		.69	
	Minimum Charge	Job		254		254	

Carpeting

Carpeting		Unit	Material	Labor	Equip.	Total	Specification
Indoor/Outdoor							
	Demolish	S.Y.		1.48		1.48	Includes material and labor to install
	Install	S.Y.	10.10	6.80		16.90	indoor-outdoor carpet.
	Demolish and Install	S.Y.	10.10	8.28		18.38	
	Reinstall	S.Y.		5.42		5.42	
	Clean	S.Y.		6.25		6.25	
	Minimum Charge	Job		254		254	
Economy Grade							
	Demolish	S.Y.		1.48		1.48	Includes material and labor to install
	Install	S.Y.	12.40	6.80		19.20	carpet including tack strips and hot
	Demolish and Install	S.Y.	12.40	8.28		20.68	melt tape on seams.
	Reinstall	S.Y.		5.42		5.42	
	Clean	S.Y.		6.25		6.25	
	Minimum Charge	Job		254		254	
Average Grade							
	Demolish	S.Y.		1.48		1.48	Includes material and labor to install
	Install	S.Y.	12.45	6.80		19.25	carpet including tack strips and hot
	Demolish and Install	S.Y.	12.45	8.28		20.73	melt tape on seams.
	Reinstall	S.Y.		5.42		5.42	
	Clean	S.Y.		6.25		6.25	
	Minimum Charge	Job		254		254	
Premium Grade							
	Demolish	S.Y.		1.48		1.48	Includes material and labor to install
	Install	S.Y.	19.60	6.80		26.40	carpet including tack strips and hot
	Demolish and Install	S.Y.	19.60	8.28		27.88	melt tape on seams.
	Reinstall	S.Y.		5.42		5.42	
	Clean	S.Y.		6.25		6.25	
	Minimum Charge	Job		254		254	

For customer support on your Contractor's Pricing Guide: Residential Repair & Remodeling, call 800.448.8182.

Flooring

Carpeting

	Unit	Material	Labor	Equip.	Total	Specification
100% Wool						
Average Grade						
Demolish	S.Y.		1.48		1.48	Includes material and labor to install
Install	S.Y.	77	9.10		86.10	wool carpet including tack strips and
Demolish and Install	S.Y.	77	10.58		87.58	hot melt tape on seams.
Reinstall	S.Y.		7.26		7.26	
Clean	S.Y.		6.25		6.25	
Minimum Charge	Job		254		254	
Premium Grade						
Demolish	S.Y.		1.48		1.48	Includes material and labor to install
Install	S.Y.	111	9.10		120.10	carpet including tack strips and hot
Demolish and Install	S.Y.	111	10.58		121.58	melt tape on seams.
Reinstall	S.Y.		7.26		7.26	
Clean	S.Y.		6.25		6.25	
Minimum Charge	Job		254		254	
Luxury Grade						
Demolish	S.Y.		1.48		1.48	Includes material and labor to install
Install	S.Y.	114	9.10		123.10	carpet including tack strips and hot
Demolish and Install	S.Y.	114	10.58		124.58	melt tape on seams.
Reinstall	S.Y.		7.26		7.26	
Clean	S.Y.		6.25		6.25	
Minimum Charge	Job		254		254	
Berber						
Average Grade						
Demolish	S.Y.		1.48		1.48	Includes material and labor to install
Install	S.Y.	68	10.80		78.80	berber carpet including tack strips and
Demolish and Install	S.Y.	68	12.28		80.28	hot melt tape on seams.
Reinstall	S.Y.		8.65		8.65	
Clean	S.Y.		6.25		6.25	
Minimum Charge	Job		254		254	
Premium Grade						
Demolish	S.Y.		1.48		1.48	Includes material and labor to install
Install	S.Y.	77	10.80		87.80	carpet including tack strips and hot
Demolish and Install	S.Y.	77	12.28		89.28	melt tape on seams.
Reinstall	S.Y.		8.65		8.65	
Clean	S.Y.		6.25		6.25	
Minimum Charge	Job		254		254	

Carpet Pad Per S.F.

	Unit	Material	Labor	Equip.	Total	Specification
Urethane						
Demolish	S.F.		.03		.03	Includes material and labor to install
Install	S.F.	.51	.38		.89	urethane carpet pad.
Demolish and Install	S.F.	.51	.41		.92	
Minimum Charge	Job		254		254	
Foam Rubber Slab						
Demolish	S.F.		.03		.03	Includes material and labor to install
Install	S.F.	.75	.38		1.13	carpet pad, 3/8" thick foam rubber.
Demolish and Install	S.F.	.75	.41		1.16	
Minimum Charge	Job		254		254	
Waffle						
Demolish	S.F.		.03		.03	Includes material and labor to install
Install	S.F.	.91	.04		.95	rubber waffle carpet pad.
Demolish and Install	S.F.	.91	.07		.98	
Minimum Charge	Job		254		254	

For customer support on your Contractor's Pricing Guide: Residential Repair & Remodeling, call 800.448.8182.

Flooring

Carpet Pad Per S.F.

Carpet Pad Per S.F.	Unit	Material	Labor	Equip.	Total	Specification
Jute						
Demolish	S.F.		.03		.03	Includes material and labor to install
Install	S.F.	.87	.33		1.20	jute hair carpet pad.
Demolish and Install	S.F.	.87	.36		1.23	
Minimum Charge	Job		254		254	
Rebound						
Demolish	S.F.		.03		.03	Includes material and labor to install
Install	S.F.	.50	.28		.78	rebound carpet pad.
Demolish and Install	S.F.	.50	.31		.81	
Minimum Charge	Job		254		254	

Carpet Pad

Carpet Pad	Unit	Material	Labor	Equip.	Total	Specification
Urethane						
Demolish	S.Y.		.25		.25	Includes material and labor to install
Install	S.Y.	4.55	3.39		7.94	urethane carpet pad.
Demolish and Install	S.Y.	4.55	3.64		8.19	
Reinstall	S.Y.		2.71		2.71	
Minimum Charge	Job		254		254	
Foam Rubber Slab						
Demolish	S.Y.		.25		.25	Includes material and labor to install
Install	S.Y.	6.70	3.39		10.09	rubber slab carpet pad.
Demolish and Install	S.Y.	6.70	3.64		10.34	
Reinstall	S.Y.		2.71		2.71	
Minimum Charge	Job		254		254	
Waffle						
Demolish	S.Y.		.25		.25	Includes material and labor to install
Install	S.Y.	8.20	.38		8.58	rubber waffle carpet pad.
Demolish and Install	S.Y.	8.20	.63		8.83	
Reinstall	S.Y.		.30		.30	
Minimum Charge	Job		254		254	
Jute						
Demolish	S.Y.		.25		.25	Includes material and labor to install
Install	S.Y.	7.80	2.99		10.79	jute hair carpet pad.
Demolish and Install	S.Y.	7.80	3.24		11.04	
Reinstall	S.Y.		2.39		2.39	
Minimum Charge	Job		254		254	
Rebound						
Demolish	S.Y.		.25		.25	Includes material and labor to install
Install	S.Y.	4.45	2.53		6.98	rebound carpet pad.
Demolish and Install	S.Y.	4.45	2.78		7.23	
Reinstall	S.Y.		2.03		2.03	
Minimum Charge	Job		254		254	

Stair Components

Stair Components	Unit	Material	Labor	Equip.	Total	Specification
Carpeting						
Demolish	Ea.		2.69		2.69	Includes material and labor to install
Install	Riser	61	13.40		74.40	carpet and pad on stairs.
Demolish and Install	Riser	61	16.09		77.09	
Reinstall	Riser		10.70		10.70	
Clean	Ea.		6.25		6.25	
Minimum Charge	Job		254		254	

For customer support on your Contractor's Pricing Guide: Residential Repair & Remodeling, call 800.448.8182.

Finish Mechanical

Tub/Shower

Stainless Steel Double Sink

Water Closet

Commode	Unit	Material	Labor	Equip.	Total	Specification
Floor Mounted w / Tank						
Demolish	Ea.		58.50		58.50	Includes material and labor to install a toilet with valve, seat and cover.
Install	Ea.	288	199		487	
Demolish and Install	Ea.	288	257.50		545.50	
Reinstall	Ea.		159.52		159.52	
Clean	Ea.	.32	20		20.32	
Minimum Charge	Job		294		294	
Wall w / Flush Valve						
Demolish	Ea.		58.50		58.50	Includes material and labor to install a wall hung toilet with flush valve, seat and cover.
Install	Ea.	1200	182		1382	
Demolish and Install	Ea.	1200	240.50		1440.50	
Reinstall	Ea.		145.77		145.77	
Clean	Ea.	.32	20		20.32	
Minimum Charge	Job		294		294	
Designer Floor Mount						
Demolish	Ea.		58.50		58.50	Includes material and labor to install a toilet with valve, seat and cover.
Install	Ea.	1075	199		1274	
Demolish and Install	Ea.	1075	257.50		1332.50	
Reinstall	Ea.		159.52		159.52	
Clean	Ea.	.32	20		20.32	
Minimum Charge	Job		294		294	
Urinal						
Demolish	Ea.		58.50		58.50	Includes material and labor to install a wall hung urinal with flush valve.
Install	Ea.	505	350		855	
Demolish and Install	Ea.	505	408.50		913.50	
Reinstall	Ea.		281.81		281.81	
Clean	Ea.	.32	20		20.32	
Minimum Charge	Job		294		294	
Bidet						
Vitreous China						
Demolish	Ea.		58.50		58.50	Includes material and labor to install a vitreous china bidet complete with trim.
Install	Ea.	840	211		1051	
Demolish and Install	Ea.	840	269.50		1109.50	
Reinstall	Ea.		169.09		169.09	
Clean	Ea.	.32	20		20.32	
Minimum Charge	Job		294		294	

Finish Mechanical

Commode		Unit	Material	Labor	Equip.	Total	Specification
Chrome Fittings							
	Install	Ea.	300	73.50		373.50	Includes labor and materials to install chrome fittings for a bidet.
Brass Fittings							
	Install	Ea.	330	73.50		403.50	Includes labor and materials to install brass fittings for a bidet.
Seat							
	Demolish	Ea.		24.50		24.50	Includes material and labor to install a toilet seat and cover.
	Install	Ea.	43.50	24.50		68	
	Demolish and Install	Ea.	43.50	49		92.50	
	Reinstall	Ea.		19.57		19.57	
	Clean	Ea.	.15	8.35		8.50	
	Minimum Charge	Job		294		294	
Rough-in							
	Install	Ea.	430	370		800	Includes labor and materials to install water supply pipe with valves and drain, waste and vent pipe with all couplings, hangers and fasteners necessary.

Sink (assembly)		Unit	Material	Labor	Equip.	Total	Specification
Single							
Porcelain							
	Demolish	Ea.		36.50		36.50	Includes material and labor to install a single bowl enamel finished cast iron kitchen sink with faucet and drain.
	Install	Ea.	1200	189		1389	
	Demolish and Install	Ea.	1200	225.50		1425.50	
	Reinstall	Ea.		150.97		150.97	
	Clean	Ea.	.01	1.57		1.58	
	Minimum Charge	Job		294		294	
Stainless Steel							
	Demolish	Ea.		36.50		36.50	Includes material and labor to install a single bowl stainless steel kitchen sink with faucet and drain.
	Install	Ea.	885	189		1074	
	Demolish and Install	Ea.	885	225.50		1110.50	
	Reinstall	Ea.		150.97		150.97	
	Clean	Ea.	.01	1.57		1.58	
	Minimum Charge	Job		294		294	
Double							
Porcelain							
	Demolish	Ea.		42		42	Includes material and labor to install a double bowl enamel finished cast iron kitchen sink with faucet and drain.
	Install	Ea.	500	220		720	
	Demolish and Install	Ea.	500	262		762	
	Reinstall	Ea.		176.13		176.13	
	Clean	Ea.	.32	20		20.32	
	Minimum Charge	Job		294		294	
Stainless Steel							
	Demolish	Ea.		42		42	Includes material and labor to install a double bowl stainless steel kitchen sink with faucet and drain.
	Install	Ea.	555	220		775	
	Demolish and Install	Ea.	555	262		817	
	Reinstall	Ea.		176.13		176.13	
	Clean	Ea.	.32	20		20.32	
	Minimum Charge	Job		294		294	

For customer support on your Contractor's Pricing Guide: Residential Repair & Remodeling, call 800.448.8182.

Finish Mechanical

Sink (assembly)

Sink (assembly)	Unit	Material	Labor	Equip.	Total	Specification
Bar						Includes material and labor to install a small single bowl stainless steel bar / vegetable / kitchen sink with faucet and drain.
Demolish	Ea.		36.50		36.50	
Install	Ea.	252	147		399	
Demolish and Install	Ea.	252	183.50		435.50	
Reinstall	Ea.		117.44		117.44	
Clean	Ea.	.01	1.57		1.58	
Minimum Charge	Job		294		294	
Floor						Includes material and labor to install an enameled cast iron floor mounted corner service sink with faucet and drain.
Demolish	Ea.		117		117	
Install	Ea.	1750	240		1990	
Demolish and Install	Ea.	1750	357		2107	
Reinstall	Ea.		192.15		192.15	
Clean	Ea.	.32	10		10.32	
Minimum Charge	Job		294		294	
Pedestal						Includes material and labor to install a vitreous china pedestal lavatory with faucet set and pop-up drain.
Demolish	Ea.		36.50		36.50	
Install	Ea.	425	165		590	
Demolish and Install	Ea.	425	201.50		626.50	
Reinstall	Ea.		132.10		132.10	
Clean	Ea.	.01	1.57		1.58	
Minimum Charge	Job		294		294	
Vanity Lavatory						Includes material and labor to install a vitreous china lavatory with faucet set and pop-up drain.
Demolish	Ea.		36.50		36.50	
Install	Ea.	246	196		442	
Demolish and Install	Ea.	246	232.50		478.50	
Reinstall	Ea.		156.56		156.56	
Clean	Ea.	.01	1.57		1.58	
Minimum Charge	Job		294		294	
Rough-in						Includes labor and materials to install copper supply pipe with valves and drain, waste and vent pipe with all couplings, hangers and fasteners necessary.
Install	Ea.	440	495		935	
Laundry						Includes material and labor to install a laundry sink (wall mounted or with legs) including faucet and pop up drain.
Demolish	Ea.		40		40	
Install	Ea.	186	163		349	
Demolish and Install	Ea.	186	203		389	
Clean	Ea.	.01	1.57		1.58	
Minimum Charge	Job		294		294	

Sink Only

Sink Only	Unit	Material	Labor	Equip.	Total	Specification
Single						
Porcelain						Includes material and labor to install a single bowl enamel finished cast iron kitchen sink.
Demolish	Ea.		36.50		36.50	
Install	Ea.	1075	210		1285	
Demolish and Install	Ea.	1075	246.50		1321.50	
Reinstall	Ea.		167.77		167.77	
Clean	Ea.	.01	1.57		1.58	
Minimum Charge	Job		294		294	

For customer support on your Contractor's Pricing Guide: Residential Repair & Remodeling, call 800.448.8182.

Finish Mechanical

Sink Only

	Unit	Material	Labor	Equip.	Total	Specification
Stainless Steel						
Demolish	Ea.		36.50		36.50	Includes material and labor to install a
Install	Ea.	750	147		897	single bowl stainless steel kitchen sink.
Demolish and Install	Ea.	750	183.50		933.50	
Reinstall	Ea.		117.44		117.44	
Clean	Ea.	.01	1.57		1.58	
Minimum Charge	Job		294		294	
Double						
Porcelain						
Demolish	Ea.		42		42	Includes material and labor to install a
Install	Ea.	340	245		585	double bowl enamel finished cast iron
Demolish and Install	Ea.	340	287		627	kitchen sink.
Reinstall	Ea.		195.73		195.73	
Clean	Ea.	.32	20		20.32	
Minimum Charge	Job		294		294	
Stainless Steel						
Demolish	Ea.		42		42	Includes material and labor to install a
Install	Ea.	1125	189		1314	double bowl stainless steel kitchen
Demolish and Install	Ea.	1125	231		1356	sink.
Reinstall	Ea.		151.54		151.54	
Clean	Ea.	.32	20		20.32	
Minimum Charge	Job		294		294	
Bar						
Demolish	Ea.		36.50		36.50	Includes material and labor to install a
Install	Ea.	101	147		248	small single bowl stainless steel bar /
Demolish and Install	Ea.	101	183.50		284.50	vegetable / kitchen sink.
Reinstall	Ea.		117.44		117.44	
Clean	Ea.	.01	1.57		1.58	
Minimum Charge	Job		294		294	
Floor						
Demolish	Ea.		117		117	Includes material and labor to install
Install	Ea.	1600	267		1867	an enameled cast iron floor mounted
Demolish and Install	Ea.	1600	384		1984	corner service sink.
Reinstall	Ea.		213.53		213.53	
Clean	Ea.	.32	10		10.32	
Minimum Charge	Job		294		294	
Pedestal						
Demolish	Ea.		36.50		36.50	Includes material and labor to install a
Install	Ea.	305	184		489	vitreous china pedestal lavatory.
Demolish and Install	Ea.	305	220.50		525.50	
Reinstall	Ea.		146.80		146.80	
Clean	Ea.	.01	1.57		1.58	
Minimum Charge	Job		294		294	
Vanity Lavatory						
Demolish	Ea.		36.50		36.50	Includes material and labor to install a
Install	Ea.	111	184		295	vitreous china lavatory.
Demolish and Install	Ea.	111	220.50		331.50	
Reinstall	Ea.		146.80		146.80	
Clean	Ea.	.01	1.57		1.58	
Minimum Charge	Job		294		294	
Rough-in						
Install	Ea.	405	460		865	Includes labor and materials to install copper supply pipe with valves and drain, waste and vent pipe with all couplings, hangers and fasteners necessary.

For customer support on your Contractor's Pricing Guide: Residential Repair & Remodeling, call 800.448.8182.

Finish Mechanical

Sink Only		Unit	Material	Labor	Equip.	Total	Specification
Faucet Sink							
Good Grade							
	Demolish	Ea.		29.50		29.50	Includes material and labor to install a
	Install	Ea.	114	40.50		154.50	sink faucet and fittings.
	Demolish and Install	Ea.	114	70		184	
	Reinstall	Ea.		32.42		32.42	
	Clean	Ea.	.13	12.50		12.63	
	Minimum Charge	Job		294		294	
Better Grade							
	Demolish	Ea.		29.50		29.50	Includes material and labor to install a
	Install	Ea.	193	40.50		233.50	sink faucet and fittings.
	Demolish and Install	Ea.	193	70		263	
	Reinstall	Ea.		32.42		32.42	
	Clean	Ea.	.13	12.50		12.63	
	Minimum Charge	Job		294		294	
Premium Grade							
	Demolish	Ea.		29.50		29.50	Includes material and labor to install a
	Install	Ea.	375	50.50		425.50	sink faucet and fittings.
	Demolish and Install	Ea.	375	80		455	
	Reinstall	Ea.		40.43		40.43	
	Clean	Ea.	.13	12.50		12.63	
	Minimum Charge	Job		294		294	
Drain & Basket							
	Demolish	Ea.		18.35		18.35	Includes material and labor to install a
	Install	Ea.	20	36.50		56.50	sink drain and basket.
	Demolish and Install	Ea.	20	54.85		74.85	
	Reinstall	Ea.		29.36		29.36	
	Clean	Ea.	.13	12.50		12.63	
	Minimum Charge	Job		294		294	

Bathtub		Unit	Material	Labor	Equip.	Total	Specification
Enameled							
Steel							
	Demolish	Ea.		106		106	Includes labor and material to install a
	Install	Ea.	895	192		1087	formed steel bathtub with spout,
	Demolish and Install	Ea.	895	298		1193	mixing valve, shower head and
	Reinstall	Ea.		153.72		153.72	pop-up drain.
	Clean	Ea.	.64	33.50		34.14	
	Minimum Charge	Job		294		294	
Cast Iron							
	Demolish	Ea.		106		106	Includes material and labor to install a
	Install	Ea.	2075	240		2315	cast iron bathtub with spout, mixing
	Demolish and Install	Ea.	2075	346		2421	valve, shower head and pop-up drain.
	Reinstall	Ea.		192.15		192.15	
	Clean	Ea.	.64	33.50		34.14	
	Minimum Charge	Job		294		294	
Fiberglass							
	Demolish	Ea.		106		106	Includes material and labor to install
	Install	Ea.	1800	192		1992	an acrylic soaking tub with spout,
	Demolish and Install	Ea.	1800	298		2098	mixing valve, shower head and
	Reinstall	Ea.		153.72		153.72	pop-up drain.
	Clean	Ea.	.64	33.50		34.14	
	Minimum Charge	Job		294		294	

Finish Mechanical

Bathtub	Unit	Material	Labor	Equip.	Total	Specification
Institutional						
Demolish	Ea.		106		106	Includes material and labor to install a
Install	Ea.	2175	350		2525	hospital / institutional type bathtub
Demolish and Install	Ea.	2175	456		2631	with spout, mixing valve, shower head
Reinstall	Ea.		281.81		281.81	and pop-up drain.
Clean	Ea.	.64	33.50		34.14	
Minimum Charge	Job		294		294	
Whirlpool (Acrylic)						
Demolish	Ea.		325		325	Includes material and labor to install a
Install	Ea.	1725	1050		2775	molded fiberglass whirlpool tub.
Demolish and Install	Ea.	1725	1375		3100	
Reinstall	Ea.		845.44		845.44	
Clean	Ea.	.64	33.50		34.14	
Minimum Charge	Job		294		294	
Pump / Motor						
Demolish	Ea.		81.50		81.50	Includes material and labor to install a
Install	Ea.	405	176		581	whirlpool tub pump / motor.
Demolish and Install	Ea.	405	257.50		662.50	
Reinstall	Ea.		141.07		141.07	
Clean	Ea.	.15	25		25.15	
Minimum Charge	Job		294		294	
Heater / Motor						
Demolish	Ea.		81.50		81.50	Includes material and labor to install a
Install	Ea.	251	261		512	whirlpool tub heater.
Demolish and Install	Ea.	251	342.50		593.50	
Reinstall	Ea.		208.78		208.78	
Clean	Ea.	.15	25		25.15	
Minimum Charge	Job		294		294	
Thermal Cover						
Demolish	Ea.		13.05		13.05	Includes material and labor to install a
Install	Ea.	635	36.50		671.50	whirlpool tub thermal cover.
Demolish and Install	Ea.	635	49.55		684.55	
Reinstall	Ea.		29.36		29.36	
Clean	Ea.	1.58	25		26.58	
Minimum Charge	Job		294		294	
Tub / Shower Combination						
Demolish	Ea.		93		93	Includes material and labor to install a
Install	Ea.	1275	264		1539	fiberglass module tub with shower
Demolish and Install	Ea.	1275	357		1632	surround with spout, mixing valve,
Reinstall	Ea.		211.36		211.36	shower head and pop-up drain.
Clean	Ea.	.79	50		50.79	
Minimum Charge	Job		294		294	
Accessories						
Sliding Door						
Demolish	Ea.		28		28	Includes material and labor to install a
Install	Ea.	350	120		470	48" aluminum framed tempered glass
Demolish and Install	Ea.	350	148		498	shower door.
Reinstall	Ea.		95.66		95.66	
Clean	Ea.	1.32	16.65		17.97	
Minimum Charge	Job		294		294	
Shower Head						
Demolish	Ea.		7.35		7.35	Includes material and labor to install a
Install	Ea.	35	24.50		59.50	water saving shower head.
Demolish and Install	Ea.	35	31.85		66.85	
Reinstall	Ea.		19.57		19.57	
Clean	Ea.	.01	5		5.01	
Minimum Charge	Job		294		294	

For customer support on your Contractor's Pricing Guide: Residential Repair & Remodeling, call 800.448.8182.

Finish Mechanical

Bathtub

Bathtub		Unit	Material	Labor	Equip.	Total	Specification
Faucet Set							Includes material and labor to install a
	Demolish	Ea.		36.50		36.50	combination spout / diverter for a
	Install	Ea.	114	73.50		187.50	bathtub.
	Demolish and Install	Ea.	114	110		224	
	Reinstall	Ea.		58.72		58.72	
	Clean	Ea.	.13	12.50		12.63	
	Minimum Charge	Job		294		294	
Shower Rod							Includes material and labor to install a
	Demolish	Ea.		4.62		4.62	chrome plated shower rod.
	Install	Ea.	13.95	40		53.95	
	Demolish and Install	Ea.	13.95	44.62		58.57	
	Reinstall	Ea.		32.05		32.05	
	Clean	Ea.		5		5	
	Minimum Charge	Job		294		294	
Rough-in							Includes labor and materials to install
	Install	Ea.	430	510		940	copper supply pipe with valves and
							drain, waste and vent pipe with all
							couplings, hangers and fasteners
							necessary.

Shower

Shower		Unit	Material	Labor	Equip.	Total	Specification
Fiberglass							
32″ x 32″							
	Demolish	Ea.		220		220	Includes material and labor to install a
	Install	Ea.	645	192		837	fiberglass shower stall with door,
	Demolish and Install	Ea.	645	412		1057	mixing valve and shower head and
	Reinstall	Ea.		153.72		153.72	drain fitting.
	Clean	Ea.	.79	50		50.79	
	Minimum Charge	Job		294		294	
36″ x 36″							
	Demolish	Ea.		220		220	Includes material and labor to install a
	Install	Ea.	705	192		897	fiberglass shower stall with door,
	Demolish and Install	Ea.	705	412		1117	mixing valve and shower head and
	Reinstall	Ea.		153.72		153.72	drain fitting.
	Clean	Ea.	.79	50		50.79	
	Minimum Charge	Job		294		294	
35″ x 60″							
	Demolish	Ea.		220		220	Includes material and labor to install a
	Install	Ea.	1075	264		1339	handicap fiberglass shower stall with
	Demolish and Install	Ea.	1075	484		1559	door & seat, mixing valve and shower
	Reinstall	Ea.		211.36		211.36	head and drain fitting.
	Clean	Ea.	.79	50		50.79	
	Minimum Charge	Job		294		294	
Accessories							
Single Door							
	Demolish	Ea.		28		28	Includes material and labor to install a
	Install	Ea.	183	63		246	shower door.
	Demolish and Install	Ea.	183	91		274	
	Reinstall	Ea.		50.49		50.49	
	Clean	Ea.	1.32	16.65		17.97	
	Minimum Charge	Job		294		294	

For customer support on your Contractor's Pricing Guide: Residential Repair & Remodeling, call 800.448.8182.

Finish Mechanical

Shower

Shower		Unit	Material	Labor	Equip.	Total	Specification
Shower Head							
	Demolish	Ea.		7.35		7.35	Includes material and labor to install a
	Install	Ea.	35	24.50		59.50	water saving shower head.
	Demolish and Install	Ea.	35	31.85		66.85	
	Reinstall	Ea.		19.57		19.57	
	Clean	Ea.	.01	5		5.01	
	Minimum Charge	Job		294		294	
Faucet Set							
	Demolish	Ea.		36.50		36.50	Includes material and labor to install a
	Install	Ea.	114	73.50		187.50	combination spout / diverter for a
	Demolish and Install	Ea.	114	110		224	bathtub.
	Reinstall	Ea.		58.72		58.72	
	Clean	Ea.	.13	12.50		12.63	
	Minimum Charge	Job		294		294	
Shower Pan							
	Demolish	Ea.		32.50		32.50	Includes material and labor to install
	Install	Ea.	172	73.50		245.50	the base portion of a fiberglass
	Demolish and Install	Ea.	172	106		278	shower unit.
	Reinstall	Ea.		58.72		58.72	
	Clean	Ea.	.64	33.50		34.14	
	Minimum Charge	Job		294		294	
Rough-in							
	Install	Ea.	525	515		1040	Includes labor and materials to install copper supply pipe with valves and drain, waste and vent pipe with all couplings, hangers and fasteners necessary.

Ductwork

Ductwork		Unit	Material	Labor	Equip.	Total	Specification
Return Grill							
Under 10" Wide							
	Demolish	Ea.		12.60		12.60	Includes material and labor to install a
	Install	Ea.	38.50	31.50		70	10" x 10" air return register.
	Demolish and Install	Ea.	38.50	44.10		82.60	
	Reinstall	Ea.		25.24		25.24	
	Clean	Ea.	.15	11.10		11.25	
	Minimum Charge	Job		284		284	
12" to 20" Wide							
	Demolish	Ea.		12.60		12.60	Includes material and labor to install a
	Install	Ea.	99	33.50		132.50	16" x 16" air return register.
	Demolish and Install	Ea.	99	46.10		145.10	
	Reinstall	Ea.		26.73		26.73	
	Clean	Ea.	.15	11.10		11.25	
	Minimum Charge	Job		284		284	
Over 20" Wide							
	Demolish	Ea.		19.10		19.10	Includes material and labor to install a
	Install	Ea.	82	51.50		133.50	24" x 24" air return register.
	Demolish and Install	Ea.	82	70.60		152.60	
	Reinstall	Ea.		41.31		41.31	
	Clean	Ea.	.15	12.50		12.65	
	Minimum Charge	Job		284		284	

Finish Mechanical

Ductwork

	Unit	Material	Labor	Equip.	Total	Specification
Supply Grill						
Baseboard Type						
Demolish	Ea.		9		9	Includes material and labor to install a
Install	Ea.	27	28.50		55.50	14" x 6" baseboard air supply
Demolish and Install	Ea.	27	37.50		64.50	register.
Reinstall	Ea.		22.72		22.72	
Clean	Ea.	.15	11.10		11.25	
Minimum Charge	Job		284		284	
10" Wide						
Demolish	Ea.		10.50		10.50	Includes material and labor to install a
Install	Ea.	40.50	28.50		69	10" x 6" air supply register.
Demolish and Install	Ea.	40.50	39		79.50	
Reinstall	Ea.		22.72		22.72	
Clean	Ea.	.15	11.10		11.25	
Minimum Charge	Job		284		284	
12" to 15" Wide						
Demolish	Ea.		10.50		10.50	Includes material and labor to install a
Install	Ea.	39	33.50		72.50	14" x 8" air supply register.
Demolish and Install	Ea.	39	44		83	
Reinstall	Ea.		26.73		26.73	
Clean	Ea.	.15	11.10		11.25	
Minimum Charge	Job		284		284	
18" to 24" Wide						
Demolish	Ea.		12.60		12.60	Includes material and labor to install a
Install	Ea.	78	43.50		121.50	24" x 8" air supply register.
Demolish and Install	Ea.	78	56.10		134.10	
Reinstall	Ea.		34.95		34.95	
Clean	Ea.	.15	12.50		12.65	
Minimum Charge	Job		284		284	
30" to 36" Wide						
Demolish	Ea.		15.80		15.80	Includes material and labor to install a
Install	Ea.	102	40.50		142.50	30" x 8" air supply register.
Demolish and Install	Ea.	102	56.30		158.30	
Reinstall	Ea.		32.46		32.46	
Clean	Ea.	.15	25		25.15	
Minimum Charge	Job		284		284	
Diffuser						
12" x 12"						
Demolish	Ea.		12.60		12.60	Includes material and labor to install a
Install	Ea.	135	47.50		182.50	12" x 12" aluminum diffuser.
Demolish and Install	Ea.	135	60.10		195.10	
Reinstall	Ea.		37.87		37.87	
Clean	Ea.	.15	11.10		11.25	
Minimum Charge	Job		284		284	
14" x 14"						
Demolish	Ea.		12.60		12.60	Includes material and labor to install a
Install	Ea.	215	47.50		262.50	14" x 14" aluminum diffuser.
Demolish and Install	Ea.	215	60.10		275.10	
Reinstall	Ea.		37.87		37.87	
Clean	Ea.	.15	12.50		12.65	
Minimum Charge	Job		284		284	
18" x 18"						
Demolish	Ea.		19.10		19.10	Includes material and labor to install a
Install	Ea.	208	63		271	18" x 18" aluminum diffuser.
Demolish and Install	Ea.	208	82.10		290.10	
Reinstall	Ea.		50.49		50.49	
Clean	Ea.	.15	16.65		16.80	
Minimum Charge	Job		284		284	

For customer support on your Contractor's Pricing Guide: Residential Repair & Remodeling, call 800.448.8182.

Finish Mechanical

Ductwork

	Unit	Material	Labor	Equip.	Total	Specification
12" Round						
Demolish	Ea.		17.55		17.55	Includes material and labor to install a
Install	Ea.	52.50	47.50		100	12" diameter aluminum diffuser with a
Demolish and Install	Ea.	52.50	65.05		117.55	butterfly damper.
Reinstall	Ea.		37.87		37.87	
Clean	Ea.	.15	11.10		11.25	
Minimum Charge	Job		284		284	
14" to 20" Round						
Demolish	Ea.		23.50		23.50	Includes material and labor to install a
Install	Ea.	235	63		298	20" diameter aluminum diffuser with a
Demolish and Install	Ea.	235	86.50		321.50	butterfly damper.
Reinstall	Ea.		50.49		50.49	
Clean	Ea.	.15	12.50		12.65	
Minimum Charge	Job		284		284	
Flush-out / Sanitize						
Install	L.F.	.64	5		5.64	Includes labor and material costs to
Minimum Charge	Job		284		284	clean ductwork or flue.

Finish Electrical

Lighting Fixture

Duplex Receptacle

Electrical Per S.F.	Unit	Material	Labor	Equip.	Total	Specification
Residential						
Light Fixtures						
Install	S.F.	.28	.31		.59	Includes labor and material to replace house light fixtures. Apply S.F. cost to floor area including garages but not basements.

Circuits	Unit	Material	Labor	Equip.	Total	Specification
120V						
Outlet with Wiring						
Demolish	Ea.		34.50		34.50	Includes material and labor to install general purpose outlet with cover, up to 20 feet of #12/2 wire.
Install	Ea.	11.55	42.50		54.05	
Demolish and Install	Ea.	11.55	77		88.55	
Reinstall	Ea.		34.13		34.13	
Clean	Ea.		2.50		2.50	
Minimum Charge	Job		310		310	
GFCI Outlet w / Wiring						
Demolish	Ea.		34.50		34.50	Cost includes material and labor to install 120V GFCI receptacle and wall plate, with up to 20 feet of #12/2 wire.
Install	Ea.	33	50.50		83.50	
Demolish and Install	Ea.	33	85		118	
Reinstall	Ea.		40.34		40.34	
Clean	Ea.		2.50		2.50	
Minimum Charge	Job		310		310	
Exterior Outlet w / Wiring						
Demolish	Ea.		34.50		34.50	Cost includes material and labor to install 110V waterproof external receptacle and cover plate with up to 20' of #12/2 wire.
Install	Ea.	62.50	51.50		114	
Demolish and Install	Ea.	62.50	86		148.50	
Reinstall	Ea.		41.39		41.39	
Clean	Ea.		2.50		2.50	
Minimum Charge	Job		310		310	
240V Outlet w / Wiring						
Demolish	Ea.		34.50		34.50	Cost includes material and labor to install 240V receptacle and cover plate with up to 20' of #12/2 wire.
Install	Ea.	90	97		187	
Demolish and Install	Ea.	90	131.50		221.50	
Reinstall	Ea.		77.48		77.48	
Clean	Ea.		2.50		2.50	
Minimum Charge	Job		310		310	

Finish Electrical

Circuits	Unit	Material	Labor	Equip.	Total	Specification
TV / Cable Jack w / Wiring						
Demolish	Ea.		34.50		34.50	Includes material and labor to install
Install	Ea.	20	39		59	TV/cable outlet box and cover plate
Demolish and Install	Ea.	20	73.50		93.50	including 20' of wiring.
Reinstall	Ea.		31.04		31.04	
Clean	Ea.		2.50		2.50	
Minimum Charge	Job		310		310	
Phone Jack with Wiring						
Demolish	Ea.		34.50		34.50	Includes material and labor to install
Install	Ea.	12.65	24		36.65	phone jack and cover plate including
Demolish and Install	Ea.	12.65	58.50		71.15	up to 20' of telephone cable.
Reinstall	Ea.		19.10		19.10	
Clean	Ea.		1.56		1.56	
Minimum Charge	Job		310		310	

Light Switch	Unit	Material	Labor	Equip.	Total	Specification
Single						
Switch						
Demolish	Ea.		6.90		6.90	Includes material and labor to install
Install	Ea.	4.13	29.50		33.63	single switch and wall plate. Wiring is
Demolish and Install	Ea.	4.13	36.40		40.53	not included.
Reinstall	Ea.		23.65		23.65	
Clean	Ea.		2.50		2.50	
Minimum Charge	Job		310		310	
Plate						
Demolish	Ea.		3.45		3.45	Includes material and labor to install a
Install	Ea.	.44	7.75		8.19	one gang brown wall switch plate.
Demolish and Install	Ea.	.44	11.20		11.64	
Reinstall	Ea.		6.21		6.21	
Clean	Ea.	.07	.50		.57	
Minimum Charge	Job		310		310	
Double						
Switch						
Demolish	Ea.		13.80		13.80	Includes material and labor to install
Install	Ea.	5.90	44.50		50.40	double switch and wall plate. Wiring
Demolish and Install	Ea.	5.90	58.30		64.20	is not included.
Reinstall	Ea.		35.47		35.47	
Clean	Ea.	.07	8		8.07	
Minimum Charge	Job		310		310	
Plate						
Demolish	Ea.		3.45		3.45	Includes material and labor to install a
Install	Ea.	.90	11.70		12.60	two gang brown wall switch plate.
Demolish and Install	Ea.	.90	15.15		16.05	
Reinstall	Ea.		9.37		9.37	
Clean	Ea.	.07	.50		.57	
Minimum Charge	Job		310		310	
Triple						
Switch						
Demolish	Ea.		21		21	Includes material and labor to install
Install	Ea.	18.15	51.50		69.65	triple switch and wall plate. Wiring is
Demolish and Install	Ea.	18.15	72.50		90.65	not included.
Reinstall	Ea.		41.39		41.39	
Clean	Ea.	.07	12		12.07	
Minimum Charge	Job		310		310	

Finish Electrical

Light Switch

Light Switch	Unit	Material	Labor	Equip.	Total	Specification
Plate						
Demolish	Ea.		3.45		3.45	Includes material and labor to install a
Install	Ea.	2.09	19.40		21.49	three gang brown wall switch plate.
Demolish and Install	Ea.	2.09	22.85		24.94	
Reinstall	Ea.		15.52		15.52	
Clean	Ea.	.07	.50		.57	
Minimum Charge	Job		310		310	
Dimmer						
Switch						
Demolish	Ea.		6.90		6.90	Includes material and labor to install
Install	Ea.	40	36.50		76.50	dimmer switch. Wiring is not included.
Demolish and Install	Ea.	40	43.40		83.40	
Reinstall	Ea.		29.21		29.21	
Clean	Ea.		2.50		2.50	
Minimum Charge	Job		310		310	
Plate						
Demolish	Ea.		3.45		3.45	Includes material and labor to install a
Install	Ea.	.44	7.75		8.19	one gang brown wall switch plate.
Demolish and Install	Ea.	.44	11.20		11.64	
Reinstall	Ea.		6.21		6.21	
Clean	Ea.	.07	.50		.57	
Minimum Charge	Job		310		310	
With Wiring						
Single						
Demolish	Ea.		34.50		34.50	Includes material and labor to install
Install	Ea.	10.75	36.50		47.25	single switch and plate including up to
Demolish and Install	Ea.	10.75	71		81.75	20' of romex cable.
Reinstall	Ea.		29.04		29.04	
Clean	Ea.		2.50		2.50	
Minimum Charge	Job		310		310	
Double						
Demolish	Ea.		34.50		34.50	Includes material and labor to install
Install	Ea.	31.50	62		93.50	double switch and plate including up to
Demolish and Install	Ea.	31.50	96.50		128	to 20' of romex cable.
Reinstall	Ea.		49.66		49.66	
Clean	Ea.	.07	8		8.07	
Minimum Charge	Job		310		310	
Triple						
Demolish	Ea.		52		52	Includes material and labor to install
Install	Ea.	28	70		98	triple switch and plate including up to
Demolish and Install	Ea.	28	122		150	20' of romex cable.
Reinstall	Ea.		55.87		55.87	
Clean	Ea.	.07	12		12.07	
Minimum Charge	Job		310		310	
Dimmer						
Demolish	Ea.		34.50		34.50	Includes material and labor to install
Install	Ea.	47	42.50		89.50	dimmer switch and cover plate
Demolish and Install	Ea.	47	77		124	including up to 20' of romex cable.
Reinstall	Ea.		34.13		34.13	
Clean	Ea.		2.50		2.50	
Minimum Charge	Job		310		310	

For customer support on your Contractor's Pricing Guide: Residential Repair & Remodeling, call 800.448.8182.

Finish Electrical

Cover for Outlet / Switch

	Unit	Material	Labor	Equip.	Total	Specification
High Grade						
Demolish	Ea.		3.45		3.45	Includes material and labor to install
Install	Ea.	2.30	7.75		10.05	wall plate, stainless steel, 1 gang.
Demolish and Install	Ea.	2.30	11.20		13.50	
Reinstall	Ea.		6.21		6.21	
Clean	Ea.	.07	.50		.57	
Minimum Charge	Job		310		310	
Deluxe Grade						
Demolish	Ea.		3.45		3.45	Includes material and labor to install
Install	Ea.	8.15	7.75		15.90	wall plate, brushed brass, 1 gang.
Demolish and Install	Ea.	8.15	11.20		19.35	
Reinstall	Ea.		6.21		6.21	
Clean	Ea.	.07	.50		.57	
Minimum Charge	Job		310		310	

Residential Light Fixture

	Unit	Material	Labor	Equip.	Total	Specification
Ceiling						
Good Quality						
Demolish	Ea.		14.35		14.35	Includes material and labor to install
Install	Ea.	28	15.50		43.50	standard quality incandescent ceiling
Demolish and Install	Ea.	28	29.85		57.85	light fixture. Wiring is not included.
Reinstall	Ea.		12.42		12.42	
Clean	Ea.	.08	33.50		33.58	
Minimum Charge	Job		310		310	
Better Quality						
Demolish	Ea.		14.35		14.35	Includes material and labor to install
Install	Ea.	99.50	32.50		132	custom quality incandescent ceiling
Demolish and Install	Ea.	99.50	46.85		146.35	light fixture. Wiring is not included.
Reinstall	Ea.		26.14		26.14	
Clean	Ea.	.08	33.50		33.58	
Minimum Charge	Job		310		310	
Premium Quality						
Demolish	Ea.		14.35		14.35	Includes material and labor to install
Install	Ea.	440	32.50		472.50	deluxe quality incandescent ceiling
Demolish and Install	Ea.	440	46.85		486.85	light fixture. Wiring is not included.
Reinstall	Ea.		26.14		26.14	
Clean	Ea.	.08	33.50		33.58	
Minimum Charge	Job		310		310	
Recessed Ceiling						
Demolish	Ea.		28.50		28.50	Includes material and labor to install
Install	Ea.	65	20.50		85.50	standard quality recessed eyeball
Demolish and Install	Ea.	65	49		114	spotlight with housing. Wiring is not
Reinstall	Ea.		16.55		16.55	included.
Clean	Ea.	.03	4.16		4.19	
Minimum Charge	Job		310		310	
Recessed Eyeball						
Demolish	Ea.		28.50		28.50	Includes material and labor to install
Install	Ea.	48.50	22		70.50	custom quality recessed eyeball
Demolish and Install	Ea.	48.50	50.50		99	spotlight with housing. Wiring is not
Reinstall	Ea.		17.74		17.74	included.
Clean	Ea.	.03	4.16		4.19	
Minimum Charge	Job		310		310	

Finish Electrical

Residential Light Fixture

	Unit	Material	Labor	Equip.	Total	Specification
Recessed Shower Type						
Demolish	Ea.		28.50		28.50	Includes material and labor to install
Install	Ea.	99.50	20.50		120	deluxe quality recessed eyeball
Demolish and Install	Ea.	99.50	49		148.50	spotlight with housing. Wiring is not
Reinstall	Ea.		16.55		16.55	included.
Clean	Ea.	.03	4.16		4.19	
Minimum Charge	Job		310		310	
Run Wiring						
Install	Ea.	21.50	25		46.50	Includes labor and material to install
Minimum Charge	Job		310		310	an electric fixture utility box with
						non-metallic cable.
Fluorescent						
Good Quality						
Demolish	Ea.		31.50		31.50	Includes material and labor to install
Install	Ea.	121	88.50		209.50	standard quality 12″ wide by 48″ long
Demolish and Install	Ea.	121	120		241	decorative surface mounted fluorescent
Reinstall	Ea.		70.95		70.95	fixture w/acrylic diffuser.
Clean	Ea.	.03	4.16		4.19	
Minimum Charge	Job		310		310	
Better Quality						
Demolish	Ea.		31.50		31.50	Includes material and labor to install
Install	Ea.	229	100		329	an interior fluorescent lighting fixture.
Demolish and Install	Ea.	229	131.50		360.50	
Reinstall	Ea.		80.10		80.10	
Clean	Ea.	.03	4.16		4.19	
Minimum Charge	Job		310		310	
Premium Quality						
Demolish	Ea.		31.50		31.50	Includes material and labor to install
Install	Ea.	172	117		289	deluxe quality 24″ wide by 48″ long
Demolish and Install	Ea.	172	148.50		320.50	decorative surface mounted fluorescent
Reinstall	Ea.		93.71		93.71	fixture w/acrylic diffuser.
Clean	Ea.	.03	4.16		4.19	
Minimum Charge	Job		310		310	
Run Wiring						
Install	Ea.	21.50	25		46.50	Includes labor and material to install
Minimum Charge	Job		310		310	an electric fixture utility box with
						non-metallic cable.
Exterior						
Good Quality						
Demolish	Ea.		27.50		27.50	Includes material and labor to install
Install	Ea.	81.50	39		120.50	standard quality incandescent outdoor
Demolish and Install	Ea.	81.50	66.50		148	light fixture. Wiring is not included.
Reinstall	Ea.		31.04		31.04	
Clean	Ea.	.15	16.65		16.80	
Minimum Charge	Job		310		310	
Better Quality						
Demolish	Ea.		27.50		27.50	Includes material and labor to install
Install	Ea.	128	39		167	custom quality incandescent outdoor
Demolish and Install	Ea.	128	66.50		194.50	light fixture. Wiring is not included.
Reinstall	Ea.		31.04		31.04	
Clean	Ea.	.15	16.65		16.80	
Minimum Charge	Job		310		310	
Premium Quality						
Demolish	Ea.		27.50		27.50	Includes material and labor to install
Install	Ea.	365	155		520	deluxe quality incandescent outdoor
Demolish and Install	Ea.	365	182.50		547.50	light fixture. Wiring is not included.
Reinstall	Ea.		124.16		124.16	
Clean	Ea.	.15	16.65		16.80	
Minimum Charge	Job		310		310	

For customer support on your Contractor's Pricing Guide: Residential Repair & Remodeling, call 800.448.8182.

Finish Electrical

Residential Light Fixture		Unit	Material	Labor	Equip.	Total	Specification
Outdoor w / Pole							Includes material and labor to install
	Demolish	Ea.		27.50		27.50	post lantern type incandescent outdoor
	Install	Ea.	395	155		550	light fixture. Wiring is not included.
	Demolish and Install	Ea.	395	182.50		577.50	
	Reinstall	Ea.		124.16		124.16	
	Clean	Ea.	.15	16.65		16.80	
	Minimum Charge	Job		310		310	
Run Wiring							Includes labor and material to install
	Install	Ea.	21.50	25		46.50	an electric fixture utility box with
	Minimum Charge	Job		310		310	non-metallic cable.
Track Lighting							Includes material and labor to install
	Demolish	L.F.		9.60		9.60	track lighting.
	Install	L.F.	35	26		61	
	Demolish and Install	L.F.	35	35.60		70.60	
	Reinstall	L.F.		20.69		20.69	
	Clean	L.F.	.15	8		8.15	
	Minimum Charge	Job		310		310	

Commercial Light Fixture		Unit	Material	Labor	Equip.	Total	Specification
Fluorescent							
2 Bulb, 2'							Includes material and labor to install a
	Demolish	Ea.		27.50		27.50	2' long surface mounted fluorescent
	Install	Ea.	39.50	77.50		117	light fixture with two 40 watt bulbs.
	Demolish and Install	Ea.	39.50	105		144.50	
	Reinstall	Ea.		62.08		62.08	
	Clean	Ea.	.79	28.50		29.29	
	Minimum Charge	Job		310		310	
2 Bulb, 4'							Includes material and labor to install a
	Demolish	Ea.		27.50		27.50	4' long surface mounted fluorescent
	Install	Ea.	158	88.50		246.50	light fixture with two 40 watt bulbs.
	Demolish and Install	Ea.	158	116		274	
	Reinstall	Ea.		70.95		70.95	
	Clean	Ea.	.79	28.50		29.29	
	Minimum Charge	Job		310		310	
3 Bulb, 4'							Includes material and labor to install a
	Demolish	Ea.		27.50		27.50	4' long pendant mounted fluorescent
	Install	Ea.	113	88.50		201.50	light fixture with three 34 watt bulbs.
	Demolish and Install	Ea.	113	116		229	
	Reinstall	Ea.		70.95		70.95	
	Clean	Ea.	.79	28.50		29.29	
	Minimum Charge	Job		310		310	
4 Bulb, 4'							Includes material and labor to install a
	Demolish	Ea.		27.50		27.50	4' long pendant mounted fluorescent
	Install	Ea.	102	95.50		197.50	light fixture with four 34 watt bulbs.
	Demolish and Install	Ea.	102	123		225	
	Reinstall	Ea.		76.41		76.41	
	Clean	Ea.	.79	28.50		29.29	
	Minimum Charge	Job		310		310	
2 Bulb, 8'							Includes material and labor to install
	Demolish	Ea.		34.50		34.50	an 8' long surface mounted
	Install	Ea.	128	117		245	fluorescent light fixture with two 110
	Demolish and Install	Ea.	128	151.50		279.50	watt bulbs.
	Reinstall	Ea.		93.71		93.71	
	Clean	Ea.	.79	28.50		29.29	
	Minimum Charge	Job		310		310	

For customer support on your Contractor's Pricing Guide: Residential Repair & Remodeling, call 800.448.8182.

Finish Electrical

Commercial Light Fixture

	Unit	Material	Labor	Equip.	Total	Specification
Run Wiring						
Install	Ea.	21.50	25		46.50	Includes labor and material to install
Minimum Charge	Job		310		310	an electric fixture utility box with
						non-metallic cable.
Ceiling, Recessed						
Eyeball Type						
Demolish	Ea.		28.50		28.50	Includes material and labor to install
Install	Ea.	48.50	22		70.50	custom quality recessed eyeball
Demolish and Install	Ea.	48.50	50.50		99	spotlight with housing. Wiring is not
Reinstall	Ea.		17.74		17.74	included.
Clean	Ea.	.03	4.16		4.19	
Minimum Charge	Job		310		310	
Shower Type						
Demolish	Ea.		28.50		28.50	Includes material and labor to install
Install	Ea.	99.50	20.50		120	deluxe quality recessed eyeball
Demolish and Install	Ea.	99.50	49		148.50	spotlight with housing. Wiring is not
Reinstall	Ea.		16.55		16.55	included.
Clean	Ea.	.03	4.16		4.19	
Minimum Charge	Job		310		310	
Run Wiring						
Install	Ea.	21.50	25		46.50	Includes labor and material to install
Minimum Charge	Job		310		310	an electric fixture utility box with
						non-metallic cable.
Ceiling Grid Type						
2 Tube						
Demolish	Ea.		40.50		40.50	Cost includes material and labor to
Install	Ea.	104	117		221	install 2 lamp fluorescent light fixture
Demolish and Install	Ea.	104	157.50		261.50	in 2' x 4' ceiling grid system including
Reinstall	Ea.		93.71		93.71	lens and tubes.
Clean	Ea.	.79	28.50		29.29	
Minimum Charge	Job		310		310	
4 Tube						
Demolish	Ea.		46		46	Cost includes material and labor to
Install	Ea.	105	132		237	install 4 lamp fluorescent light fixture
Demolish and Install	Ea.	105	178		283	in 2' x 4' ceiling grid system including
Reinstall	Ea.		105.67		105.67	lens and tubes.
Clean	Ea.	.79	28.50		29.29	
Minimum Charge	Job		310		310	
Acrylic Diffuser						
Demolish	Ea.		7.20		7.20	Includes material and labor to install
Install	Ea.	27.50	18.80		46.30	fluorescent light fixture w/acrylic
Demolish and Install	Ea.	27.50	26		53.50	diffuser.
Reinstall	Ea.		15.05		15.05	
Clean	Ea.	.51	5		5.51	
Minimum Charge	Job		310		310	
Louver Diffuser						
Demolish	Ea.		8.60		8.60	Includes material and labor to install
Install	Ea.	110	18.80		128.80	fluorescent light fixture w/louver
Demolish and Install	Ea.	110	27.40		137.40	diffuser.
Reinstall	Ea.		15.05		15.05	
Clean	Ea.	.51	5		5.51	
Minimum Charge	Job		310		310	
Run Wiring						
Install	Ea.	21.50	25		46.50	Includes labor and material to install
Minimum Charge	Job		310		310	an electric fixture utility box with
						non-metallic cable.

For customer support on your Contractor's Pricing Guide: Residential Repair & Remodeling, call 800.448.8182.

Finish Electrical

Commercial Light Fixture

Commercial Light Fixture	Unit	Material	Labor	Equip.	Total	Specification
Track Lighting						
Demolish	L.F.		9.60		9.60	Includes material and labor to install
Install	L.F.	35	26		61	track lighting.
Demolish and Install	L.F.	35	35.60		70.60	
Reinstall	L.F.		20.69		20.69	
Clean	L.F.	.15	8		8.15	
Minimum Charge	Job		310		310	
Emergency						
Demolish	Ea.		49.50		49.50	Includes material and labor to install
Install	Ea.	164	155		319	battery-powered emergency lighting
Demolish and Install	Ea.	164	204.50		368.50	unit with two (2) lights, including
Reinstall	Ea.		124.16		124.16	wiring.
Clean	Ea.	.40	28.50		28.90	
Minimum Charge	Job		310		310	
Run Wiring						
Install	Ea.	21.50	25		46.50	Includes labor and material to install
Minimum Charge	Job		310		310	an electric fixture utility box with
						non-metallic cable.

Fan

Fan	Unit	Material	Labor	Equip.	Total	Specification
Ceiling Paddle						
Good Quality						
Demolish	Ea.		38.50		38.50	Includes material and labor to install
Install	Ea.	135	62		197	standard ceiling fan.
Demolish and Install	Ea.	135	100.50		235.50	
Reinstall	Ea.		49.66		49.66	
Clean	Ea.	.32	25		25.32	
Minimum Charge	Job		310		310	
Better Quality						
Demolish	Ea.		38.50		38.50	Includes material and labor to install
Install	Ea.	315	62		377	better ceiling fan.
Demolish and Install	Ea.	315	100.50		415.50	
Reinstall	Ea.		49.66		49.66	
Clean	Ea.	.32	25		25.32	
Minimum Charge	Job		310		310	
Premium Quality						
Demolish	Ea.		38.50		38.50	Includes material and labor to install
Install	Ea.	284	77.50		361.50	deluxe ceiling fan.
Demolish and Install	Ea.	284	116		400	
Reinstall	Ea.		62.08		62.08	
Clean	Ea.	.32	25		25.32	
Minimum Charge	Job		310		310	
Light Kit						
Install	Ea.	52.50	77.50		130	Includes labor and material to install a
Minimum Charge	Job		310		310	ceiling fan light kit.
Bathroom Exhaust						
Ceiling Mounted						
Demolish	Ea.		29		29	Cost includes material and labor to
Install	Ea.	59.50	62		121.50	install 60 CFM ceiling mounted
Demolish and Install	Ea.	59.50	91		150.50	bathroom exhaust fan, up to 10' of
Reinstall	Ea.		49.52		49.52	wiring with one wall switch.
Clean	Ea.	.15	10		10.15	
Minimum Charge	Job		310		310	

Finish Electrical

Fan

Fan	Unit	Material	Labor	Equip.	Total	Specification
Wall Mounted						
Demolish	Ea.		29		29	Cost includes material and labor to
Install	Ea.	76.50	62		138.50	install 110 CFM wall mounted
Demolish and Install	Ea.	76.50	91		167.50	bathroom exhaust fan, up to 10' of
Reinstall	Ea.		49.52		49.52	wiring with one wall switch.
Clean	Ea.	.15	10		10.15	
Minimum Charge	Job		310		310	
Lighted						
Demolish	Ea.		29		29	Cost includes material and labor to
Install	Ea.	77.50	62		139.50	install 50 CFM ceiling mounted lighted
Demolish and Install	Ea.	77.50	91		168.50	exhaust fan, up to 10' of wiring with
Reinstall	Ea.		49.52		49.52	one wall switch..
Clean	Ea.	.15	10		10.15	
Minimum Charge	Job		310		310	
Blower Heater w / Light						
Demolish	Ea.		29		29	Includes material and labor to install
Install	Ea.	106	103		209	ceiling blower type ventilator with
Demolish and Install	Ea.	106	132		238	switch, light and snap-on grill. Includes
Reinstall	Ea.		82.53		82.53	1300 watt/120 volt heater.
Clean	Ea.	.15	10		10.15	
Minimum Charge	Job		310		310	
Fan Heater w / Light						
Demolish	Ea.		29		29	Includes material and labor to install
Install	Ea.	129	103		232	circular ceiling fan heater unit, 1500
Demolish and Install	Ea.	129	132		261	watt, surface mounted.
Reinstall	Ea.		82.53		82.53	
Clean	Ea.	.15	10		10.15	
Minimum Charge	Job		310		310	
Whole House Exhaust						
36" Ceiling Mounted						
Install	Ea.	1500	155		1655	Includes material and labor to install a
Clean	Ea.	.04	8.10		8.14	whole house exhaust fan.
Minimum Charge	Job		310		310	
Kitchen Exhaust						
Standard Model						
Install	Ea.	86.50	41.50		128	Includes labor and materials to install
Clean	Ea.	.04	8.10		8.14	a bathroom or kitchen vent fan.
Minimum Charge	Job		310		310	
Low Noise Model						
Install	Ea.	139	41.50		180.50	Includes labor and materials to install
Clean	Ea.	.04	8.10		8.14	a bathroom or kitchen vent fan, low
Minimum Charge	Job		310		310	noise model.

Low Voltage Systems

Low Voltage Systems	Unit	Material	Labor	Equip.	Total	Specification
Smoke Detector						
Pre-wired						
Demolish	Ea.		14.35		14.35	Includes material and labor to install
Install	Ea.	27.50	29		56.50	AC type smoke detector including
Demolish and Install	Ea.	27.50	43.35		70.85	wiring.
Reinstall	Ea.		23.28		23.28	
Clean	Ea.	.07	12.50		12.57	
Minimum Charge	Job		310		310	

For customer support on your Contractor's Pricing Guide: Residential Repair & Remodeling, call 800.448.8182.

Finish Electrical

Low Voltage Systems

	Unit	Material	Labor	Equip.	Total	Specification
Run Wiring						
Install	Ea.	21.50	25		46.50	Includes labor and material to install
Minimum Charge	Job		310		310	an electric fixture utility box with
						non-metallic cable.
Door Bell / Chimes						
Good Quality						
Demolish	Ea.		43		43	Includes material and labor to install
Install	Ea.	62.50	155		217.50	surface mounted 2 note door bell.
Demolish and Install	Ea.	62.50	198		260.50	Wiring is not included.
Reinstall	Ea.		124.16		124.16	
Clean	Ea.	.15	12.50		12.65	
Minimum Charge	Job		310		310	
Better Quality						
Demolish	Ea.		43		43	Includes material and labor to install
Install	Ea.	112	155		267	standard quality surface mounted 2
Demolish and Install	Ea.	112	198		310	note door bell. Wiring is not included.
Reinstall	Ea.		124.16		124.16	
Clean	Ea.	.15	12.50		12.65	
Minimum Charge	Job		310		310	
Premium Quality						
Demolish	Ea.		43		43	Includes material and labor to install
Install	Ea.	117	155		272	deluxe quality surface mounted 2 note
Demolish and Install	Ea.	117	198		315	door bell. Wiring is not included.
Reinstall	Ea.		124.16		124.16	
Clean	Ea.	.15	12.50		12.65	
Minimum Charge	Job		310		310	
Run Wiring						
Install	Ea.	21.50	25		46.50	Includes labor and material to install
Minimum Charge	Job		310		310	an electric fixture utility box with
						non-metallic cable.
Intercom System						
Intercom (master)						
Demolish	Ea.		115		115	Includes material and labor to install
Install	Ea.	930	310		1240	master intercom station with AM/FM
Demolish and Install	Ea.	930	425		1355	music and room monitoring. Wiring is
Reinstall	Ea.		248.32		248.32	not included.
Clean	Ea.	.15	33.50		33.65	
Minimum Charge	Job		310		310	
Intercom (remote)						
Demolish	Ea.		69		69	Includes material and labor to install
Install	Ea.	84.50	77.50		162	remote station for master intercom
Demolish and Install	Ea.	84.50	146.50		231	system. Wiring is not included.
Reinstall	Ea.		62.08		62.08	
Clean	Ea.	.15	12.50		12.65	
Minimum Charge	Job		310		310	
Security System						
Alarm Panel						
Demolish	Ea.		138		138	Includes material and labor to install a
Install	Ea.	91.50	310		401.50	6 zone burglar alarm panel.
Demolish and Install	Ea.	91.50	448		539.50	
Reinstall	Ea.		248.32		248.32	
Clean	Ea.	.15	12.50		12.65	
Minimum Charge	Job		310		310	

For customer support on your Contractor's Pricing Guide: Residential Repair & Remodeling, call 800.448.8182.

Finish Electrical

Low Voltage Systems

Low Voltage Systems		Unit	Material	Labor	Equip.	Total	Specification
Motion Detector							Includes material and labor to install
	Demolish	Ea.		98.50		98.50	an ultrasonic motion detector for a
	Install	Ea.	191	270		461	security system.
	Demolish and Install	Ea.	191	368.50		559.50	
	Reinstall	Ea.		215.93		215.93	
	Clean	Ea.	.15	12.50		12.65	
	Minimum Charge	Job		310		310	
Door Switch							Includes material and labor to install a
	Demolish	Ea.		43		43	door / window contact for a security
	Install	Ea.	67	117		184	system.
	Demolish and Install	Ea.	67	160		227	
	Reinstall	Ea.		93.71		93.71	
	Clean	Ea.	.15	4		4.15	
	Minimum Charge	Job		310		310	
Window Switch							Includes material and labor to install a
	Demolish	Ea.		43		43	door / window contact for a security
	Install	Ea.	67	117		184	system.
	Demolish and Install	Ea.	67	160		227	
	Reinstall	Ea.		93.71		93.71	
	Clean	Ea.	.15	4		4.15	
	Minimum Charge	Job		310		310	
Satellite System							
Dish							Includes material, labor and
	Demolish	Ea.		192		192	equipment to install 10' mesh dish.
	Install	Ea.	1750	515		2265	
	Demolish and Install	Ea.	1750	707		2457	
	Reinstall	Ea.		413.87		413.87	
	Clean	Ea.	3.17	50		53.17	
	Minimum Charge	Job		310		310	
Motor							Includes material and labor to install a
	Demolish	Ea.		96		96	television satellite dish motor unit.
	Install	Ea.	255	259		514	
	Demolish and Install	Ea.	255	355		610	
	Reinstall	Ea.		206.93		206.93	
	Clean	Ea.	.15	12.50		12.65	
	Minimum Charge	Job		310		310	
Television Antenna							
40' Pole							Includes material and labor to install
	Demolish	Ea.		108		108	stand-alone 1"-2" metal pole, 10' long.
	Install	Ea.	23.50	194		217.50	Cost does not include antenna.
	Demolish and Install	Ea.	23.50	302		325.50	
	Reinstall	Ea.		155.20		155.20	
	Clean	Ea.	.64	50		50.64	
	Paint	Ea.	24	54		78	
	Minimum Charge	Job		310		310	
Rotor Unit							Includes material and labor to install a
	Demolish	Ea.		86		86	television antenna rotor.
	Install	Ea.	62.50	77.50		140	
	Demolish and Install	Ea.	62.50	163.50		226	
	Reinstall	Ea.		62.08		62.08	
	Clean	Ea.	.64	50		50.64	
	Minimum Charge	Job		310		310	
Single Booster							Includes material and labor to install
	Demolish	Ea.		86		86	standard grade VHF television signal
	Install	Ea.	31	77.50		108.50	booster unit and all hardware required
	Demolish and Install	Ea.	31	163.50		194.50	for connection to TV cable.
	Reinstall	Ea.		62.08		62.08	
	Clean	Ea.	.03	6.25		6.28	
	Minimum Charge	Job		310		310	

For customer support on your Contractor's Pricing Guide: Residential Repair & Remodeling, call 800.448.8182.

Improvements / Appliances / Treatments

	Gazebo		Gutter			Shutter	

Window Treatment		Unit	Material	Labor	Equip.	Total	Specification
Drapery							
Good Grade							
	Install	L.F.	81	8		89	Includes material and labor to install
	Reinstall	L.F.		6.39		6.39	good quality drapery.
	Clean	L.F.	2.33	.20		2.53	
	Minimum Charge	Job		200		200	
Premium Grade							
	Install	L.F.	96.50	8		104.50	Includes material and labor to install
	Reinstall	L.F.		6.39		6.39	premium quality drapery.
	Clean	L.F.	2.33	.22		2.55	
	Minimum Charge	Job		200		200	
Lining							
	Install	L.F.	47	8		55	Includes material and labor to install
	Reinstall	L.F.		6.39		6.39	drapery lining.
	Clean	L.F.	1.53	.11		1.64	
	Minimum Charge	Job		200		200	
Valance Board							
	Demolish	L.F.		2.22		2.22	Includes material and labor to install a
	Install	L.F.	12.55	16		28.55	drapery valance board.
	Demolish and Install	L.F.	12.55	18.22		30.77	
	Reinstall	L.F.		12.79		12.79	
	Clean	L.F.	1.53	.40		1.93	
	Minimum Charge	Job		200		200	
Rod							
	Demolish	L.F.		5.55		5.55	Includes material and labor to install
	Install	L.F.	19.35	8.85		28.20	standard traverse drapery hardware.
	Demolish and Install	L.F.	19.35	14.40		33.75	
	Reinstall	L.F.		7.06		7.06	
	Clean	L.F.	.06	.40		.46	
	Paint	L.F.	.18	.83		1.01	
	Minimum Charge	Job		200		200	
Blinds							
Mini-type							
	Demolish	S.F.		1.13		1.13	Includes material and labor to install
	Install	S.F.	9	.88		9.88	horizontal 1" slat blinds.
	Demolish and Install	S.F.	9	2.01		11.01	
	Reinstall	S.F.		.71		.71	
	Clean	S.F.	1.89	.04		1.93	
	Minimum Charge	Job		200		200	

Improvements / Appliances / Treatments

Window Treatment

Window Treatment	Unit	Material	Labor	Equip.	Total	Specification
Vertical Type						
Demolish	S.F.		1.13		1.13	Includes material and labor to install
Install	S.F.	5.30	1.13		6.43	vertical 3" - 5" cloth or PVC blinds.
Demolish and Install	S.F.	5.30	2.26		7.56	
Reinstall	S.F.		.91		.91	
Clean	S.F.	1.98	.04		2.02	
Minimum Charge	Job		200		200	
Vertical Type per L.F.						
Demolish	L.F.		1.85		1.85	Includes material and labor to install
Install	L.F.	32	5.35		37.35	vertical 3" - 5" cloth or PVC blinds x
Demolish and Install	L.F.	32	7.20		39.20	72" long.
Reinstall	L.F.		4.26		4.26	
Clean	L.F.	12.30	.28		12.58	
Minimum Charge	Job		200		200	
Venetian Type						
Demolish	S.F.		1.13		1.13	Includes material and labor to install
Install	S.F.	8.20	.88		9.08	horizontal 2" slat blinds.
Demolish and Install	S.F.	8.20	2.01		10.21	
Reinstall	S.F.		.71		.71	
Clean	S.F.	1.98	.04		2.02	
Minimum Charge	Job		200		200	

Wood Shutter

27" x 36", 4 Panel

	Unit	Material	Labor	Equip.	Total	Specification
Demolish	Ea.		36		36	Includes material and labor to install
Install	Pair	143	30.50		173.50	movable louver, up to 33" x 36", 4
Demolish and Install	Pair	143	66.50		209.50	panel shutter.
Reinstall	Pair		24.51		24.51	
Clean	Ea.	.06	13.50		13.56	
Paint	Ea.	3.87	31		34.87	
Minimum Charge	Job		260		260	

33" x 36", 4 Panel

	Unit	Material	Labor	Equip.	Total	Specification
Demolish	Ea.		36		36	Includes material and labor to install
Install	Pair	143	30.50		173.50	movable louver, up to 33" x 36", 4
Demolish and Install	Pair	143	66.50		209.50	panel shutter.
Reinstall	Pair		24.51		24.51	
Clean	Ea.	.06	13.50		13.56	
Paint	Ea.	4.04	33		37.04	
Minimum Charge	Job		260		260	

39" x 36", 4 Panel

	Unit	Material	Labor	Equip.	Total	Specification
Demolish	Ea.		36		36	Includes material and labor to install
Install	Pair	156	30.50		186.50	movable louver, up to 47" x 36", 4
Demolish and Install	Pair	156	66.50		222.50	panel shutter.
Reinstall	Pair		24.51		24.51	
Clean	Ea.	.06	13.50		13.56	
Paint	Ea.	4.04	33		37.04	
Minimum Charge	Job		260		260	

47" x 36", 4 Panel

	Unit	Material	Labor	Equip.	Total	Specification
Demolish	Ea.		36		36	Includes material and labor to install
Install	Pair	156	30.50		186.50	movable louver, up to 47" x 36", 4
Demolish and Install	Pair	156	66.50		222.50	panel shutter.
Reinstall	Pair		24.51		24.51	
Clean	Ea.	.06	13.50		13.56	
Paint	Ea.	4.21	36		40.21	
Minimum Charge	Job		260		260	

Improvements / Appliances / Treatments

Stairs

Stairs	Unit	Material	Labor	Equip.	Total	Specification
Concrete Steps						
Precast						
Demolish	Ea.		77	54.50	131.50	Includes material and labor to install
Install	Flight	900	181	150	1231	precast concrete front entrance stairs,
Demolish and Install	Flight	900	258	204.50	1362.50	4' wide, with 48" platform and 2
Clean	Ea.		5.55		5.55	risers.
Paint	Ea.	16.55	54		70.55	
Minimum Charge	Job		475		475	
Metal						
Concrete Filled						
Demolish	Ea.		22		22	Includes material and labor to install
Install	Riser	805	79.50	6.25	890.75	steel, cement filled metal pan stairs
Demolish and Install	Riser	805	101.50	6.25	912.75	with picket rail.
Clean	Ea.	.10	3.39		3.49	
Paint	Ea.	1.24	3.20		4.44	
Minimum Charge	Job		294		294	
Landing						
Demolish	S.F.		5.55		5.55	Includes material and labor to install
Install	S.F.	158	9.35	.74	168.09	pre-erected conventional steel pan
Demolish and Install	S.F.	158	14.90	.74	173.64	landing.
Clean	S.F.	.10	.71		.81	
Paint	S.F.	.43	.92		1.35	
Minimum Charge	Job		294		294	
Railing						
Demolish	L.F.		3.06		3.06	Includes material and labor to install
Install	L.F.	22	13.55	1.07	36.62	primed steel pipe.
Demolish and Install	L.F.	22	16.61	1.07	39.68	
Clean	L.F.	.64	2.50		3.14	
Paint	L.F.	2.32	7.20		9.52	
Minimum Charge	Job		294		294	

Elevator

Elevator	Unit	Material	Labor	Equip.	Total	Specification
2 Stop Residential						
Install	Ea.	21000	8125		29125	Includes labor, material and
Clean	Ea.	1.12	25		26.12	equipment to install 2 stop elevator for
Minimum Charge	Job		815		815	residential use.
2 Stop Commercial						
Install	Ea.	55000	16300		71300	Includes labor, material and
Clean	Ea.	1.12	25		26.12	equipment to install 2 stop elevator for
Minimum Charge	Job		815		815	apartment or commercial building with
3 Stop Commercial						2,000 lb/13 passenger capacity.
Install	Ea.	66500	25000		91500	Includes labor, material and
Clean	Ea.	1.12	25		26.12	equipment to install 3 stop elevator for
Minimum Charge	Job		815		815	apartment or commercial building with
4 Stop Commercial						2,000 lb/13 passenger capacity.
Install	Ea.	78500	33800		112300	Includes labor, material and
Clean	Ea.	1.12	25		26.12	equipment to install 4 stop elevator for
Minimum Charge	Job		815		815	apartment or commercial building with
						2,000 lb/13 passenger capacity.

For customer support on your Contractor's Pricing Guide: Residential Repair & Remodeling, call 800.448.8182.

Improvements / Appliances / Treatments

Elevator		Unit	Material	Labor	Equip.	Total	Specification
5 Stop Commercial							Includes labor, material and
	Install	Ea.	90000	42800		132800	equipment to install 5 stop elevator for
	Clean	Ea.	1.12	25		26.12	apartment or commercial building with
	Minimum Charge	Job		815		815	2,000 lb/13 passenger capacity.

Rangetop		Unit	Material	Labor	Equip.	Total	Specification
4-burner							
	Demolish	Ea.		40.50		40.50	Cost includes material and labor to
	Install	Ea.	330	103		433	install 4-burner electric surface unit.
	Demolish and Install	Ea.	330	143.50		473.50	
	Reinstall	Ea.		82.77		82.77	
	Clean	Ea.	.04	40.50		40.54	
	Minimum Charge	Job		100		100	
6-burner							
	Demolish	Ea.		40.50		40.50	Includes material and labor to install a
	Install	Ea.	3750	124		3874	6-burner cooktop unit.
	Demolish and Install	Ea.	3750	164.50		3914.50	
	Reinstall	Ea.		99.33		99.33	
	Clean	Ea.	.04	40.50		40.54	
	Minimum Charge	Job		100		100	
Premium Brand w / Grill							
	Demolish	Ea.		40.50		40.50	Includes material and labor to install
	Install	Ea.	2475	124		2599	electric downdraft cooktop with grill.
	Demolish and Install	Ea.	2475	164.50		2639.50	
	Reinstall	Ea.		99.33		99.33	
	Clean	Ea.	.04	40.50		40.54	
	Minimum Charge	Job		100		100	

Rangehood		Unit	Material	Labor	Equip.	Total	Specification
Vented							
30″							
	Demolish	Ea.		37		37	Includes material and labor to install a
	Install	Ea.	98.50	124		222.50	vented range hood.
	Demolish and Install	Ea.	98.50	161		259.50	
	Reinstall	Ea.		99.33		99.33	
	Clean	Ea.	.04	11.55		11.59	
	Minimum Charge	Job		100		100	
36″							
	Demolish	Ea.		37		37	Includes material and labor to install a
	Install	Ea.	197	124		321	vented range hood.
	Demolish and Install	Ea.	197	161		358	
	Reinstall	Ea.		99.33		99.33	
	Clean	Ea.	.04	11.55		11.59	
	Minimum Charge	Job		100		100	
42″							
	Demolish	Ea.		37		37	Includes material and labor to install a
	Install	Ea.	150	135		285	vented range hood.
	Demolish and Install	Ea.	150	172		322	
	Reinstall	Ea.		108.16		108.16	
	Clean	Ea.	.04	11.55		11.59	
	Minimum Charge	Job		100		100	

Improvements / Appliances / Treatments

Rangehood

Rangehood	Unit	Material	Labor	Equip.	Total	Specification
Ventless						
30"						
Demolish	Ea.		37		37	Includes material and labor to install a
Install	Ea.	109	77.50		186.50	ventless range hood.
Demolish and Install	Ea.	109	114.50		223.50	
Reinstall	Ea.		62.08		62.08	
Clean	Ea.	.04	11.55		11.59	
Minimum Charge	Job		100		100	
36"						
Demolish	Ea.		37		37	Includes material and labor to install a
Install	Ea.	267	124		391	ventless range hood.
Demolish and Install	Ea.	267	161		428	
Reinstall	Ea.		99.33		99.33	
Clean	Ea.	.04	11.55		11.59	
Minimum Charge	Job		100		100	
42"						
Demolish	Ea.		37		37	Includes material and labor to install a
Install	Ea.	167	103		270	ventless range hood.
Demolish and Install	Ea.	167	140		307	
Reinstall	Ea.		82.77		82.77	
Clean	Ea.	.04	11.55		11.59	
Minimum Charge	Job		100		100	

Oven

Oven	Unit	Material	Labor	Equip.	Total	Specification
Free Standing						
Demolish	Ea.		40.50		40.50	Cost includes material and labor to
Install	Ea.	820	80		900	install 30" free standing electric range
Demolish and Install	Ea.	820	120.50		940.50	with self-cleaning oven.
Reinstall	Ea.		63.94		63.94	
Clean	Ea.	.04	40.50		40.54	
Minimum Charge	Job		100		100	
Single Wall Oven						
Demolish	Ea.		40.50		40.50	Includes material and labor to install a
Install	Ea.	1050	155		1205	built-in oven.
Demolish and Install	Ea.	1050	195.50		1245.50	
Reinstall	Ea.		124.16		124.16	
Clean	Ea.	.04	40.50		40.54	
Minimum Charge	Job		100		100	
Double Wall Oven						
Demolish	Ea.		40.50		40.50	Includes material and labor to install a
Install	Ea.	2850	260		3110	conventional double oven.
Demolish and Install	Ea.	2850	300.50		3150.50	
Reinstall	Ea.		208.32		208.32	
Clean	Ea.	.04	40.50		40.54	
Minimum Charge	Job		100		100	
Drop-in Type w / Range						
Demolish	Ea.		40.50		40.50	Includes material and labor to install a
Install	Ea.	1425	103		1528	built-in cooking range with oven.
Demolish and Install	Ea.	1425	143.50		1568.50	
Reinstall	Ea.		82.77		82.77	
Clean	Ea.	.04	40.50		40.54	
Minimum Charge	Job		100		100	

For customer support on your Contractor's Pricing Guide: Residential Repair & Remodeling, call 800.448.8182.

Improvements / Appliances / Treatments

Oven	Unit	Material	Labor	Equip.	Total	Specification
Hi/Lo w / Range, Microwave						
Demolish	Ea.		40.50		40.50	Includes material and labor to install a
Install	Ea.	3175	520		3695	double oven, one conventional, one
Demolish and Install	Ea.	3175	560.50		3735.50	microwave.
Reinstall	Ea.		416.64		416.64	
Clean	Ea.	.04	40.50		40.54	
Minimum Charge	Job		100		100	
Hi/Lo w/ Range / Double Oven						
Demolish	Ea.		40.50		40.50	Includes material and labor to install a
Install	Ea.	2850	260		3110	conventional double oven.
Demolish and Install	Ea.	2850	300.50		3150.50	
Reinstall	Ea.		208.32		208.32	
Clean	Ea.	.04	40.50		40.54	
Minimum Charge	Job		100		100	
Countertop Microwave						
Demolish	Ea.		13.90		13.90	Includes material and labor to install a
Install	Ea.	186	50		236	1.0 cubic foot microwave oven.
Demolish and Install	Ea.	186	63.90		249.90	
Reinstall	Ea.		39.96		39.96	
Clean	Ea.	.04	20		20.04	
Minimum Charge	Job		100		100	
Cabinet / Wall Mounted Microwave						
Demolish	Ea.		28		28	Includes material and labor to install a
Install	Ea.	1175	155		1330	space saver microwave oven.
Demolish and Install	Ea.	1175	183		1358	
Reinstall	Ea.		124.16		124.16	
Clean	Ea.	.04	20		20.04	
Minimum Charge	Job		100		100	

Dryer	Unit	Material	Labor	Equip.	Total	Specification
Basic Grade						
Demolish	Ea.		18.50		18.50	Includes material and labor to install
Install	Ea.	880	310		1190	electric or gas dryer. Vent kit not
Demolish and Install	Ea.	880	328.50		1208.50	included.
Reinstall	Ea.		247.38		247.38	
Clean	Ea.	.04	40.50		40.54	
Minimum Charge	Job		100		100	
Good Grade						
Demolish	Ea.		18.50		18.50	Includes material and labor to install
Install	Ea.	1100	310		1410	electric or gas dryer. Vent kit not
Demolish and Install	Ea.	1100	328.50		1428.50	included.
Reinstall	Ea.		247.38		247.38	
Clean	Ea.	.04	40.50		40.54	
Minimum Charge	Job		100		100	
Premium Grade						
Demolish	Ea.		18.50		18.50	Includes material and labor to install
Install	Ea.	1575	465		2040	electric or gas dryer. Vent kit not
Demolish and Install	Ea.	1575	483.50		2058.50	included.
Reinstall	Ea.		371.07		371.07	
Clean	Ea.	.04	40.50		40.54	
Minimum Charge	Job		100		100	

For customer support on your Contractor's Pricing Guide: Residential Repair & Remodeling, call 800.448.8182.

Improvements / Appliances / Treatments

Dryer	Unit	Material	Labor	Equip.	Total	Specification
Washer / Dryer Combination Unit						Includes material and labor to install
Demolish	Ea.		18.50		18.50	electric washer/dryer combination
Install	Ea.	1750	80		1830	unit, 27" wide, stackable or unitized.
Demolish and Install	Ea.	1750	98.50		1848.50	
Reinstall	Ea.		63.94		63.94	
Clean	Ea.	.04	40.50		40.54	
Minimum Charge	Job		100		100	

Washer	Unit	Material	Labor	Equip.	Total	Specification
Basic Grade						
Demolish	Ea.		18.50		18.50	Includes material and labor to install
Install	Ea.	680	196		876	washing machine. New hoses not
Demolish and Install	Ea.	680	214.50		894.50	included.
Reinstall	Ea.		156.59		156.59	
Clean	Ea.	.04	40.50		40.54	
Minimum Charge	Job		100		100	
Good Grade						
Demolish	Ea.		18.50		18.50	Includes material and labor to install
Install	Ea.	1125	294		1419	washing machine. New hoses not
Demolish and Install	Ea.	1125	312.50		1437.50	included.
Reinstall	Ea.		234.88		234.88	
Clean	Ea.	.04	40.50		40.54	
Minimum Charge	Job		100		100	
Premium Grade						
Demolish	Ea.		18.50		18.50	Includes material and labor to install
Install	Ea.	1675	585		2260	washing machine with digital
Demolish and Install	Ea.	1675	603.50		2278.50	computer readouts and large capacity.
Reinstall	Ea.		469.76		469.76	New hoses not included.
Clean	Ea.	.04	40.50		40.54	
Minimum Charge	Job		100		100	
Washer / Dryer Combination Unit						
Demolish	Ea.		18.50		18.50	Includes material and labor to install
Install	Ea.	1750	80		1830	electric washer/dryer combination
Demolish and Install	Ea.	1750	98.50		1848.50	unit, 27" wide, stackable or unitized.
Reinstall	Ea.		63.94		63.94	
Clean	Ea.	.04	40.50		40.54	
Minimum Charge	Job		100		100	

Refrigerator	Unit	Material	Labor	Equip.	Total	Specification
12 Cubic Foot						
Demolish	Ea.		55.50		55.50	Includes material and labor to install a
Install	Ea.	680	80		760	refrigerator.
Demolish and Install	Ea.	680	135.50		815.50	
Reinstall	Ea.		63.94		63.94	
Clean	Ea.	.04	40.50		40.54	
Minimum Charge	Job		100		100	
16 Cubic Foot						
Demolish	Ea.		55.50		55.50	Includes material and labor to install a
Install	Ea.	870	89		959	refrigerator.
Demolish and Install	Ea.	870	144.50		1014.50	
Reinstall	Ea.		71.04		71.04	
Clean	Ea.	.04	40.50		40.54	
Minimum Charge	Job		100		100	

For customer support on your Contractor's Pricing Guide: Residential Repair & Remodeling, call 800.448.8182.

Improvements / Appliances / Treatments

Refrigerator	Unit	Material	Labor	Equip.	Total	Specification
18 Cubic Foot						
Demolish	Ea.		55.50		55.50	Includes material and labor to install a refrigerator.
Install	Ea.	915	100		1015	
Demolish and Install	Ea.	915	155.50		1070.50	
Reinstall	Ea.		79.92		79.92	
Clean	Ea.	.04	40.50		40.54	
Minimum Charge	Job		100		100	
21 Cubic Foot						
Demolish	Ea.		55.50		55.50	Includes material and labor to install a refrigerator.
Install	Ea.	2250	114		2364	
Demolish and Install	Ea.	2250	169.50		2419.50	
Reinstall	Ea.		91.34		91.34	
Clean	Ea.	.04	40.50		40.54	
Minimum Charge	Job		100		100	
24 Cubic Foot						
Demolish	Ea.		55.50		55.50	Includes material and labor to install a refrigerator.
Install	Ea.	3000	114		3114	
Demolish and Install	Ea.	3000	169.50		3169.50	
Reinstall	Ea.		91.34		91.34	
Clean	Ea.	.04	40.50		40.54	
Minimum Charge	Job		100		100	
27 Cubic Foot						
Demolish	Ea.		55.50		55.50	Includes material and labor to install a refrigerator.
Install	Ea.	2250	80		2330	
Demolish and Install	Ea.	2250	135.50		2385.50	
Reinstall	Ea.		63.94		63.94	
Clean	Ea.	.04	40.50		40.54	
Minimum Charge	Job		100		100	
Ice Maker						
Install	Ea.	127	73.50		200.50	Includes labor and material to install an automatic ice maker in a refrigerator.
Minimum Charge	Job		100		100	

Freezer	Unit	Material	Labor	Equip.	Total	Specification
15 Cubic Foot						
Demolish	Ea.		55.50		55.50	Cost includes material and labor to install 15 cubic foot freezer.
Install	Ea.	970	80		1050	
Demolish and Install	Ea.	970	135.50		1105.50	
Reinstall	Ea.		63.94		63.94	
Clean	Ea.	.04	40.50		40.54	
Minimum Charge	Job		100		100	
18 Cubic Foot						
Demolish	Ea.		55.50		55.50	Cost includes material and labor to install 18 cubic foot freezer.
Install	Ea.	1300	80		1380	
Demolish and Install	Ea.	1300	135.50		1435.50	
Reinstall	Ea.		63.94		63.94	
Clean	Ea.	.04	40.50		40.54	
Minimum Charge	Job		100		100	
21 Cubic Foot						
Demolish	Ea.		55.50		55.50	Includes material and labor to install a 15 to 23 cubic foot deep freezer.
Install	Ea.	595	80		675	
Demolish and Install	Ea.	595	135.50		730.50	
Reinstall	Ea.		63.94		63.94	
Clean	Ea.	.04	40.50		40.54	
Minimum Charge	Job		100		100	

For customer support on your Contractor's Pricing Guide: Residential Repair & Remodeling, call 800.448.8182.

Improvements / Appliances / Treatments

Freezer

Freezer	Unit	Material	Labor	Equip.	Total	Specification
24 Cubic Foot						
Demolish	Ea.		55.50		55.50	Includes material and labor to install a
Install	Ea.	660	160		820	24 cubic foot deep freezer.
Demolish and Install	Ea.	660	215.50		875.50	
Reinstall	Ea.		127.87		127.87	
Clean	Ea.	.04	40.50		40.54	
Minimum Charge	Job		100		100	
Chest-type						
Demolish	Ea.		55.50		55.50	Cost includes material and labor to
Install	Ea.	970	80		1050	install 15 cubic foot freezer.
Demolish and Install	Ea.	970	135.50		1105.50	
Clean	Ea.	.04	40.50		40.54	
Minimum Charge	Job		100		100	

Dishwasher

Dishwasher	Unit	Material	Labor	Equip.	Total	Specification
4 Cycle						
Demolish	Ea.		44.50		44.50	Includes material and labor to install
Install	Ea.	700	186		886	good grade automatic dishwasher.
Demolish and Install	Ea.	700	230.50		930.50	
Reinstall	Ea.		148.48		148.48	
Clean	Ea.	.04	40.50		40.54	
Minimum Charge	Job		100		100	

Garbage Disposal

Garbage Disposal	Unit	Material	Labor	Equip.	Total	Specification
In Sink Unit						
Demolish	Ea.		41		41	Cost includes material and labor to
Install	Ea.	320	74		394	install 1/2 HP custom disposal unit.
Demolish and Install	Ea.	320	115		435	
Reinstall	Ea.		59.39		59.39	
Clean	Ea.	.02	11.55		11.57	
Minimum Charge	Job		100		100	

Trash Compactor

Trash Compactor	Unit	Material	Labor	Equip.	Total	Specification
4 to 1 Compaction						
Demolish	Ea.		43		43	Includes material and labor to install a
Install	Ea.	1750	104		1854	trash compactor.
Demolish and Install	Ea.	1750	147		1897	
Reinstall	Ea.		83.33		83.33	
Clean	Ea.	.04	25		25.04	
Minimum Charge	Job		100		100	

Kitchenette Unit

Kitchenette Unit	Unit	Material	Labor	Equip.	Total	Specification
Range, Refrigerator and Sink						
Demolish	Ea.		172		172	Includes material and labor to install
Install	Ea.	3350	620		3970	range/refrigerator/sink unit, 72"
Demolish and Install	Ea.	3350	792		4142	wide, base unit only.
Reinstall	Ea.		494.93		494.93	
Clean	Ea.	.79	20		20.79	
Minimum Charge	Job		100		100	

Improvements / Appliances / Treatments

Bath Accessories	Unit	Material	Labor	Equip.	Total	Specification
Towel Bar						
Demolish	Ea.		9.25		9.25	Includes material and labor to install
Install	Ea.	22	21.50		43.50	towel bar up to 24" long.
Demolish and Install	Ea.	22	30.75		52.75	
Reinstall	Ea.		17.36		17.36	
Clean	Ea.		5		5	
Minimum Charge	Job		260		260	
Toothbrush Holder						
Demolish	Ea.		9.25		9.25	Includes material and labor to install a
Install	Ea.	16.60	26		42.60	surface mounted tumbler and
Demolish and Install	Ea.	16.60	35.25		51.85	toothbrush holder.
Reinstall	Ea.		20.83		20.83	
Clean	Ea.		5		5	
Minimum Charge	Job		260		260	
Grab Rail						
Demolish	Ea.		13.90		13.90	Includes material and labor to install a
Install	Ea.	39.50	22.50		62	24" long grab bar. Blocking is not
Demolish and Install	Ea.	39.50	36.40		75.90	included.
Reinstall	Ea.		18.11		18.11	
Clean	Ea.	.32	8.35		8.67	
Minimum Charge	Job		260		260	
Soap Dish						
Demolish	Ea.		9.25		9.25	Includes material and labor to install a
Install	Ea.	7.80	26		33.80	surface mounted soap dish.
Demolish and Install	Ea.	7.80	35.25		43.05	
Reinstall	Ea.		20.83		20.83	
Clean	Ea.		5		5	
Minimum Charge	Job		260		260	
Toilet Paper Roller						
Demolish	Ea.		9.25		9.25	Includes material and labor to install a
Install	Ea.	22.50	17.35		39.85	surface mounted toilet tissue dispenser.
Demolish and Install	Ea.	22.50	26.60		49.10	
Reinstall	Ea.		13.89		13.89	
Clean	Ea.	.32	12.50		12.82	
Minimum Charge	Job		260		260	
Dispenser						
Soap						
Demolish	Ea.		12.35		12.35	Includes material and labor to install a
Install	Ea.	66	26		92	surface mounted soap dispenser.
Demolish and Install	Ea.	66	38.35		104.35	
Reinstall	Ea.		20.83		20.83	
Clean	Ea.	.32	12.50		12.82	
Minimum Charge	Job		260		260	
Towel						
Demolish	Ea.		13.90		13.90	Includes material and labor to install a
Install	Ea.	72	32.50		104.50	surface mounted towel dispenser.
Demolish and Install	Ea.	72	46.40		118.40	
Reinstall	Ea.		26.04		26.04	
Clean	Ea.	.32	16.65		16.97	
Minimum Charge	Job		260		260	
Seat Cover						
Demolish	Ea.		9.25		9.25	Includes material and labor to install a
Install	Ea.	50	34.50		84.50	surface mounted toilet seat cover
Demolish and Install	Ea.	50	43.75		93.75	dispenser.
Reinstall	Ea.		27.78		27.78	
Clean	Ea.	.32	12.50		12.82	
Minimum Charge	Job		260		260	

For customer support on your **Contractor's Pricing Guide: Residential Repair & Remodeling**, call 800.448.8182.

Improvements / Appliances / Treatments

Bath Accessories

		Unit	Material	Labor	Equip.	Total	Specification
Sanitary Napkin							Includes material and labor to install a
	Demolish	Ea.		23.50		23.50	surface mounted sanitary napkin
	Install	Ea.	290	55		345	dispenser.
	Demolish and Install	Ea.	290	78.50		368.50	
	Reinstall	Ea.		43.86		43.86	
	Clean	Ea.	.15	16.65		16.80	
	Minimum Charge	Job		260		260	
Partition							
Urinal							
	Demolish	Ea.		28		28	Includes material and labor to install a
	Install	Ea.	365	130		495	urinal screen.
	Demolish and Install	Ea.	365	158		523	
	Reinstall	Ea.		104.16		104.16	
	Clean	Ea.	.47	11.10		11.57	
	Minimum Charge	Job		260		260	
Toilet							
	Demolish	Ea.		82		82	Includes material and labor to install
	Install	Ea.	1175	208		1383	baked enamel standard size partition,
	Demolish and Install	Ea.	1175	290		1465	floor or ceiling mounted with one wall
	Reinstall	Ea.		166.66		166.66	and one door.
	Clean	Ea.	3.17	25		28.17	
	Minimum Charge	Job		260		260	
Handicap							
	Demolish	Ea.		82		82	Includes material and labor to install
	Install	Ea.	1350	208		1558	baked enamel handicap partition,
	Demolish and Install	Ea.	1350	290		1640	floor or ceiling mounted with one wall
	Reinstall	Ea.		166.66		166.66	and one door.
	Clean	Ea.	3.17	25		28.17	
	Minimum Charge	Job		260		260	

Fire Prevention / Protection

		Unit	Material	Labor	Equip.	Total	Specification
Fire Extinguisher							
5# Carbon Dioxide							Includes labor and material to install
	Install	Ea.	320			320	one wall mounted 5 lb. carbon dioxide factory charged unit, complete with hose, horn and wall mounting bracket.
10# Carbon Dioxide							Includes labor and material to install
	Install	Ea.	440			440	one wall mounted 10 lb. carbon dioxide factory charged unit, complete with hose, horn and wall mounting bracket.
15# Carbon Dioxide							Includes labor and material to install
	Install	Ea.	585			585	one wall mounted 15 lb. carbon dioxide factory charged unit, complete with hose, horn and wall mounting bracket.

For customer support on your Contractor's Pricing Guide: Residential Repair & Remodeling, call 800.448.8182.

Improvements / Appliances / Treatments

Fire Prevention / Protection		Unit	Material	Labor	Equip.	Total	Specification
5# Dry Chemical							
	Install	Ea.	73.50			73.50	Includes labor and material to install one wall mounted 5 lb. dry chemical factory charged unit, complete with hose, horn and wall mounting bracket.
20# Dry Chemical							
	Install	Ea.	259			259	Includes labor and material to install one wall mounted 20 lb. dry chemical factory charged unit, complete with hose, horn and wall mounting bracket.

Gutters		Unit	Material	Labor	Equip.	Total	Specification
Galvanized							
	Demolish	L.F.		1.85		1.85	Cost includes material and labor to install 5" galvanized steel gutter with enamel finish, fittings, hangers, and corners.
	Install	L.F.	4.84	4.54		9.38	
	Demolish and Install	L.F.	4.84	6.39		11.23	
	Reinstall	L.F.		3.64		3.64	
	Clean	L.F.		.53		.53	
	Paint	L.F.	.65	1.33		1.98	
	Minimum Charge	Job		284		284	
Aluminum							
	Demolish	L.F.		1.85		1.85	Cost includes material and labor to install 5" aluminum gutter with enamel finish, fittings, hangers, and corners.
	Install	L.F.	3.41	4.54		7.95	
	Demolish and Install	L.F.	3.41	6.39		9.80	
	Reinstall	L.F.		3.64		3.64	
	Clean	L.F.		.53		.53	
	Paint	L.F.	.65	1.33		1.98	
	Minimum Charge	Job		284		284	
Copper							
	Demolish	L.F.		1.85		1.85	Cost includes material and labor to install 5" half-round copper gutter with fittings, hangers, and corners.
	Install	L.F.	11.60	4.54		16.14	
	Demolish and Install	L.F.	11.60	6.39		17.99	
	Reinstall	L.F.		3.64		3.64	
	Clean	L.F.		.53		.53	
	Minimum Charge	Job		284		284	
Plastic							
	Demolish	L.F.		1.85		1.85	Cost includes material and labor to install 4" white vinyl gutter fittings, hangers, and corners.
	Install	L.F.	2.82	4.53		7.35	
	Demolish and Install	L.F.	2.82	6.38		9.20	
	Reinstall	L.F.		3.62		3.62	
	Clean	L.F.		.53		.53	
	Paint	L.F.	.65	1.33		1.98	
	Minimum Charge	Job		284		284	

Downspouts		Unit	Material	Labor	Equip.	Total	Specification
Galvanized							
	Demolish	L.F.		1.27		1.27	Cost includes material and labor to install 4" diameter downspout with enamel finish.
	Install	L.F.	6.25	3.92		10.17	
	Demolish and Install	L.F.	6.25	5.19		11.44	
	Reinstall	L.F.		3.13		3.13	
	Clean	L.F.		.53		.53	
	Paint	L.F.	.65	1.33		1.98	
	Minimum Charge	Job		284		284	

For customer support on your Contractor's Pricing Guide: Residential Repair & Remodeling, call 800.448.8182.

Improvements / Appliances / Treatments

Downspouts

Downspouts	Unit	Material	Labor	Equip.	Total	Specification
Aluminum						
Demolish	L.F.		1.27		1.27	Includes material and labor to install
Install	L.F.	4.73	4.06		8.79	aluminum enameled downspout.
Demolish and Install	L.F.	4.73	5.33		10.06	
Reinstall	L.F.		3.25		3.25	
Clean	L.F.		.53		.53	
Paint	L.F.	.65	1.33		1.98	
Minimum Charge	Job		284		284	
Copper						
Demolish	L.F.		1.27		1.27	Includes material and labor to install
Install	L.F.	13.55	2.99		16.54	round copper downspout.
Demolish and Install	L.F.	13.55	4.26		17.81	
Reinstall	L.F.		2.39		2.39	
Clean	L.F.		.53		.53	
Paint	L.F.	.65	1.33		1.98	
Minimum Charge	Job		284		284	
Plastic						
Demolish	L.F.		1.27		1.27	Cost includes material and labor to
Install	L.F.	3.23	2.70		5.93	install 2" x 3" vinyl rectangular
Demolish and Install	L.F.	3.23	3.97		7.20	downspouts.
Reinstall	L.F.		2.16		2.16	
Clean	L.F.		.53		.53	
Paint	L.F.	.65	1.33		1.98	
Minimum Charge	Job		284		284	

Shutters

Shutters	Unit	Material	Labor	Equip.	Total	Specification
Exterior						
25" High						
Demolish	Ea.		13.90		13.90	Includes material and labor to install
Install	Pair	240	52		292	pair of exterior shutters 25" x 16",
Demolish and Install	Pair	240	65.90		305.90	unpainted, installed as fixed shutters
Reinstall	Pair		52.08		52.08	on concrete or wood frame.
Clean	Ea.	.06	13.50		13.56	
Paint	Ea.	5.15	23		28.15	
Minimum Charge	Job		260		260	
39" High						
Demolish	Ea.		13.90		13.90	Includes material and labor to install
Install	Pair	335	52		387	pair of shutters 39" x 16", unpainted,
Demolish and Install	Pair	335	65.90		400.90	installed as fixed shutters on concrete
Reinstall	Pair		52.08		52.08	or wood frame.
Clean	Ea.	.06	13.50		13.56	
Paint	Ea.	8	31.50		39.50	
Minimum Charge	Job		260		260	
51" High						
Demolish	Ea.		13.90		13.90	Includes material and labor to install
Install	Pair	430	52		482	pair of shutters 51" x 16", unpainted,
Demolish and Install	Pair	430	65.90		495.90	installed as fixed shutters on concrete
Reinstall	Pair		52.08		52.08	or wood frame.
Clean	Ea.	.06	13.50		13.56	
Paint	Ea.	10.40	32.50		42.90	
Minimum Charge	Job		260		260	

Improvements / Appliances / Treatments

Shutters		Unit	Material	Labor	Equip.	Total	Specification
59" High							
	Demolish	Ea.		13.90		13.90	Includes material and labor to install
	Install	Pair	585	52		637	pair of shutters 59" x 16", unpainted,
	Demolish and Install	Pair	585	65.90		650.90	installed as fixed shutters on concrete
	Reinstall	Pair		52.08		52.08	or wood frame.
	Clean	Ea.	.06	13.50		13.56	
	Paint	Ea.	12	35.50		47.50	
	Minimum Charge	Job		260		260	
67" High							
	Demolish	Ea.		13.90		13.90	Includes material and labor to install
	Install	Pair	620	58		678	pair of shutters 67" x 16", unpainted,
	Demolish and Install	Pair	620	71.90		691.90	installed as fixed shutters on concrete
	Reinstall	Pair		57.87		57.87	or wood frame.
	Clean	Ea.	.06	13.50		13.56	
	Paint	Ea.	13.70	40		53.70	
	Minimum Charge	Job		260		260	
Interior Wood							
23" x 24", 2 Panel							
	Demolish	Ea.		13.90		13.90	Cost includes material and labor to
	Install	Ea.	227	30.50		257.50	install 23" x 24", 4 panel interior
	Demolish and Install	Ea.	227	44.40		271.40	movable pine shutter, 1-1/4" louver,
	Reinstall	Ea.		24.51		24.51	knobs and hooks and hinges.
	Clean	Ea.	.06	13.50		13.56	
	Paint	Ea.	3.87	31		34.87	
	Minimum Charge	Job		260		260	
27" x 24", 4 Panel							
	Demolish	Ea.		13.90		13.90	Cost includes material and labor to
	Install	Ea.	230	30.50		260.50	install 27" x 24", 4 panel interior
	Demolish and Install	Ea.	230	44.40		274.40	movable pine shutter, 1-1/4" louver,
	Reinstall	Ea.		24.51		24.51	knobs and hooks and hinges.
	Clean	Ea.	.06	13.50		13.56	
	Paint	Ea.	3.87	31		34.87	
	Minimum Charge	Job		260		260	
31" x 24", 4 Panel							
	Demolish	Ea.		13.90		13.90	Cost includes material and labor to
	Install	Ea.	253	30.50		283.50	install 31" x 24", 4 panel interior
	Demolish and Install	Ea.	253	44.40		297.40	movable pine shutter, 1-1/4" louver,
	Reinstall	Ea.		24.51		24.51	knobs and hooks and hinges.
	Clean	Ea.	.06	13.50		13.56	
	Paint	Ea.	3.87	31		34.87	
	Minimum Charge	Job		260		260	
35" x 24", 4 Panel							
	Demolish	Ea.		13.90		13.90	Cost includes material and labor to
	Install	Ea.	272	30.50		302.50	install 35" x 24", 4 panel interior
	Demolish and Install	Ea.	272	44.40		316.40	movable pine shutter, 1-1/4" louver,
	Reinstall	Ea.		24.51		24.51	knobs and hooks and hinges.
	Clean	Ea.	.06	13.50		13.56	
	Paint	Ea.	3.87	31		34.87	
	Minimum Charge	Job		260		260	
39" x 24", 4 Panel							
	Demolish	Ea.		13.90		13.90	Cost includes material and labor to
	Install	Ea.	300	30.50		330.50	install 39" x 24", 4 panel interior
	Demolish and Install	Ea.	300	44.40		344.40	movable pine shutter, 1-1/4" louver,
	Reinstall	Ea.		24.51		24.51	knobs and hooks and hinges.
	Clean	Ea.	.06	13.50		13.56	
	Paint	Ea.	3.87	31		34.87	
	Minimum Charge	Job		260		260	

For customer support on your Contractor's Pricing Guide: Residential Repair & Remodeling, call 800.448.8182.

Improvements / Appliances / Treatments

Shutters

Shutters	Unit	Material	Labor	Equip.	Total	Specification
47" x 24", 4 Panel						
Demolish	Ea.		13.90		13.90	Cost includes material and labor to
Install	Ea.	340	30.50		370.50	install 47" x 24", 4 panel interior
Demolish and Install	Ea.	340	44.40		384.40	movable pine shutter, 1-1/4" louver,
Reinstall	Ea.		24.51		24.51	knobs and hooks and hinges.
Clean	Ea.	.06	13.50		13.56	
Paint	Ea.	4.04	33		37.04	
Minimum Charge	Job		260		260	

Awning

Awning	Unit	Material	Labor	Equip.	Total	Specification
Standard Aluminum						
36" Wide x 30" Deep						
Demolish	Ea.		13.90		13.90	Includes material and labor to install
Install	Ea.	310	43.50		353.50	one aluminum awning with
Demolish and Install	Ea.	310	57.40		367.40	weather-resistant finish.
Reinstall	Ea.		34.72		34.72	
Clean	Ea.		16.65		16.65	
Minimum Charge	Job		260		260	
48" Wide x 30" Deep						
Demolish	Ea.		13.90		13.90	Includes material and labor to install
Install	Ea.	360	52		412	one aluminum awning with
Demolish and Install	Ea.	360	65.90		425.90	weather-resistant finish.
Reinstall	Ea.		41.66		41.66	
Clean	Ea.		16.65		16.65	
Minimum Charge	Job		260		260	
60" Wide x 30" Deep						
Demolish	Ea.		28		28	Includes material and labor to install
Install	Ea.	385	58		443	one aluminum awning with
Demolish and Install	Ea.	385	86		471	weather-resistant finish.
Reinstall	Ea.		46.29		46.29	
Clean	Ea.		16.65		16.65	
Minimum Charge	Job		260		260	
72" Wide x 30" Deep						
Demolish	Ea.		28		28	Includes material and labor to install
Install	Ea.	425	65		490	one aluminum awning with
Demolish and Install	Ea.	425	93		518	weather-resistant finish.
Reinstall	Ea.		52.08		52.08	
Clean	Ea.		16.65		16.65	
Minimum Charge	Job		260		260	
Roll-up Type						
Demolish	S.F.		1		1	Includes material and labor to install
Install	S.F.	30.50	5.20		35.70	an aluminum roll-up awning.
Demolish and Install	S.F.	30.50	6.20		36.70	
Reinstall	S.F.		4.17		4.17	
Clean	S.F.	.03	.83		.86	
Minimum Charge	Job		260		260	
Canvas						
30" Wide						
Demolish	Ea.		8.35		8.35	Includes material and labor to install a
Install	Ea.	117	43.50		160.50	canvas window awning.
Demolish and Install	Ea.	117	51.85		168.85	
Reinstall	Ea.		34.72		34.72	
Clean	Ea.		5		5	
Minimum Charge	Job		260		260	

For customer support on your Contractor's Pricing Guide: Residential Repair & Remodeling, call 800.448.8182.

Improvements / Appliances / Treatments

Awning

Awning	Unit	Material	Labor	Equip.	Total	Specification
36″ Wide						
Demolish	Ea.		8.35		8.35	Includes material and labor to install a
Install	Ea.	117	52		169	canvas window awning.
Demolish and Install	Ea.	117	60.35		177.35	
Reinstall	Ea.		41.66		41.66	
Clean	Ea.		5.45		5.45	
Minimum Charge	Job		260		260	
42″ Wide						
Demolish	Ea.		8.35		8.35	Includes material and labor to install a
Install	Ea.	121	65		186	canvas window awning.
Demolish and Install	Ea.	121	73.35		194.35	
Reinstall	Ea.		52.08		52.08	
Clean	Ea.		6.50		6.50	
Minimum Charge	Job		260		260	
48″ Wide						
Demolish	Ea.		8.35		8.35	Includes material and labor to install a
Install	Ea.	98	74.50		172.50	canvas window awning.
Demolish and Install	Ea.	98	82.85		180.85	
Reinstall	Ea.		59.52		59.52	
Clean	Ea.		7.55		7.55	
Minimum Charge	Job		260		260	
Cloth Canopy Cover						
8′ x 10′						
Demolish	Ea.		450		450	Includes material and labor to install
Install	Ea.	296	260		556	one 8′ x 10′ cloth canopy patio cover
Demolish and Install	Ea.	296	710		1006	with front bar and tension support
Reinstall	Ea.		208.32		208.32	rafters and 9″ valance.
Clean	Ea.	2.53	66.50		69.03	
Minimum Charge	Job		520		520	
8′ x 15′						
Demolish	Ea.		450		450	Includes material and labor to install
Install	Ea.	360	260		620	one 8′ x 15′ cloth canopy patio cover
Demolish and Install	Ea.	360	710		1070	with front bar and tension support
Reinstall	Ea.		208.32		208.32	rafters and 9″ valance.
Clean	Ea.	2.53	100		102.53	
Minimum Charge	Job		520		520	

Vent

Vent	Unit	Material	Labor	Equip.	Total	Specification
Small Roof Turbine						
Demolish	Ea.		13.90		13.90	Includes material and labor to install
Install	Ea.	120	51		171	spinner ventilator, wind driven,
Demolish and Install	Ea.	120	64.90		184.90	galvanized, 4″ neck diam, 180 CFM.
Reinstall	Ea.		40.92		40.92	
Clean	Ea.		8.90		8.90	
Minimum Charge	Job		294		294	
Large Roof Turbine						
Demolish	Ea.		11.10		11.10	Includes material and labor to install
Install	Ea.	139	73		212	spinner ventilator, wind driven,
Demolish and Install	Ea.	139	84.10		223.10	galvanized, 8″ neck diam, 360 CFM.
Reinstall	Ea.		58.45		58.45	
Clean	Ea.		10		10	
Minimum Charge	Job		294		294	

For customer support on your Contractor's Pricing Guide: Residential Repair & Remodeling, call 800.448.8182.

Improvements / Appliances / Treatments

Vent		Unit	Material	Labor	Equip.	Total	Specification
Gable Louvered							
	Demolish	Ea.		28		28	Includes material and labor to install a
	Install	Ea.	65.50	17.35		82.85	vinyl gable end vent.
	Demolish and Install	Ea.	65.50	45.35		110.85	
	Reinstall	Ea.		13.89		13.89	
	Clean	Ea.		5.35		5.35	
	Paint	Ea.	3.23	6.65		9.88	
	Minimum Charge	Job		260		260	
Ridge Vent							
	Demolish	L.F.		1.43		1.43	Includes material and labor to install a
	Install	L.F.	4.41	3.66		8.07	mill finish aluminum ridge vent strip.
	Demolish and Install	L.F.	4.41	5.09		9.50	
	Minimum Charge	Job		260		260	
Foundation							
	Demolish	Ea.		10.05		10.05	Includes material and labor to install
	Install	Ea.	223	17.25		240.25	galvanized foundation block vent.
	Demolish and Install	Ea.	223	27.30		250.30	
	Reinstall	Ea.		13.79		13.79	
	Clean	Ea.	.15	10		10.15	
	Minimum Charge	Job		260		260	
Frieze							
	Demolish	L.F.		.99		.99	Includes material and labor to install
	Install	L.F.	.90	2.08		2.98	pine frieze board.
	Demolish and Install	L.F.	.90	3.07		3.97	
	Reinstall	L.F.		1.67		1.67	
	Clean	L.F.	.03	.33		.36	
	Minimum Charge	Job		260		260	
Water Heater Cap							
	Demolish	Ea.		20.50		20.50	Includes material and labor to install a
	Install	Ea.	69.50	71		140.50	galvanized steel water heater cap.
	Demolish and Install	Ea.	69.50	91.50		161	
	Reinstall	Ea.		56.80		56.80	
	Clean	Ea.	1.11	25		26.11	

Cupolas		Unit	Material	Labor	Equip.	Total	Specification
22" x 22"							
	Demolish	Ea.		258		258	Includes material and labor to install a
	Install	Ea.	470	130		600	cedar cupola with aluminum roof
	Demolish and Install	Ea.	470	388		858	covering.
	Reinstall	Ea.		104.16		104.16	
	Minimum Charge	Job		260		260	
29" x 29"							
	Demolish	Ea.		258		258	Includes material and labor to install a
	Install	Ea.	1125	130		1255	cedar cupola with aluminum roof
	Demolish and Install	Ea.	1125	388		1513	covering.
	Reinstall	Ea.		104.16		104.16	
	Minimum Charge	Job		260		260	
35" x 35"							
	Demolish	Ea.		258		258	Includes material and labor to install a
	Install	Ea.	1350	174		1524	cedar cupola with aluminum roof
	Demolish and Install	Ea.	1350	432		1782	covering.
	Reinstall	Ea.		138.88		138.88	
	Minimum Charge	Job		260		260	

For customer support on your Contractor's Pricing Guide: Residential Repair & Remodeling, call 800.448.8182.

Improvements / Appliances / Treatments

Cupolas

	Unit	Material	Labor	Equip.	Total	Specification
47" x 47"						
Demolish	Ea.		258		258	Includes material and labor to install a
Install	Ea.	2525	260		2785	cedar cupola with aluminum roof
Demolish and Install	Ea.	2525	518		3043	covering.
Reinstall	Ea.		208.32		208.32	
Minimum Charge	Job		260		260	
Weather Vane						
Install	Ea.	252	65		317	Includes minimum labor and
Minimum Charge	Job		260		260	equipment to install residential type weathervane.

Aluminum Carport

	Unit	Material	Labor	Equip.	Total	Specification
20' x 20' Complete						
Demolish	Ea.		278		278	Includes material, labor and
Install	Ea.	5475	845	296	6616	equipment to install an aluminum
Demolish and Install	Ea.	5475	1123	296	6894	carport. Foundations are not included.
Reinstall	Ea.		677.47	236.93	914.40	
Clean	Ea.	.17	100		100.17	
Minimum Charge	Job		520		520	
24' x 24' Complete						
Demolish	Ea.		278		278	Includes material, labor and
Install	Ea.	3625	1125	395	5145	equipment to install an aluminum
Demolish and Install	Ea.	3625	1403	395	5423	carport. Foundations are not included.
Reinstall	Ea.		903.30	315.90	1219.20	
Clean	Ea.	.17	200		200.17	
Minimum Charge	Job		520		520	
Metal Support Posts						
Demolish	Ea.		3.70		3.70	Includes material and labor to install
Install	Ea.	660	65		725	metal support posts for a carport.
Demolish and Install	Ea.	660	68.70		728.70	
Reinstall	Ea.		52.08		52.08	
Clean	Ea.	.17	4.16		4.33	
Paint	Ea.	.41	8.65		9.06	
Minimum Charge	Job		520		520	

Storage Shed

	Unit	Material	Labor	Equip.	Total	Specification
Aluminum						
Pre-fab (8' x 10')						
Demolish	Ea.		80.50		80.50	Includes material and labor to install
Install	Ea.	995	775		1770	an aluminum storage shed.
Demolish and Install	Ea.	995	855.50		1850.50	
Clean	Ea.	1.06	50		51.06	
Minimum Charge	Ea.		590		590	
Pre-fab (10' x 12')						
Demolish	Ea.		150		150	Includes material and labor to install
Install	Ea.	1825	775		2600	an aluminum storage shed.
Demolish and Install	Ea.	1825	925		2750	
Clean	Ea.	1.06	55.50		56.56	
Minimum Charge	Ea.		590		590	

For customer support on your Contractor's Pricing Guide: Residential Repair & Remodeling, call 800.448.8182.

Improvements / Appliances / Treatments

Storage Shed		Unit	Material	Labor	Equip.	Total	Specification
Wood							
Pre-fab (8' x 10')							
	Demolish	Ea.		80.50		80.50	Includes material and labor to install a
	Install	Ea.	5725	775		6500	pre-fabricated wood storage shed.
	Demolish and Install	Ea.	5725	855.50		6580.50	
	Clean	Ea.	1.06	50		51.06	
	Minimum Charge	Ea.		590		590	
Pre-fab (10' x 12')							
	Demolish	Ea.		150		150	Includes material and labor to install a
	Install	Ea.	6350	775		7125	pre-fabricated wood storage shed.
	Demolish and Install	Ea.	6350	925		7275	
	Clean	Ea.	1.06	55.50		56.56	
	Minimum Charge	Ea.		590		590	

Gazebo		Unit	Material	Labor	Equip.	Total	Specification
Good Grade							
	Demolish	S.F.		4.44		4.44	Includes material and labor to install a
	Install	S.F.	17.10	39		56.10	gazebo.
	Demolish and Install	S.F.	17.10	43.44		60.54	
	Reinstall	S.F.		31.25		31.25	
	Clean	S.F.	.13	2.66		2.79	
	Minimum Charge	Job		260		260	
Custom Grade							
	Demolish	S.F.		4.44		4.44	Includes material and labor to install a
	Install	S.F.	35	39		74	gazebo.
	Demolish and Install	S.F.	35	43.44		78.44	
	Reinstall	S.F.		31.25		31.25	
	Clean	S.F.	.13	2.66		2.79	
	Minimum Charge	Job		260		260	
Bench Seating							
	Demolish	L.F.		7.40		7.40	Includes material and labor to install
	Install	L.F.	3.56	26		29.56	pressure treated bench seating.
	Demolish and Install	L.F.	3.56	33.40		36.96	
	Reinstall	L.F.		20.83		20.83	
	Clean	L.F.	.13	2		2.13	
	Paint	L.F.	1.28	3.45		4.73	
	Minimum Charge	Job		260		260	
Re-screen							
	Install	S.F.	4.74	2.08		6.82	Includes labor and material to
	Minimum Charge	Job		260		260	re-screen wood frame.

Pool		Unit	Material	Labor	Equip.	Total	Specification
Screen Enclosure							
Re-screen							
	Install	SF Wall	.06	5.90		5.96	Includes material and labor to install a
	Minimum Charge	Job		260		260	screen enclosure for a pool.
Screen Door							
	Demolish	Ea.		10.35		10.35	Includes material and labor to install
	Install	Ea.	735	87		822	wood screen door with aluminum cloth
	Demolish and Install	Ea.	735	97.35		832.35	screen. Frame and hardware not
	Reinstall	Ea.		69.44		69.44	included.
	Clean	Ea.	.12	16.20		16.32	
	Paint	Ea.	9.80	17.25		27.05	
	Minimum Charge	Job		260		260	

For customer support on your Contractor's Pricing Guide: Residential Repair & Remodeling, call 800.448.8182.

Improvements / Appliances / Treatments

Pool	Unit	Material	Labor	Equip.	Total	Specification
Heater / Motor						
Demolish	Ea.		54.50		54.50	Includes material and labor to install
Install	Ea.	4000	700		4700	swimming pool heater, not incl. base
Demolish and Install	Ea.	4000	754.50		4754.50	or pad, elec., 12 kW, 4,800 gal
Reinstall	Ea.		559.26		559.26	pool, incl. pump.
Clean	Ea.	3.17	50		53.17	
Minimum Charge	Job		315		315	
Filter System						
Demolish	Ea.		89		89	Includes material, labor and
Install	Total	6300	470		6770	equipment to install sand filter system
Demolish and Install	Total	6300	559		6859	tank including hook-up to existing
Reinstall	Total		375.81		375.81	equipment. Pump and motor not
Minimum Charge	Job		294		294	included.
Pump / Motor						
Demolish	Ea.		54.50		54.50	Includes material and labor to install a
Install	Ea.	845	117		962	22 GPM pump.
Demolish and Install	Ea.	845	171.50		1016.50	
Reinstall	Ea.		93.95		93.95	
Clean	Ea.	3.17	50		53.17	
Minimum Charge	Job		315		315	
Pump Water						
Install	Day		555	119	674	Includes labor and equipment to pump
Minimum Charge	Job		256		256	water from a swimming pool.
Replaster						
Install	SF Surf	.90	13.55		14.45	Includes labor and material to
Minimum Charge	Job		256		256	replaster gunite pool with a surface
						area of 500 to 600 S.F.
Solar Panels						
Demolish	S.F.		2.17		2.17	Includes material and labor to install
Install	S.F.	38.50	3.86		42.36	solar panel.
Demolish and Install	S.F.	38.50	6.03		44.53	
Reinstall	S.F.		3.09		3.09	
Minimum Charge	Ea.		590		590	

For customer support on your Contractor's Pricing Guide: Residential Repair & Remodeling, call 800.448.8182.

Close Out

Construction Clean-up	Unit	Material	Labor	Equip.	Total	Specification
Final						
Clean	S.F.		.35		.35	Includes labor for final site/job
Minimum Charge	Job		200		200	clean-up including detail cleaning.

Location Factors - Residential

Costs shown in *Residential Costs with RSMeans Data* are based on national averages for materials and installation. To adjust these costs to a specific location, simply multiply the base cost by the factor for that city. The data is arranged alphabetically by state and postal zip code numbers. For a city not listed, use the factor for a nearby city with similar economic characteristics.

STATE	CITY	Residential
ALABAMA		
350-352	Birmingham	.89
354	Tuscaloosa	.88
355	Jasper	.87
356	Decatur	.87
357-358	Huntsville	.89
359	Gadsden	.86
360-361	Montgomery	.89
362	Anniston	.85
363	Dothan	.88
364	Evergreen	.87
365-366	Mobile	.87
367	Selma	.86
368	Phenix City	.86
369	Butler	.85
ALASKA		
995-996	Anchorage	1.12
997	Fairbanks	1.14
998	Juneau	1.13
999	Ketchikan	1.18
ARIZONA		
850,853	Phoenix	.90
851,852	Mesa/Tempe	.88
855	Globe	.87
856-857	Tucson	.88
859	Show Low	.88
860	Flagstaff	.89
863	Prescott	.89
864	Kingman	.88
865	Chambers	.88
ARKANSAS		
716	Pine Bluff	.87
717	Camden	.84
718	Texarkana	.85
719	Hot Springs	.83
720-722	Little Rock	.85
723	West Memphis	.85
724	Jonesboro	.86
725	Batesville	.81
726	Harrison	.82
727	Fayetteville	.83
728	Russellville	.83
729	Fort Smith	.85
CALIFORNIA		
900-902	Los Angeles	1.15
903-905	Inglewood	1.13
906-908	Long Beach	1.13
910-912	Pasadena	1.12
913-916	Van Nuys	1.14
917-918	Alhambra	1.14
919-921	San Diego	1.14
922	Palm Springs	1.12
923-924	San Bernardino	1.12
925	Riverside	1.13
926-927	Santa Ana	1.13
928	Anaheim	1.12
930	Oxnard	1.12
931	Santa Barbara	1.13
932-933	Bakersfield	1.13
934	San Luis Obispo	1.14
935	Mojave	1.12
936-938	Fresno	1.18
939	Salinas	1.21
940-941	San Francisco	1.31
942,956-958	Sacramento	1.19
943	Palo Alto	1.26
944	San Mateo	1.28
945	Vallejo	1.23
946	Oakland	1.27
947	Berkeley	1.29
948	Richmond	1.26
949	San Rafael	1.27
950	Santa Cruz	1.23
951	San Jose	1.27
952	Stockton	1.20
953	Modesto	1.18

STATE	CITY	Residential
CALIFORNIA (CONT'D)		
954	Santa Rosa	1.25
955	Eureka	1.22
959	Marysville	1.20
960	Redding	1.20
961	Susanville	1.19
COLORADO		
800-802	Denver	.92
803	Boulder	.91
804	Golden	.90
805	Fort Collins	.92
806	Greeley	.91
807	Fort Morgan	.91
808-809	Colorado Springs	.90
810	Pueblo	.89
811	Alamosa	.89
812	Salida	.90
813	Durango	.90
814	Montrose	.89
815	Grand Junction	.91
816	Glenwood Springs	.91
CONNECTICUT		
060	New Britain	1.09
061	Hartford	1.10
062	Willimantic	1.09
063	New London	1.09
064	Meriden	1.09
065	New Haven	1.09
066	Bridgeport	1.09
067	Waterbury	1.09
068	Norwalk	1.10
069	Stamford	1.10
D.C.		
200-205	Washington	.95
DELAWARE		
197	Newark	1.04
198	Wilmington	1.06
199	Dover	1.06
FLORIDA		
320,322	Jacksonville	.89
321	Daytona Beach	.88
323	Tallahassee	.90
324	Panama City	.88
325	Pensacola	.88
326,344	Gainesville	.88
327-328,347	Orlando	.89
329	Melbourne	.89
330-332,340	Miami	.87
333	Fort Lauderdale	.86
334,349	West Palm Beach	.84
335-336,346	Tampa	.86
337	St. Petersburg	.85
338	Lakeland	.84
339,341	Fort Myers	.84
342	Sarasota	.87
GEORGIA		
300-303,399	Atlanta	.92
304	Statesboro	.89
305	Gainesville	.85
306	Athens	.87
307	Dalton	.89
308-309	Augusta	.90
310-312	Macon	.91
313-314	Savannah	.91
315	Waycross	.88
316	Valdosta	.90
317,398	Albany	.90
318-319	Columbus	.90
HAWAII		
967	Hilo	1.10
968	Honolulu	1.15

For customer support on your Contractor's Pricing Guide: Residential Repair & Remodeling, call 800.448.8182.

Location Factors - Residential

STATE	CITY	Residential
STATES & POSS.		
969	Guam	.89
IDAHO		
832	Pocatello	.93
833	Twin Falls	.92
834	Idaho Falls	.89
835	Lewiston	.98
836-837	Boise	.94
838	Coeur d'Alene	.96
ILLINOIS		
600-603	North Suburban	1.15
604	Joliet	1.17
605	South Suburban	1.15
606-608	Chicago	1.19
609	Kankakee	1.09
610-611	Rockford	1.10
612	Rock Island	.97
613	La Salle	1.08
614	Galesburg	1.01
615-616	Peoria	1.03
617	Bloomington	1.03
618-619	Champaign	1.01
620-622	East St. Louis	1.01
623	Quincy	1.00
624	Effingham	1.01
625	Decatur	1.01
626-627	Springfield	1.01
628	Centralia	1.00
629	Carbondale	.98
INDIANA		
460	Anderson	.89
461-462	Indianapolis	.91
463-464	Gary	1.00
465-466	South Bend	.89
467-468	Fort Wayne	.88
469	Kokomo	.87
470	Lawrenceburg	.87
471	New Albany	.86
472	Columbus	.91
473	Muncie	.92
474	Bloomington	.92
475	Washington	.89
476-477	Evansville	.90
478	Terre Haute	.89
479	Lafayette	.92
IOWA		
500-503,509	Des Moines	.93
504	Mason City	.85
505	Fort Dodge	.86
506-507	Waterloo	.87
508	Creston	.89
510-511	Sioux City	.87
512	Sibley	.78
513	Spencer	.79
514	Carroll	.87
515	Council Bluffs	.89
516	Shenandoah	.86
520	Dubuque	.88
521	Decorah	.85
522-524	Cedar Rapids	.91
525	Ottumwa	.88
526	Burlington	.90
527-528	Davenport	.94
KANSAS		
660-662	Kansas City	.95
664-666	Topeka	.87
667	Fort Scott	.87
668	Emporia	.84
669	Belleville	.81
670-672	Wichita	.83
673	Independence	.85
674	Salina	.82
675	Hutchinson	.79
676	Hays	.82
677	Colby	.83
678	Dodge City	.83
679	Liberal	.81
KENTUCKY		
400-402	Louisville	.90
403-405	Lexington	.91

STATE	CITY	Residential
KENTUCKY (CONT'D)		
406	Frankfort	.89
407-409	Corbin	.85
410	Covington	.86
411-412	Ashland	.87
413-414	Campton	.85
415-416	Pikeville	.87
417-418	Hazard	.85
420	Paducah	.86
421-422	Bowling Green	.89
423	Owensboro	.89
424	Henderson	.88
425-426	Somerset	.85
427	Elizabethtown	.85
LOUISIANA		
700-701	New Orleans	.88
703	Thibodaux	.85
704	Hammond	.83
705	Lafayette	.86
706	Lake Charles	.86
707-708	Baton Rouge	.87
710-711	Shreveport	.87
712	Monroe	.85
713-714	Alexandria	.86
MAINE		
039	Kittery	.93
040-041	Portland	.96
042	Lewiston	.95
043	Augusta	.95
044	Bangor	.94
045	Bath	.94
046	Machias	.94
047	Houlton	.94
048	Rockland	.94
049	Waterville	.95
MARYLAND		
206	Waldorf	.89
207-208	College Park	.89
209	Silver Spring	.88
210-212	Baltimore	.94
214	Annapolis	.93
215	Cumberland	.91
216	Easton	.88
217	Hagerstown	.94
218	Salisbury	.87
219	Elkton	.93
MASSACHUSETTS		
010-011	Springfield	1.07
012	Pittsfield	1.06
013	Greenfield	1.05
014	Fitchburg	1.09
015-016	Worcester	1.09
017	Framingham	1.10
018	Lowell	1.11
019	Lawrence	1.11
020-022, 024	Boston	1.15
023	Brockton	1.09
025	Buzzards Bay	1.07
026	Hyannis	1.10
027	New Bedford	1.09
MICHIGAN		
480,483	Royal Oak	.91
481	Ann Arbor	.94
482	Detroit	.99
484-485	Flint	.91
486	Saginaw	.89
487	Bay City	.89
488-489	Lansing	.92
490	Battle Creek	.90
491	Kalamazoo	.89
492	Jackson	.90
493,495	Grand Rapids	.92
494	Muskegon	.89
496	Traverse City	.89
497	Gaylord	.87
498-499	Iron Mountain	.90
MINNESOTA		
550-551	Saint Paul	1.07
553-555	Minneapolis	1.08
556-558	Duluth	1.01

Location Factors - Residential

STATE	CITY	Residential
MINNESOTA (CONT'd)		
559	Rochester	1.01
560	Mankato	1.03
561	Windom	1.00
562	Willmar	1.01
563	St. Cloud	1.07
564	Brainerd	.99
565	Detroit Lakes	.95
566	Bemidji	.96
567	Thief River Falls	.95
MISSISSIPPI		
386	Clarksdale	.82
387	Greenville	.88
388	Tupelo	.82
389	Greenwood	.83
390-392	Jackson	.86
393	Meridian	.85
394	Laurel	.82
395	Biloxi	.85
396	McComb	.81
397	Columbus	.81
MISSOURI		
630-631	St. Louis	.98
633	Bowling Green	.93
634	Hannibal	.89
635	Kirksville	.86
636	Flat River	.91
637	Cape Girardeau	.89
638	Sikeston	.88
639	Poplar Bluff	.87
640-641	Kansas City	.99
644-645	St. Joseph	.95
646	Chillicothe	.92
647	Harrisonville	.96
648	Joplin	.91
650-651	Jefferson City	.90
652	Columbia	.89
653	Sedalia	.90
654-655	Rolla	.92
656-658	Springfield	.93
MONTANA		
590-591	Billings	.94
592	Wolf Point	.89
593	Miles City	.90
594	Great Falls	.91
595	Havre	.87
596	Helena	.90
597	Butte	.91
598	Missoula	.89
599	Kalispell	.89
NEBRASKA		
680-681	Omaha	.89
683-685	Lincoln	.90
686	Columbus	.87
687	Norfolk	.87
688	Grand Island	.90
689	Hastings	.86
690	McCook	.82
691	North Platte	.87
692	Valentine	.83
693	Alliance	.81
NEVADA		
889-891	Las Vegas	1.05
893	Ely	.95
894-895	Reno	.98
897	Carson City	.98
898	Elko	.93
NEW HAMPSHIRE		
030	Nashua	.99
031	Manchester	1.00
032-033	Concord	.99
034	Keene	.94
035	Littleton	.94
036	Charleston	.94
037	Claremont	.94
038	Portsmouth	.99

STATE	CITY	Residential
NEW JERSEY		
070-071	Newark	1.19
072	Elizabeth	1.19
073	Jersey City	1.19
074-075	Paterson	1.19
076	Hackensack	1.17
077	Long Branch	1.17
078	Dover	1.17
079	Summit	1.19
080,083	Vineland	1.17
081	Camden	1.18
082,084	Atlantic City	1.18
085-086	Trenton	1.18
087	Point Pleasant	1.17
088-089	New Brunswick	1.20
NEW MEXICO		
870-872	Albuquerque	.88
873	Gallup	.87
874	Farmington	.88
875	Santa Fe	.88
877	Las Vegas	.87
878	Socorro	.87
879	Truth/Consequences	.86
880	Las Cruces	.86
881	Clovis	.87
882	Roswell	.88
883	Carrizozo	.87
884	Tucumcari	.88
NEW YORK		
100-102	New York	1.30
103	Staten Island	1.27
104	Bronx	1.26
105	Mount Vernon	1.17
106	White Plains	1.17
107	Yonkers	1.18
108	New Rochelle	1.24
109	Suffern	1.21
110	Queens	1.29
111	Long Island City	1.31
112	Brooklyn	1.31
113	Flushing	1.30
114	Jamaica	1.30
115,117,118	Hicksville	1.21
116	Far Rockaway	1.30
119	Riverhead	1.22
120-122	Albany	1.04
123	Schenectady	1.05
124	Kingston	1.16
125-126	Poughkeepsie	1.22
127	Monticello	1.15
128	Glens Falls	1.02
129	Plattsburgh	1.01
130-132	Syracuse	.99
133-135	Utica	.99
136	Watertown	.98
137-139	Binghamton	.98
140-142	Buffalo	1.07
143	Niagara Falls	1.06
144-146	Rochester	1.01
147	Jamestown	.99
148-149	Elmira	1.00
NORTH CAROLINA		
270,272-274	Greensboro	.98
271	Winston-Salem	.98
275-276	Raleigh	.98
277	Durham	.98
278	Rocky Mount	.96
279	Elizabeth City	.96
280	Gastonia	1.01
281-282	Charlotte	1.01
283	Fayetteville	.99
284	Wilmington	.98
285	Kinston	.95
286	Hickory	.97
287-288	Asheville	.98
289	Murphy	.99
NORTH DAKOTA		
580-581	Fargo	.88
582	Grand Forks	.89
583	Devils Lake	.88
584	Jamestown	.87
585	Bismarck	.90

For customer support on your Contractor's Pricing Guide: Residential Repair & Remodeling, call 800.448.8182.

Location Factors - Residential

STATE	CITY	Residential
NORTH DAKOTA (CONT'D)		
586	Dickinson	.87
587	Minot	.88
588	Williston	.88
OHIO		
430-432	Columbus	.91
433	Marion	.89
434-436	Toledo	.93
437-438	Zanesville	.88
439	Steubenville	.90
440	Lorain	.91
441	Cleveland	.94
442-443	Akron	.92
444-445	Youngstown	.92
446-447	Canton	.90
448-449	Mansfield	.90
450	Hamilton	.90
451-452	Cincinnati	.88
453-454	Dayton	.90
455	Springfield	.90
456	Chillicothe	.92
457	Athens	.92
458	Lima	.90
OKLAHOMA		
730-731	Oklahoma City	.88
734	Ardmore	.84
735	Lawton	.85
736	Clinton	.84
737	Enid	.86
738	Woodward	.84
739	Guymon	.84
740-741	Tulsa	.88
743	Miami	.88
744	Muskogee	.88
745	McAlester	.87
746	Ponca City	.86
747	Durant	.88
748	Shawnee	.87
749	Poteau	.86
OREGON		
970-972	Portland	1.04
973	Salem	1.01
974	Eugene	1.01
975	Medford	1.01
976	Klamath Falls	1.00
977	Bend	1.02
978	Pendleton	1.01
979	Vale	.97
PENNSYLVANIA		
150-152	Pittsburgh	1.00
153	Washington	.97
154	Uniontown	.97
155	Bedford	.93
156	Greensburg	.97
157	Indiana	.95
158	Dubois	.94
159	Johnstown	.94
160	Butler	.96
161	New Castle	.96
162	Kittanning	.98
163	Oil City	.92
164-165	Erie	.95
166	Altoona	.95
167	Bradford	.92
168	State College	.95
169	Wellsboro	.92
170-171	Harrisburg	.98
172	Chambersburg	.95
173-174	York	.96
175-176	Lancaster	.96
177	Williamsport	.93
178	Sunbury	.93
179	Pottsville	.93
180	Lehigh Valley	1.01
181	Allentown	1.02
182	Hazleton	.96
183	Stroudsburg	.96
184-185	Scranton	.97
186-187	Wilkes-Barre	.95
188	Montrose	.96
189	Doylestown	1.09

STATE	CITY	Residential
PENNSYLVANIA (CONT'D)		
190-191	Philadelphia	1.17
193	Westchester	1.07
194	Norristown	1.07
195-196	Reading	.99
PUERTO RICO		
009	San Juan	.79
RHODE ISLAND		
028	Newport	1.07
029	Providence	1.08
SOUTH CAROLINA		
290-292	Columbia	1.01
293	Spartanburg	1.01
294	Charleston	1.02
295	Florence	1.01
296	Greenville	1.01
297	Rock Hill	.97
298	Aiken	1.02
299	Beaufort	1.01
SOUTH DAKOTA		
570-571	Sioux Falls	.91
572	Watertown	.85
573	Mitchell	.80
574	Aberdeen	.85
575	Pierre	.90
576	Mobridge	.80
577	Rapid City	.85
TENNESSEE		
370-372	Nashville	.91
373-374	Chattanooga	.88
375,380-381	Memphis	.90
376	Johnson City	.83
377-379	Knoxville	.86
382	McKenzie	.77
383	Jackson	.87
384	Columbia	.82
385	Cookeville	.77
TEXAS		
750	McKinney	.80
751	Waxahachie	.82
752-753	Dallas	.86
754	Greenville	.81
755	Texarkana	.82
756	Longview	.81
757	Tyler	.82
758	Palestine	.81
759	Lufkin	.81
760-761	Fort Worth	.84
762	Denton	.84
763	Wichita Falls	.84
764	Eastland	.84
765	Temple	.84
766-767	Waco	.86
768	Brownwood	.84
769	San Angelo	.84
770-772	Houston	.84
773	Huntsville	.83
774	Wharton	.82
775	Galveston	.81
776-777	Beaumont	.84
778	Bryan	.84
779	Victoria	.81
780	Laredo	.84
781-782	San Antonio	.85
783-784	Corpus Christi	.84
785	McAllen	.83
786-787	Austin	.84
788	Del Rio	.82
789	Giddings	.82
790-791	Amarillo	.86
792	Childress	.86
793-794	Lubbock	.85
795-796	Abilene	.86
797	Midland	.88
798-799,885	El Paso	.86
UTAH		
840-841	Salt Lake City	.91
842,844	Ogden	.90
843	Logan	.90

298

For customer support on your Contractor's Pricing Guide: Residential Repair & Remodeling, call 800.448.8182.

Location Factors - Residential

Location Factors

STATE	CITY	Residential
UTAH (CONT'D)		
845	Price	.90
846-847	Provo	.90
VERMONT		
050	White River Jct.	.98
051	Bellows Falls	1.03
052	Bennington	1.02
053	Brattleboro	1.03
054	Burlington	.96
056	Montpelier	1.02
057	Rutland	.95
058	St. Johnsbury	.99
059	Guildhall	.98
VIRGINIA		
220-221	Fairfax	1.02
222	Arlington	1.04
223	Alexandria	1.09
224-225	Fredericksburg	1.03
226	Winchester	1.02
227	Culpeper	1.03
228	Harrisonburg	1.00
229	Charlottesville	.98
230-232	Richmond	1.00
233-235	Norfolk	.99
236	Newport News	.98
237	Portsmouth	.97
238	Petersburg	.99
239	Farmville	.97
240-241	Roanoke	.99
242	Bristol	.96
243	Pulaski	.94
244	Staunton	.94
245	Lynchburg	.97
246	Grundy	.94
WASHINGTON		
980-981,987	Seattle	1.07
982	Everett	1.05
983-984	Tacoma	1.04
985	Olympia	1.03
986	Vancouver	1.01
988	Wenatchee	.98
989	Yakima	1.01
990-992	Spokane	.97
993	Richland	.98
994	Clarkston	.94
WEST VIRGINIA		
247-248	Bluefield	.92
249	Lewisburg	.93
250-253	Charleston	.94
254	Martinsburg	.91
255-257	Huntington	.96
258-259	Beckley	.94
260	Wheeling	.94
261	Parkersburg	.92
262	Buckhannon	.92
263-264	Clarksburg	.92
265	Morgantown	.93
266	Gassaway	.93
267	Romney	.93
268	Petersburg	.93
WISCONSIN		
530,532	Milwaukee	1.03
531	Kenosha	1.03
534	Racine	1.01
535	Beloit	.97
537	Madison	.99
538	Lancaster	.93
539	Portage	.93
540	New Richmond	.95
541-543	Green Bay	.97
544	Wausau	.95
545	Rhinelander	.92
546	La Crosse	.97
547	Eau Claire	.98
548	Superior	.98
549	Oshkosh	.93
WYOMING		
820	Cheyenne	.89
821	Yellowstone Nat. Pk.	.84
822	Wheatland	.83

STATE	CITY	Residential
WYOMING (CONT'D)		
823	Rawlins	.85
824	Worland	.84
825	Riverton	.84
826	Casper	.87
827	Newcastle	.84
828	Sheridan	.87
829-831	Rock Springs	.87
CANADIAN FACTORS (reflect Canadian currency)		
ALBERTA		
	Calgary	1.05
	Edmonton	1.05
	Fort McMurray	1.06
	Lethbridge	1.03
	Lloydminster	.96
	Medicine Hat	.99
	Red Deer	.99
BRITISH COLUMBIA		
	Kamloops	.96
	Prince George	.96
	Vancouver	1.06
	Victoria	1.00
MANITOBA		
	Brandon	1.02
	Portage la Prairie	.95
	Winnipeg	.98
NEW BRUNSWICK		
	Bathurst	.88
	Dalhousie	.91
	Fredericton	.94
	Moncton	.94
	Newcastle	.89
	St. John	.99
NEWFOUNDLAND		
	Corner Brook	1.00
	St. Johns	1.05
NORTHWEST TERRITORIES		
	Yellowknife	1.07
NOVA SCOTIA		
	Bridgewater	.92
	Dartmouth	.96
	Halifax	1.00
	New Glasgow	.96
	Sydney	.95
	Truro	.91
	Yarmouth	.96
ONTARIO		
	Barrie	1.06
	Brantford	1.05
	Cornwall	1.02
	Hamilton	1.06
	Kingston	1.02
	Kitchener	1.04
	London	1.04
	North Bay	1.10
	Oshawa	1.06
	Ottawa	1.05
	Owen Sound	1.06
	Peterborough	1.03
	Sarnia	1.05
	Sault Ste. Marie	1.01
	St. Catharines	1.05
	Sudbury	1.01
	Thunder Bay	1.04
	Timmins	1.02
	Toronto	1.09
	Windsor	1.04
PRINCE EDWARD ISLAND		
	Charlottetown	.91
	Summerside	.91
QUEBEC		
	Cap-de-la-Madeleine	.99
	Charlesbourg	.99
	Chicoutimi	1.08
	Gatineau	.99
	Granby	1.02

For customer support on your Contractor's Pricing Guide: Residential Repair & Remodeling, call 800.448.8182.

Location Factors - Residential

STATE	CITY	Residential
QUEBEC (CONT'D)		
	Hull	1.02
	Joliette	1.03
	Laval	.99
	Montreal	1.03
	Quebec City	1.04
	Rimouski	1.08
	Rouyn-Noranda	1.02
	Saint-Hyacinthe	1.03
	Sherbrooke	.99
	Sorel	1.03
	Saint-Jerome	1.02
	Trois-Rivieres	1.12
SASKATCHEWAN		
	Moose Jaw	.87
	Prince Albert	.86
	Regina	1.05
	Saskatoon	1.02
YUKON		
	Whitehorse	1.01

Abbreviations

A	Area
ASTM	American Society for Testing and Materials
B.F.	Board feet
Carp.	Carpenter
C.F.	Cubic feet
CWJ	Composite wood joist
C.Y.	Cubic yard
Ea.	Each
Equip.	Equipment
Exp.	Exposure
Ext.	Exterior
F	Fahrenheit
Ft.	Foot, feet
Gal.	Gallon
Hr.	Hour
in.	Inch, inches
Inst.	Installation
Int.	Interior
Lb.	Pound
L.F.	Linear feet
LVL	Laminated veneer lumber
Mat.	Material
Max.	Maximum
MBF	Thousand board feet
MBM	Thousand feet board measure
MSF	Thousand square feet
Min.	Minimum
O.C.	On center
O&P	Overhead and profit
OWJ	Open web wood joist
Oz.	Ounce
Pr.	Pair
Quan.	Quantity
S.F.	Square foot
Sq.	Square, 100 square feet
S.Y.	Square yard
V.L.F.	Vertical linear feet
'	Foot, feet
"	Inch, inches
o	Degrees

Index

A

A/C junction box 163
  junction box furnace unit 156
  junction box heat pump 157
Accessories bathtub 256
  fireplace 131
  shower 257
  wallboard 179
Accordion type bifold door 198
Acoustical ceiling 187
Acrylic impregnated parquet
floor 242
Add for additional story 74
  for steep pitch 74
Adhesive tile acoustical ceiling .. 189
Adobe paver 140
Air compressor 4
  conditioning 159
  handler 161
  mover 11
  purifier-per day 12
  vent system negative 13
Airless sprayer 13
Allowance to repair 166
Aluminum awning standard 287
  carport 290
  downspouts 285
  fascia 67
  flashing 71
  gutters 284
  screen door 101
  sheet 81
  siding 133
  sliding patio door 99
  soffit 69
  storage shed 290
  valley flashing 72
  window 106
Analysis asbestos 12
Anodized curtain wall 122
Antenna television 271
Architectural fees 1
  shingles 73
Asbestos analysis 12
  ceiling encapsulate 12
  pipe insulation 12
  removal 12
  test 1
Asbestos-based plaster 12
  siding 12
Asphalt paving 139
  roofing 87
Average grade carpeting 247
  room 6
Awning 287
  sash aluminum window 110

B

Back bar 228
Backfill 17
  trenching 17
Bannister stair 209
Bar back 228
  joist 45
  liquor 228
  sink 253, 254
  towel 282
Barge rafter fascia 68
Base cabinets 222
  molding 201
  pad chimney 129
Baseboard carpet 204
  electric 158

hot water 153
Bath accessories 282
Bathroom dispenser 282
  exhaust fan 268
  partition 283
Bathtub 255
Batt unfaced 167, 169, 171
Bay aluminum window 111
Beam / girder 27
  collar 52
  hardware 44
Beams glue-laminated 42
Bench seating gazebo 291
  seating porch 63
  seating wood deck 57
Berber carpeting 247, 248
Bidet 251
Bifold door 196
Blinds treatment 273
Block bridge 31
  fencing 140
Blown acoustical ceiling 190
  insulation 169, 171
Board and batten siding 136
  ledger 52
  rigid 168, 170-172
Boards ledger 32, 49
Bobcat 4
Boiler 156
Bookcase built-in 229
  custom 207
Border wallcovering 186
Bowstring truss 55
Box fire 130
Bracing let-in 36
Breast (face) brick fireplace 130
Brick paver deluxe 140
  paver standard 140
  structural 126
  veneer 126
  wall 14
Bridge block 31
Bucket concrete 5
Building concrete 9
  demolition 9
  masonry 9
  steel 9
  wood-framed 9
Built-in bookcase 229
  desk 229
Built-up roofing 87
  roofing components 88
Bullet resistant window 120
Butcher block countertop 230
Bypass sliding door 200

C

Cabinet garage base 229
  hardwood base 226
  hardwood tall 218
  hardwood vanity 233
  hardwood wall 215
  island base 227
  laminated base 222
  laminated tall 218
  laminated vanity 231
  laminated wall 211
  medicine 217
  stain 221, 229
  strip refinish ... 218, 221, 229, 234
  unit heater 154
  wall garage 218
  wall mounted microwave ... 278
  wood base 223, 224

wood tall 218
wood vanity 231, 232
wood wall 212, 214
Canopy cover cloth 288
Cant strips 85
Canvas awning 287
Cap chimney metal 129
  vent 147
Carpenter 2
Carpet baseboard 204
  cleaning 11
  pad 249
  pad per s.f. 248
  stair 249
  treatment 11
Carpeting 246, 247
  average grade 247
  berber 247, 248
  economy grade 247
  indoor-outdoor 246, 247
  per s.f. 246
  premium grade 247
  wool 246, 248
Carport aluminum 290
Cased opening double 205
  opening single 205
Casement bow wood window ... 117
  wood window 116
Casing double window 206
  single window 205
  trim 91, 93, 96, 194, 198, 201
  triple window 206
  window 206
Cast iron bathtub 255
Caulk renail base molding 204
CDX plywood 45
Cedar column or post 57
  fascia 67
  railing and trim 60
  shake 75
  soffit 69
  wood shingles 75
Ceiling acoustical 187
  acoustical scrape 12
  asbestos seal 13
  clean 10
  clean seal 237
  earthquake strapping 49
  framing 47
  grid type lighting 267
  insulation 170
  lath 177
  light fixture 264
  paddle fan 268
  paint 237
  plaster 176
  porch 63
  prep 236
  recessed lighting 267
  sand 236
  seal 236
  texture 236
Ceramic tile countertop 230
  tile cove base 204
  tile flooring 243
  wall tile 180
Chain link fencing 142
Chair rail molding 204
Chemical toilet 3
Cherry parquet floor 241
Chest-type freezer 281
Chimes door bell 270
Chimney 129
  base pad 129
  install 129
  install from roof up 129

Circuits 261
Clay tile roofing 77
Clean ceiling 10
  fixtures steam 11
  seal ceiling 237
  seal wall 236
  wall 10, 235
  wax floor 241
  wax parquet 242
  wax vinyl 245
Cleaning carpet 11
  maintenance 128
Clean-up final 293
  initial 8
  progressive 8
Clear glass 120
  plate glass 121
Clips hurricane 36, 49
Closet shelf and rod 207
  shelving 207
Cloth canopy cover 288
Clothing rod 208
CMU reinforced 125
  unreinforced 125
Coat seal 139
Collar beam 52
Column 29
  concrete 29
  or post cedar 57
  or post redwood 58
  or post treated 57
Commercial elevator 275
  grade door 94
  light fixture 266
  metal door 104
  trusses 55
Commode 251
Common brick fireplace 129
Compactor trash 281
Complete panel electrical 165
Components 96
  built-up roofing 88
  plumbing 147
Composite decking 61
Composition shingles 73
Compressor air 4
  air conditioning 159
Concealed sprinkler system 163
Concrete bucket 5
  building 9
  column 29
  footing 24
  footing continuous 24
  masonry wall 14
  mixer 4
  pad 163
  paver natural 139
  piles precast 20
  pump 5
  slab 14, 25
  slab lightweight 25
  slab reinforced 25
  slab unreinforced 25
  steps 275
  tile roofing 80
Condensing unit air
conditioning 160
Construction clean-up 8, 293
  fees 1
  permits 1
Container storage 3
Contents store 6
Continuous concrete footing 24
Cooler evaporative 161
Coping wall cap 72
Copper downspouts 285

302

Index

flashing 72
gutters 284
roofing 84
Cork wall tile 181
Corner molding 205
Corrugated aluminum sheet ... 81
galvanized steel 83
Countertop 229
microwave 278
Course wearing 139
Cove molding 205
Cover and protect 6
for outlet / switch 264
outlet/switch 264
Cross bracing piles 21
Cross-bracing foundation post ... 23
Crown molding 205
CSPE type roofing 86
Cultured marble countertop ... 230
marble vanity top 234
Cupolas 289
Curtain wall anodized 122
wall system 122
Custom bookcase 207

D

Debris hauling 7
remove 10
Decking composite 61
fir 61
mahogany 62
PVC 62
treated 61
Decontamination unit 13
Dehumidifier 11
Deluxe brick paver 140
Demo haul exterior chimney ... 129
haul interior chimney 129
Deodorize disinfect 10
Designer floor mount toilet ... 251
wallcovering 186
Desk built-in 229
Detector smoke 269
Diffuser 259
Dimmer switch 263
Disappearing stair 208
Dish soap 282
Dishwasher 281
Disinfect deodorize 10
Dispenser bathroom 282
Disposal garbage 281
Dome skylight 123
Door accordion type bifold 198
aluminum screen 101
aluminum sliding patio 99
bell chimes 270
bifold louvered 197
bifold mirrored 197
bifold wood 196
commercial grade 94
entry 89
exterior 93
French style 90
full louver 199
half louver 199
hardware 96, 194
hardware metal 105
hollow core 193
insulated storm 102
jamb 90, 195, 198, 199, 201
metal garage 102
mirrored bypass sliding 200
roll up garage 103
rough-in 38

shave refit . 101, 195, 198, 200, 201
sliding screen patio 101
solid core 95
stain bifold 198
stain bypass sliding 201
stain louver 200
steel 92
storm 102
vinyl sliding patio 100
wood bypass sliding 200
wood framed sliding patio 100
wood screen 101
Double cased opening 205
hung aluminum window 108
hung vinyl window 113
hung wood window 114
oven range 278
porcelain sink 252, 254
stainless steel sink 252, 254
switch 262
wall oven 277
Douglas fir soffit 70
Downspouts 284
Drain sink basket 255
Drapery treatment 273
Drip edge 72
Dryer 278
washer 279
Drywaller 2
Ductwork 150, 258
Dump fees 7
truck 3
Dumpster load 7
rental 7

E

Earthquake strapping ceiling 49
strapping floor 33
strapping wall 36
Economy grade carpeting 247
Edge drip 72
Elastomeric roofing 85
Electric baseboard 158
forced air furnace unit 156
opener 104
water heater 148
Electrical per s.f. 165, 261
Electrician 2
Elevator 275
Emergency lighting 268
repairs roofing 87
service call 10
Encapsulate asbestos ceiling 12
Enclosure pool screen 291
End gable 54
wall 72
Engineered lumber joist 31
lumber truss / joist 55
Engineering fees 1
Entrance storefront 122
Entry door 89
EPDM type roofing 86
Epoxy inject 26
Equipment rental 3
Etched glass window 120
Evaporative cooler 161
Excavation 17
footing 17
Exhaust fan bathroom 268
fan kitchen 269
fan whole-house 269
Exposed sprinkler system 163
Exterior door 93
insulation 172

light fixture 265
OSB 46
sheathing 45
shutters 285
siding 133
trim 65, 66, 133
trim PVC 66
type door 89
vapor barrier 46, 172
Extinguisher fire 283
Extra large room 6
Extraction water 10

F

Faced batt foil 167, 169, 170
batt kraft 167, 168, 170
Fan 268
Fascia 67
aluminum 67
barge rafter 68
cedar 67
hem-fir 68
plywood 67
redwood 67
vinyl 67
Faucet sink 255
Fees architectural 1
dump 7
engineering 1
Felt underlay 34
Fencing 140
chain link 142
gate 144
picket 143
post 144
red cedar 143
redwood 143
shadowbox 143
temporary 3
wrought iron 144
Fiber cement siding 135
Fiberglass bathtub 255
ductwork 151
insulation 172
sheet 83
shingles 73
shower 257
Filter system pool 292
Final clean-up 293
Finish plywood 46
Finished wallboard 177
Fir decking 61
Fire box 130
extinguisher 283
prevention / protection 283
Fireblock 35
Fireplace 129
accessories 131
breast (face) brick 130
common brick 129
hearth 130
natural stone 130
prefabricated 130
Fixed skylight 123
Flashing aluminum 71
aluminum valley 72
copper 72
galvanized 71
Flat clay tile roofing 77
Flatbed truck 3
Flood clean-up 10
Floor acrylic impregnated
parquet 242
asbestos seal 13

ceramic tile 243
cherry parquet 241
clean wax 241
earthquake strapping 33
framing system 29
gas-fired furnace unit 155
granite 244
insulation 167
maple 239
marble 243
mounted w/tank toilet 251
oak 239
oak parquet 241
pine 240
prefinished oak 240
protect/cover 6, 235, 236
refinish existing 240
sand finish new 240
sink 253, 254
slate 244
teak parquet 242
terrazzo 244
thick set paver 243
tile 243
underlayment 241
urethane coat 241
vapor barrier 168
vinyl sheet 245
walnut 240
walnut parquet 242
Fluorescent light fixture ... 265, 266
Fly rafter 53
Foam rubber pad 248, 249
Fog area thermal 11
Foil faced batt 167, 169, 170
wallcovering 186
Footing excavation 17
Forced air furnace unit electric .. 156
air furnace unit gas 155
Forklift 4
Foundation post 23
post cross-bracing 23
post wood 23
vent 289
Framing ceiling 47
roof 49
system floor 29
wall 34
Free standing oven 277
Freezer 280
chest-type 281
French style door 90
Frieze 289
Full louver door 199
Furnace unit 154
Furring 128
strips 179, 183, 190

G

Gable end 54
louvered vent 289
Galvanized downspouts 284
ductwork 150
flashing 71
gutters 284
steel 83
Garage base cabinet 229
cabinet wall 218
door 102
Garbage disposal 281
Gas forced air furnace unit 155
Gas-fired boiler 156
furnace unit floor 155
furnace unit wall 154

303

Index

Gas-fired water heater 149
Gate fencing 144
Gazebo 291
General clean-up 10
   laborer 2
Generator 5
Girder treated 58
Glass block wall tile 181
   clear 120
   curtain wall 122
   shelving 207
   window 119
   window etched 120
   window insulated 119
   window obscure/Florentine . . . 119
   window stained 120
   window tempered 120
   window tinted 119
   window wired 120
Glue-laminated beams 42
Grab rail 282
Grading 139
   hand 139
   machine 139
Granite floor 244
Grasscloth string wallcovering . . . 186
Gravel stop 72
Grid acoustical ceiling t-bar 188
   type lighting ceiling 267
Grill return 258
   supply 259
GRM membrane roofing 85
Gunite 26
Gutters 284
Gypsum wallboard 177

H

Half louver door 199
Hand backfill 17
   grading 139
Handler air 161
Handrail and baluster PVC 62
   stair 209
Hard tile flooring 243
Hardboard wall paneling 182
Hardware aluminum window 113
   beam 44
   door 96, 194
   metal door 105
   truss 55
Hardwood base cabinet 226
   oak 207
   stair spiral 208
   stair straight 208
   tall cabinet 218
   vanity cabinet 233
   wall cabinet 215
Haul exterior chimney demo 129
   interior chimney demo 129
Hauling debris 7
Headers wall 37
Hearth fireplace 130
Heat pump 157
Heater cabinet unit 154
   cap water 289
   space 6
Heater/motor pool 292
Heating 153, 154, 158
   control 159
Hem-fir fascia 68
HEPA vacuum cleaner 13
Holder toothbrush 282
Hole in gyp. wallboard patch . . . 179
Hollow core door 191

Hot water baseboard 153
   water system solar 150
Hurricane clips 36, 49

I

Ice maker 280
Impregnated walk 88
Indoor-outdoor carpeting . . . 246, 247
Industrial steel window 124
Initial clean-up 8
Inject epoxy 26
Install chimney 129
   from roof up chimney 129
Institutional bathtub 256
Insulated glass window 119
   storm door 102
Insulating wrap water heater 150
Insulation blown 169, 171
   fiberglass 172
   perlite 173
   polystyrene 173
   urethane 174
Intercom system 270
Interior ceiling plaster 176
   wall plaster 175
   wood shutters 286
Island base cabinet 227
I-type steel 28

J

Jack and relevel 7
   relevel 7
   roof 74
   with wiring phone 262
Jackhammer 4
Jamb door 90, 195, 198, 199, 201
   metal 105
Job permit 1
Joist bar 45
   treated 59
Junction box A/C 163
   box furnace unit A/C 156
   box heat pump A/C 157
Jute carpet pad 249

K

Kitchen exhaust fan 269
Kitchenette unit 281
Kraft faced batt 167, 168, 170

L

Laborer general 2
Laminated base cabinet 222
   countertop 229
   safety window 119
   tall cabinet 218
   vanity cabinet 231
   vanity top 234
   wall cabinet 211
Large room 6
   room extra 6
Lath ceiling 177
   wall 176
Laundry sink 253
Lavatory sink vanity 253, 254
Ledger board 52
   boards 32, 49
   strips 32, 48
   treated 59

Let-in bracing 36
Light fixture ceiling 264
   fixture exterior 265
   fixture fluorescent 265, 266
   stand 13
   switch 262
Lighting emergency 268
   track 266, 268
Lightweight concrete slab 25
Liquor bar 228
Load dumpster 7
Lookout rafter 53
Louver door 199
Louvered door bifold 197
   vent gable 289
Low voltage systems 269
Lumber joist engineered 31
   stock 65, 66
   truss / joist engineered 55

M

Machine backfill 17
   grading 139
Mahogany decking 62
Main service 165
Maintenance cleaning 128
   plumbing 147
Maker ice 280
Maple floor 239
Marble countertop cultured 230
   floor 243
   vanity top cultured 234
   wall tile 181
   window sill 206
Marlite 182
Mason 2
Masonite door pre-hung 192
Masonry building 9
   pavers 139
   wall 125
   wall concrete 14
Masthead 166
Medicine cabinet 217
Metal cap chimney 129
   door 92
   flashing 71
   garage door 102
   jamb 105
   joist 45
   pipe piles 21
   stairs 275
   stud framing 44
   support posts carport 290
Microwave cabinet wall
mounted 278
   countertop 278
   range 278
Mildicide wall 11
Mirror wall 120
   wall tile 181
Mirrored bypass sliding door . . . 200
   door bifold 197
Mission tile roofing 77
Mixer concrete 4
Modified bitumen roofing 87
Molding 183, 204
   shingle 74
Motor pool pump 292
Move / reset contents 6
Mover air 11
Muck-out 10
Mural 185

N

Natural concrete paver 139
   stone fireplace 130
Negative air vent system 13
Neoprene roofing 85

O

Oak base molding 202, 204
   floor 239
   hardwood 207
   parquet floor 241
Obscure/Florentine glass wind. . . 119
Office trailer 3
Opener electric 104
Operable skylight 123
OSB exterior 46
   roofing 56
Outlet/switch cover 264
Outriggers 53
Oven 277
   cabinet tall 219
   range 277

P

Pad concrete 163
   foam rubber 248, 249
   jute carpet 249
   rebound carpet 249
   urethane carpet 248, 249
   waffle carpet 248, 249
Paddle fan ceiling 268
Paint / texture, ceilings 236
   / texture, walls 236
   ceiling 237
   preparation, ceilings 236
   preparation, walls 235
   scrape 26
   wall 236
Painter 2
Panel acoustical ceiling tile 188
   electrical complete 165
   plywood 135
   pool solar 292
Paper roller toilet 282
Parquet clean wax 242
Particle board shelving 207
   board subflooring 34
Partition bathroom 283
Patch hole in gypsum
wallboard 179
   wall 184
Paver adobe 140
Pedestal sink 253, 254
Perlite insulation 173
Permit job 1
Phone jack with wiring 262
Picket fencing 143
Picture molding 205
   wood window 118
Piles 19
   cross bracing 21
   metal pipe 21
   steel H section 22
Pine base molding 201
   floor 240
   shelving 206
Pipe insulation asbestos 12
Plank board subflooring 34
   roofing 56
Plaster asbestos-based 12
   interior ceiling 176

304

Index

interior wall 175
repairs 176, 177
thin coat ceiling 177
thin coat wall 175
Plastic downspouts 285
gutters 284
Plate glass clear 121
glass polished 121
sill 33
wall 37
Plumber 2
Plumbing 147, 148
Plywood cdx 45
fascia 67
finish 46
panel 135
roofing 56
sheathing 46
shelving 207
soffit 69
subflooring 33
wall paneling 182
Pole power 3
Polished plate glass 121
Polystyrene insulation 173
Polyvinyl chloride (PVC) roofing . . 85
Pool 291
Porcelain sink double 252, 254
sink single 252, 253
Porch 63
ceiling 63
privacy wall 63
re-screen 63
stair 63
Post fencing 144
sleeve PVC 62
Power pole 3
Precast concrete piles 20
Prefabricated fireplace 130
Prefinished base molding 203
oak floor 240
Pre-hung masonite door 192
steel door 104
steel-clad door 91
wood door 191, 195
Premium grade carpeting 247
Prep ceiling 236
for flooring 34
wall 235
Pressure wash sewage 11
Privacy wall porch 63
wall wood deck 57
Progressive clean-up 8
Protect/cover floor 6, 235, 236
wall 6, 235, 236
Pump concrete 5
heat 157
motor pool 292
pool water 292
sump 148
Purifier-per day air 12
Purlins 52
PVC decking 62
exterior trim 66
handrail and baluster 62
post sleeve 62

Q

Quarry tile base molding 204
tile countertop 230
Quarter round molding 204

R

Radiant wall type 159
Rafter 50
fly 53
lookout 53
Rail grab 282
Railing and trim cedar 60
and trim redwood 60
and trim treated 60
temporary 3
Range double oven 278
microwave 278
oven 277
Rangehood 276
Rangetop 276
Rebound carpet pad 249
Recessed lighting ceiling 267
Red cedar fencing 143
Redwood column or post 58
fascia 67
fencing 143
railing and trim 60
soffit 69
Refinish existing floor 240
Refrigerator 279
Re-grout tile floor 243, 244
wall tile 180, 181
Reinforced CMU 125
concrete slab 25
Relevel jack 7
Removal tree 15
Remove debris 10
Repairs plaster 176, 177
Replace slate tiles 76
Replaster pool 292
Re-screen gazebo 291
porch 63
Residential electrical 165, 261
elevator 275
light fixture 264
metal door 91
trusses 54
Return grill 258
Ribbed aluminum sheet 82
galvanized steel 84
Ridge vent 74, 289
Ridgeboard 51
Rigid board 168, 170-172
Rod clothing 208
Roll up garage door 103
Rolled roofing 73
Roof framing 49
insulation board 172
jack 74
sheathing 56
shingles 73
Roofer 2
Roofing asphalt 87
clay tile 77
copper 84
CSPE type 86
emergency repairs 87
EPDM type 86
GRM membrane 85
modified bitumen 87
neoprene 85
OSB 56
plank 56
plywood 56
polyvinyl chloride (PVC) 85
rolled 73
skip-type 56
stainless steel 84

tile 76
Room average 6
demolition 13
large 6
small 6
Rough frame stair 47
Rough-in bathtub 257
commode 252
door 38
opening 38, 40
shower 258
sink 253
sink only 254
window 40
Run wiring 268

S

Safety window laminated 119
Sand ceiling 236
finish new floor 240
wall 235
Sandblaster 5
Sanitize flush-out ductwork 260
Satellite system 271
Scaffolding 7
Scrape ceiling acoustical 12
paint 26
Screen door 101
enclosure pool 291
window 118
Seal ceiling 236
ceiling asbestos 13
coat 139
floor asbestos 13
wall 235
wall asbestos 13
Seat toilet 252
Seating gazebo bench 291
porch bench 63
wood deck bench 57
Security system 270
Selective demolition 14
Self-contained air conditioning . . 159
Service call emergency 10
water 3
Shadowbox fencing 143
Shake cedar 75
Shave refit door . . 101, 195, 198, 200, 201
Shed aluminum storage 290
wood storage 291
Sheet fiberglass 83
vinyl 245
vinyl per s.f. 245
Shelf and rod closet 207
Shelving 206
Shim support piers 6
Shingle molding 74
siding wood 136
Shingles architectural 73
fiberglass 73
Shoe base molding 203
molding 204
Shower 257
combination tub 256
Shutter wood 274
Shutters 285
Side wall 72
Siding aluminum 133
asbestos-based 12
board and batten 136
fiber cement 135

tongue and groove 135
vinyl 134
Sill marble window 206
plate 33
wood window 206
Single cased opening 205
hung window 106
porcelain sink 252, 253
stainless steel sink 252, 254
switch 262
wall oven 277
Sink (assembly) 252
faucet 255
only 253
Skip-type roofing 56
Skylight 123
Slab concrete 14
vapor barrier 26
Slabjacking 26
Slate floor 244
tile roofing 76
tiles replace 76
Sliding patio door 99
sash aluminum window 109
screen patio door 101
Slumpstone 127
veneer 130
Small room 6
Smoke detector 269
Soap dish 282
Soffit 69
aluminum 69
cedar 69
douglas fir 70
plywood 69
redwood 69
vinyl 69
Softener water 150
Solar hot water system 150
panel pool 292
Solid core door 93, 95, 195
surface countertop 230
surface vanity top 234
vinyl wallcovering 185
wood beam 27
Space heater 6
Spanish tile roofing 78
Spinner ventilator 288
Spiral hardwood stair 208
Sprayer airless 13
Springs garage door 104
Sprinkler system 163
Stain bifold door 198
bypass sliding door 201
cabinet 221, 229
louver door 200
Stained glass window 120
Stainless steel roofing 84
steel sink double 252, 254
steel sink single 252, 254
Stair assemblies 208
bannister 209
carpet 249
components 209, 249
handrail 209
porch 63
rough frame 47
wood deck 57
Stairs 47, 208, 275
metal 275
Stand light 13
Standard aluminum awning 287
brick paver 140
Standing seam 84

305

Index

Stapled tile acoustical ceiling . . . 189
Steam clean fixtures 11
Steel bathtub 255
   building 9
   door 92
   door pre-hung 104
   H section piles 22
   I-type 28
   window 124
   window vented 124
Steel-clad door pre-hung 91
Stem wall 24
Steps concrete 275
Stock lumber 65, 66
Stop gravel 72
Storage container 3
   shed 290
Store contents 6
Storefront entrance 122
   system 120
Storm aluminum window 112
   door 102
Straight hardwood stair 208
Strip refinish cabinet . . 218, 221, 229,
                              234
   typical bathroom 13
   typical kitchen 13
   typical laundry 13
   typical room 13
Strips cant 85
   furring 179, 183, 190
   ledger 32, 48
Structural brick 126
Stucco 128
   siding synthetic 137
Stud framing metal 44
   wall 36
Subflooring 33
   particle board 34
   plank board 34
   plywood 33
Sump pump 148
Supply grill 259
Support piers shim 6
   posts carport metal 290
   wire 168
Switch dimmer 263
   double 262
   single 262
   triple 262
Synthetic stucco siding 137
System intercom 270
   satellite 271
   security 270

T

Tall cabinets 218
T-bar grid acoustical ceiling 188
Teak parquet floor 242
Telephone 3
Television antenna 271
Tempered glass window 120
Temporary bracing 6
   fencing 3
   railing 3
   utilities 3
Terrazzo floor 244
Test asbestos 1
Texture ceiling 236
   wall 236
   wallboard 179
Thermal fog area 11
Thermostat air conditioning 163
   heat pump 157

heating control 159
Thick set paver floor 243
Thin coat ceiling plaster 177
   coat wall plaster 175
Tile acoustical ceiling adhesive . . 189
   acoustical ceiling stapled 189
   floor 243
   floor ceramic 243
   floor re-grout 243, 244
   panel acoustical ceiling 188
   roofing 76
   thick set wall 180
   thin set wall 180
   vinyl 245
Tinted glass window 119
Toilet chemical 3
   designer floor mount 251
   floor mounted w/tank 251
   paper roller 282
   seat 252
   wall hung 251
Tongue and groove siding 135
Toothbrush holder 282
Towel bar 282
Track lighting 266, 268
Trade labor 2
Trailer office 3
Trash compactor 281
Treated column or post 57
   decking 61
   girder 58
   joist 59
   ledger 59
   railing and trim 60
   wood piles 19
Treatment blinds 273
   carpet 11
   drapery 273
Tree removal 15
Trenching backfill 17
Trim casing . 91, 93, 96, 194, 198, 201
   exterior 133
   set aluminum window 113
   set wood window 118
Triple switch 262
Truck dump 3
   flatbed 3
Truss bowstring 55
   hardware 55
   w or fink 54
Tub shower combination 256
TV / cable jack w / wiring 262
Typical bathroom strip 13
   kitchen strip 13
   laundry strip 13
   room strip 13

U

Underlay felt 34
Underlayment 73
   floor 241
Unfaced batt 167, 169, 171
Unfinished wallboard 178
Unit decontamination 13
   kitchenette 281
Unreinforced CMU 125
   concrete slab 25
Untreated wood piles 19
Urethane carpet pad 248, 249
   coat floor 241
   insulation 174
Urinal 251
Utility cabinet tall 220

V

Vacuum cleaner HEPA 13
Vane weather 290
Vanity cabinet tops 234
   cabinets 231
   lavatory sink 253, 254
Vapor barrier exterior 46, 172
   barrier floor 168
   barrier slab 26
   barrier wall 170
Veneer brick 126
   slumpstone 130
Vent 288
   cap 147
   ridge 74, 289
Vented rangehood 276
Ventilator spinner 288
Ventless rangehood 277
Vinyl base molding 203
   clean wax 245
   fascia 67
   sheet 245
   sheet floor 245
   siding 134
   sliding patio door 100
   soffit 69
   tile 245
   wallcovering solid 185
   window 113
Vinyl-coated wallcovering . . 184, 185

W

W or fink truss 54
Waffle carpet pad 248, 249
Walk impregnated 88
Wall asbestos seal 13
   brick 14
   cabinets 211
   cap coping 72
   clean 10, 235
   clean seal 236
   earthquake strapping 36
   end 72
   framing 34
   framing system 34
   gas-fired furnace unit 154
   headers 37
   hung toilet 251
   insulation 168
   lath 176
   mildicide 11
   mirror 120
   oven double 277
   oven single 277
   paint 236
   paneling 182
   patch 184
   plaster 175
   plate 37
   prep 235
   protect/cover 6, 235, 236
   sand 235
   seal 235
   side 72
   stem 24
   stud 36
   texture 236
   tile 181
   tile re-grout 180, 181
   tile thick set 180
   tile thin set 180
   type radiant 159
   vapor barrier 170

Wallboard accessories 179
   finished 177
   texture 179
   unfinished 178
Wallcovering 184
   border 186
Walnut floor 240
   parquet floor 242
Wash sewage pressure 11
Washer 279
   dryer 279
Water extraction 10
   heater 148
   heater cap 289
   pump pool 292
   service 3
   softener 150
Wearing course 139
Weather vane 290
Welder 5
Wet fog / ulv area 11
Whirlpool bathtub 256
Whole-house exhaust fan 269
Window awning sash aluminum . . 110
   bay aluminum 111
   bullet resistant 120
   casement bow wood 117
   casement wood 116
   casing 206
   casing double 206
   casing single 205
   casing triple 206
   double hung aluminum 108
   double hung vinyl 113
   double hung wood 114
   glass 119
   hardware aluminum 113
   mounted air conditioning 162
   picture wood 118
   rough-in 40
   screens 118
   sill 206
   single hung 106
   sliding sash aluminum 109
   steel 124
   storm aluminum 112
   treatment 273
   trim set 205
   trim set aluminum 113
   trim set wood 118
   vented steel 124
Wire support 168
Wired glass window 120
Wiring 120v 261
   240v 261
   run 268
Wood base cabinet 223, 224
   beam solid 27
   bypass sliding door 200
   deck 57
   deck stair 57
   door bifold 196
   door pre-hung 191, 195
   foundation post 23
   framed sliding patio door 100
   parquet tile floor 241
   piles treated 19
   piles untreated 19
   plank flooring 239
   screen door 101
   shingle siding 136
   shingles 75
   shingles cedar 75
   shutter 274
   shutters interior 286
   storage shed 291

306

Index

    tall cabinet 218
    vanity cabinet 231, 232
    wall cabinet 212, 214
    window 114
    window sill 206
Wood-framed building 9
Wool carpeting 246, 248
Wrap water heater insulating ... 150
Wrought iron fencing 144

Notes

Notes

Notes

Notes

Notes

Notes

Notes

Notes

Notes

Notes

2025 Training Class Options
A tradition of excellence in construction cost information and services since 1942

RSMeans Data Training

Virtual and in person classes offered monthly. Call for the schedule. ☎ **877.620.6245**

Gordian offers educational content in a variety of formats — including self-paced training modules, live virtual seminars, and in-person classes. We also offer customized private training solutions to meet your organization's unique continuing education needs. Please visit our website at https://www.rsmeans.com/products/services/training for our current schedule of offerings.

Seminars include: Introduction to Cost Estimating, Advanced Cost Estimating, Facilities Estimating, Maintenance & Repair Estimating, Mechanical & Electrical Estimating, and RSMeans Online Training.

Registration Information

How to register

By Phone
Register by phone at 877.620.6245

Online
Register online at rsmeans.com/products/services/training

Note: Purchase orders or credits cards are required to register.

Instructor-led Government Pricing
All federal government employees are eligible for discounted training. Other promotional discounts cannot be combined with the government discount. Call 877.620.6245 for government pricing.

If you have any questions or you would like to register for any class or purchase a training module, call us at 877.620.6245.

Become an RSMeans™ Data Certified Professional

Gordian's RSMeans Data Certification program allows you to differentiate yourself and your organization when it comes to demonstrating your mastery of RSMeans Data, the construction industry standard for reliable costs. Visit https://www.rsmeans.com/landing-pages/rsmeans-data-certification to learn more.

Introduction to Cost Estimating

This one-day introductory course to improve estimating skills and effectiveness starts with the details of interpreting bid documents and ends with the summary of the estimate and bid submission.

Topics include:
- Using plans and specifications to create estimates
- The takeoff process — deriving all tasks with correct quantities
- Summarizing the estimate to arrive at the final number
- Formulas for area and cubic measure, adding waste and adjusting productivity to specific projects
- Evaluating subcontractors' proposals and prices
- Adding insurance and bonds
- Understanding how labor costs are calculated
- Submitting bids and proposals

Who should attend: novice estimators, project managers, architects, engineers, owners' representatives, contractors, and anyone responsible for budgeting or estimating construction projects.

Advanced Cost Estimating

This three-day practical training program addresses the vital importance of understanding the scope of projects in order to produce accurate cost estimates for facilities repair and remodeling.

Topics include:
- Discussions of site visits, plans/specs, record drawings of facilities and site-specific lists
- Review of CSI divisions, including means, methods, materials and the challenges of scoping each topic
- Exercises in scope identification and scope writing for accurate estimating of projects
- Hands-on exercises that require scope, takeoff and pricing

Who should attend: corporate and government estimators, planners, facility managers and others who need to produce accurate project estimates.

Visit rsmeans.com for more details.

Mechanical & Electrical Estimating

This three-day course teaches attendees how to prepare more accurate and complete mechanical/electrical estimates, avoid the pitfalls of omission and double-counting, and understand the composition and rationale within the RSMeans Data mechanical/electrical database.

Topics include:
- The unique way mechanical and electrical systems are interrelated
- M&E estimates — conceptual, planning, budgeting, and bidding stages
- Order of magnitude, square foot, assemblies and unit price estimating
- Comparative cost analysis of equipment and design alternatives

Who should attend: architects, engineers, facilities managers, mechanical and electrical contractors, and others who need a highly reliable method for developing, understanding, and evaluating mechanical and electrical contracts.

Facilities Estimating

In this three-day course, professionals working in facilities management can get help with their daily challenges to establish budgets for all phases of a project.

Topics include:
- Determining the full scope of a project
- Understanding of RSMeans Data and what is included in prices
- Identifying appropriate factors to be included in your estimate
- Creative solutions to estimating issues
- Organizing estimates for presentation and discussion
- Special estimating techniques for repair/remodel and maintenance projects
- Appropriate use of contingency, city cost indexes and reference notes
- Techniques to get to the correct estimate quickly

Who should attend: facility managers, engineers, contractors, facility tradespeople, planners and project managers.

Maintenance & Repair Estimating

This three-day course teaches attendees how to plan, budget and estimate the cost of ongoing and preventive maintenance and repair for existing buildings and grounds.

Topics include:
- The most financially favorable maintenance, repair, and replacement scheduling and estimating
- Preventive planning and facilities upgrading
- Determining both in-house and contract-out service costs
- Annual asset-protecting M&R plan

Who should attend: facility managers, maintenance supervisors, buildings and grounds superintendents, plant managers, planners, estimators, and others involved in facilities planning and budgeting.

RSMeans Online Training

Construction estimating is vital to the decision-making process at each state of every project. Our online solution works the way you do. It's systematic, flexible and intuitive. In this two-day course you will see how you can estimate any phase of any project faster and better.

Topics include:
- Customizing our online estimating solution
- Making the most of RSMeans Data "Circle Reference" numbers
- How to integrate your cost data
- Generating reports, exporting estimates to MS Excel, sharing, collaborating and more

Also offered as a self-paced or on-site training program!

RSMeans Data Training

RSMeans Data Training Cancellation Policy

Virtual Training (Individual)
If you are unable to attend a training class, a colleague may substitute at any time before the session starts by notifying the training registrar at 877.620.6245 or your sales representative. Gordian is not responsible for sending duplicate materials to the substitute colleague.

If you are unable to find a substitute and cancel fewer than ten (10) business days prior to the training class, you will either forfeit the registration fee or be allowed to apply your registration fee to the same class only once within the same calendar year. No-shows will forfeit their registration fees and no rescheduling is permitted. In the unfortunate circumstance that Gordian must cancel your event, Gordian will work with you to reschedule your attendance in the same training class at a later date or will fully refund your registration fee.

Any on-demand (self-paced) training modules are not eligible for cancellation, substitution, transfer, return or refund.

Virtual Training (Private Group)
If you provide notice more than ten (10) business days prior to the training class, Gordian will work to reschedule your event. If you cancel your event fewer than ten (10) business days prior to the training class, you are responsible for paying for the trainer's time for the cancelled training, as well as the full cost of materials and shipping. In the unfortunate circumstance that Gordian must cancel your event, Gordian will work with you to reschedule your attendance in the same training class at a later date or will fully refund your registration fee.

Public Training Events (Individual)
If you are unable to attend a training conference, a colleague may substitute no fewer than ten (10) business days before the session starts by notifying the training registrar at 877.620.6245 or your sales representative. Gordian is not responsible for sending duplicate materials to the substitute colleague.

If you are unable to find a substitute and cancel fewer than ten (10) business days prior to the conference, you will forfeit the registration fee. No-shows will forfeit their registration fees and no rescheduling is permitted. In the unfortunate circumstance that Gordian must cancel your event, Gordian will fully refund your registration fee.

On-Site Training (Private Group)
If you provide notice more than ten (10) business days prior to the training class, Gordian will work with you to reschedule your event, but you are responsible for rescheduling fees associated with the trainer's time, airfare and lodging. If you cancel fewer than ten (10) business days before your event, you are responsible for the full cost of the event. In the unfortunate event that Gordian must cancel your event, Gordian will work with you to reschedule your attendance at a later date or will fully refund your registration fee.

AACE Approved Courses
The AACE recognizes .1 CEUs per hour of training attended for all of Gordian's virtual training courses and public training events.

Important
Class sizes are limited, so please register as soon as possible.